# Administration of Recreation, Parks, and Leisure Services

# Administration of Recreation, Parks, and Leisure Services

SECOND EDITION

**Lynn S. Rodney**
University of Oregon

**Robert F. Toalson**
Champaign Park District
Champaign, Illinois

**John Wiley & Sons**

New York • Chichester • Brisbane • Toronto

**Library of Congress Cataloging in Publication Data:**

Rodney, Lynn Smith, 1915–
Administration of recreation, parks, and leisure services.

First ed. (1964) by L.S. Rodney, published under title: Administration of public recreation.
Includes index.
1. Recreation—Management. 2. Parks—Management.
I. Toalson, Robert F., joint author. II. Title.

GV181.5.R62 1981 352.7 80-17723
ISBN 0-471-05806-8

Printed in the United States of America

10 9 8 7 6 5 4 3 2 1

# Preface

Many books have been written in recent years on recreation and park administration. Each has its own style and conveys its reason for the occurrence of major changes in our society that affect the park, recreation, and leisure services field. The quality and success of our park and recreation programs depend primarily on the skill and insight of the leadership that understands administrative concepts as a result of proven practices and recent research.

This book is an introduction to the general field of recreation and park administration and an aid to members of park and recreation boards, supervisors, administrators, park and recreation leaders, and others involved with relationships that call for better understanding of the problems and responsibilities facing the administrative staff of a recreation and park system. Major emphasis is given to the administration of local park and recreation systems rather than to the state or federal, but the administrative principles, practices, and concepts are not restricted to this level.

This volume deals with the subject areas of administrative relationships, leadership, theory, and practices that provide a conceptual framework for the administration of a park and recreation enterprise. The core of operational procedures that provide the foundation for sound and proven practices are discussed.

One of the distinctive features of this book is the manner in which it probes into the subject areas. We have tried to provide practical guidelines for dealing with situations that face park and recreation leaders. Separate texts could be written on each of the topics discussed. However, our intent is to provide a more general panoramic view that is understandable and useful to recreation and park officials and to students of this field.

Special appreciation goes to our professional colleagues, our students, and the many park and recreation organizations that contributed to the material presented herein, and to the authors and publishers that granted permission to use copyrighted material.

*Lynn S. Rodney*
*Robert F. Toalson*

# Contents

# CHAPTER 1

# Recreation: Purposes and Administrative Foundations

Any time a person is required to work in a cooperative organizational venture to attain institutional objectives the need for administration is evident. People cannot achieve organizational goals without administration.

Although administrative activity can be traced back to biblical times, its study gained impetus in modern times when scientific management came to the fore and the complexity of our modern institutions made demands for formal study of this unique area.

Public recreation and park administration as an area of formal study is a phenomenon of the present century. It is a relatively new field of endeavor, and much of what has been written was taken from the experiences of those who are practitioners in the field. However, the practitioners need direction in new concepts, as do the students who are entering this professional field. Our complex society requires institutions to provide the means by which people can effectively use their expanding leisure time, and the public recreation organizations provide the means by which this can be accomplished. Thus these institutions require a pattern of administration that is focused toward specific goals.

## THE SETTING FOR ADMINISTRATION

To understand administration and the extent of its function in American society, one must look at the setting in which it works. Many changes in the way Americans live have become the setting for administrative action. Actually, more changes have taken place in the American scene during the past 20 years than during the past thousands of years of human's existence. Such changes are both awesome and challenging. In reviewing these changes, one should place them in the setting for administration, inasmuch as patterns of administration must be adaptive to changing life styles.

## Growth in Population

There is every reason to believe that the population of this country will continue to show a phenomenal growth, even though current percentages indicate a drop in the rate of growth. Population growth during the twentieth century has had its ups and downs. Declining birthrates in the 1930s were erased in the 1940s. In turn, the growth rate peaked in the 1960s, fell somewhat in the early 1970s, and will be increasing in the 1980s. Looking at the awesome world figures, we see that it took millions of years to reach a billion population in 1830, whereas it took 100 years to reach two billion in 1930, and only 30 years to reach three billion in 1960 and four billion in 1975. Although such great growth is more evident in the underdeveloped countries, it still touches our country where we now see the growth of massive strip cities, greater number of metropolitan areas and major population shifts. Leisure service institutions must plan for growth and the changing social composition of our cities and rural areas.

Furthermore, a definite shift in population has taken place between the rural farm sector and urban centers. Since 1900 the percentage of population has shifted from a predominantly rural population to one that is primarily urban. In 1900, 60.3 percent of the population lived in rural areas; in 1970, this was reduced to 26.5 percent, and the 1980s will indicate further reductions.

In other words, the growth of our cities and the decline in the rural farm population have created problems that have great significance for the administration of parks, recreation, and leisure services in our cities, as well as in the rural sector of our nation.

## The Metropolitan Scene

Since our populations have become more urbanized, special attention needs to be focused toward the metropolitan scene.

During the latter part of the eighteenth century, only 5 percent of the population of America lived in cities. By 1900, 35 percent lived in urban centers. This increased to 64 percent in 1970 and to 75 percent by 1975. The forecast is that 80 percent of the population, or four out of five persons, will be living in these metropolitan centers by 1985. This means that current population growth has largely taken place in what can be identified as metropolitan areas.

A metropolitan area is one that comprises a core city of 50,000 or more people, plus the surrounding urban and suburban fringe. The formation of these large urban areas creates problems for recreation planners and administrators alike, since the population shift is a continuing one. As the central core of the city becomes densely built up, the population pressure shifts to the suburbs and the areas immediately surrounding the city. This in turn creates what is known as the "urban fringe." In other words, population pressure and the availability of modern means of transportation provide people an opportunity to flee the congestion of the core city for the amenities of the countryside. However, this growth at the peripheries soon reaches a dimension that takes on the configuration of the urban community, which in turn nullifies the purpose of the outmigration from the central city. The city and surrounding areas soon take on the cloak of highly dense urban communities in which city boundaries are no longer physically identifiable.

People living in the outer unincorporated fringe of the central city benefit from the amenities provided them by the core city in which most are gainfully employed or come for pleasure, but they are not required to pay taxes for these services since they live outside the taxing jurisdiction of the city. This so-called "no-man's land" has created what is known as the "urban fringe" problem. Use of golf courses, swimming pools, community centers, parks, tennis courts, and other park and recreation facilities and areas that are financed by the taxpayers of a city are available for use by people in the "urban fringe" who pay no city taxes. Park and recreation systems serve these people who cannot be identified, thus creating pressures and problems that call for solutions. Further, the flight to the suburbs creates pockets of neighborhood decay in the core city that soon become identified as slums or ghettos.

## Expanding Educational Opportunities

During the latter part of the nineteenth century, educational opportunities were limited and the work week, as we know it today, did not exist. People were largely concerned with work, and education was limited because of the desire of children and youth to enter the work market.

The education scene has changed. More and more children, youth, and adults have been and are being exposed to educational opportunities. At least $100 billion were spent for public education alone by 1980. This means that additional millions of people will have been exposed to recreation and cultural activities and in turn will have built-in expectations of the types and kinds of activity in which they wish to engage.

The educational attainment of people in this nation is high and will continue to go even higher. In fact, the years of schooling for our people have climbed to a point where graduation from high school and even from community colleges has become a common goal. Since administration operates in the culture in which it is located and this culture is one where people are vastly better educated than in the past, public parks and recreation programs and services must keep pace with this educational phenomenon. What people in our nation wanted in the 1940s is not necessarily what they want in the 1980s.

### Changing Needs of People

As recreation, park, and leisure service organizations have grown in number and complexity during the past 50 years, new program shifts as well as priorities have emerged. These program shifts are the result of changes brought about by great technical advances that have taken place in recent years. Since such changes affect the setting for administration, it is important to identify them as much as possible. All scientific and technological advances create hazards in addition to comforts and in turn create degrees of stress as well as concerns.

### Environmental Concerns

The present generation wants broader and more meaningful directions given to protect the physical environment through outdoor recreational planning for wilderness areas, park preserves, and recreation lands that make life more livable. No longer can society turn its back to the need for conservation, environmental, and land-use planning in the name of progress when a badly needed partnership relationship between people and their physical environment must be established to provide the human amenities for a growing population. It is encouraging that our recreation institutions and organizations are beginning to recognize the environmental movement as one that has a close link to the leisure, recreation, and park movement.

### Concept of Leisure and Recreation

Society is beginning to understand leisure and recreation as a shaper of life-styles and a value in itself. The concept of leisure now includes a larger view and a broader concept of its meaning to people and its place in the value structure of society.

Actually, leisure has come of age; it is a symbol of an American life-style. Whether it be a concept relating to personal and social values or a symbol that expresses participation in an experience that provides self-gratification or self-fulfillment, leisure has implications to the American way of life. We must accept the broader context of leisure and its impact on our society, just as we must examine the concept of recreation, its relation to leisure, to the individual, and to the community of which it is a part.

## The Cultural Arts

An increasing awareness of the need to encompass the cultural arts as a viable part of recreation and leisure services programs is becoming evident, as more and more programs are directed toward the creative, social, and cultural arts.

Since one of the major purposes of recreation is the improvement in the quality of living and self-fulfillment of the individual, it seems only logical and expeditious that the many and varied cultural arts opportunities that enrich people's lives, making them more meaningful, should be encompassed within the recreation program context. Programs in the cultural arts—music, drama, art, dance, and the like—provide a dimension for people to respond to aesthetic experiences that give them feelings of personal achievement, well-being, and self-satisfaction.

## Minority and Other Special Problems

A close look at the recreation programs provided for the minorities, disabled, and handicapped is necessary. These special groups have, in some instances, been discriminated against, although no overt attempt has been made to do so. However, today efforts are being focused on their special needs. Recreation organizations are beginning to make social equality meaningful by integrating programs toward the goal of meeting the needs of all regardless of disability, race, or ethnic origin. No longer can our recreation and park organizations fail to recognize the necessity to share with all special groups the opportunity for full participation and involvement in services and democratic processes that have been identified with the middle class, the well-to-do, and the very rich. Further, making provision for equality of the sexes in all activity involvement is becoming a reality, and the relationships between the sexes are taking on new meanings. The same can be said in making special provision for the disabled, poor, and ethnic minorities. Actually, the changing needs of people are influenced by one group with another, each providing its own enrichment to the totality of the group.

## Lifelong Learning

Our society is now facing the desire and need for more people to be involved in lifelong learning situations. Lifelong learning is not merely programs involved with public schools, although these institutions are not excluded, but does involve programs where individuals can find a greater degree of life enrichment through a learning process that is tied closer to leisure service organizations, as well as to schools, churches, and other social institutions.

One of the indications of the movement toward lifelong learning has been the tremendous expansion in the field of community education, which is now being supported by our local, state, and national governments. Although much evidence is being focused on lifelong learning experiences, still more needs to be done, and our recreation and leisure service agencies should be at the forefront of this expanding movement.

### Movement of People

Another primary reason for the changing needs of people relates to transportation and the speed of modern living. Actually, today we can almost consider the world as a playground and the community, as a backyard. While at one time we felt that the neighborhood playground was absolutely essential to the recreation needs of people living in such a setting, we know today, that speed has changed the urban scene and has enhanced the need of facilities beyond the close confines of the neighborhood and community.

### Rising Standard of Living

One cannot discuss the setting for recreation services without considering the growth of the American economy and the rising affluence of people in this country. No longer is consideration given to only the rich and the poor. Today we have an affluent middle class that is not only seeking but demanding opportunities for expanded leisure services. Furthermore, the work week has rapidly been reduced from the standard 10-hour day, which was prevalent at the turn of the twentieth century, to one of eight, and in some instances to six or less hours. With technological advances one can predict that the future will see a man or woman accomplishing in a mere hour what it takes eight hours of work to do at present. But this also means that people have more money to spend in an expanding age of leisure. A rising standard of living makes possible more discretionary spending. Americans spending for leisure pursuits soared to an estimated $180 billion in 1978, which is about one dollar out of every eight of a consumer's personal spending.[1] Leisure spending is continuing to boom, and there is every reason to believe it will double in the next decade as it has done in the past.

### Energy Problem

The American public, for the first time in its history, is facing an energy problem that will have a dramatic effect on the way of life for all citizens. The leisure industry, in particular, will be drastically affected because people will be unable to have the freedom of movement due to lack of oil and gasoline resources, as well as to the heavy cost imposed on the limited available resources.

[1] *U.S. News & World Report* (January 15, 1979): 41.

In the past, part of the American way of life was to use the automobile for recreation or leisure purposes. However, the situation has changed and more and more people are forced to drive less. Therefore they will make greater and heavier use of neighborhood and community parks, of regional and state recreational resources, and of all other leisure services that involve a minimum of travel. There is no question that people will make greater use of public transportation, which may offset the cutback in automobile travel, but even this will not completely eliminate the problem. The American dream of unlimited automobile travel for all classes of people is gone. The recreation and park professionals will need to adapt their programs to this situation as we see the shift of emphasis of activity in involvement.

## RECREATION DEFINED

"Recreation" is a popular word. It has different meanings to different people. Generally, however, all meanings at the community level are focused toward the attainment of an ideal, toward the contribution in one form or another to a better way of life. Hence the word "recreation" as used here has a positive connotation—it is community and goal oriented, and provides a means for the improvement of society.

Before discussing the purposes of recreation, it is important to look at definitions and point out that the word "recreation," in its popular usage, is related to recreating, refreshment, diversion, or play. In other words, it is commonly held to be any activity that is engaged in for the satisfaction it brings to the individual. Meyer and Brightbill state that "recreation is activity engaged in during leisure and primarily motivated by the satisfaction derived from it."[2] Sapora and Mitchell express the concept of recreation as "the leisure activities of adults as well as children; all types of activities, active as well as passive, sought by both children or adults for personal expression."[3] Butler states that "recreation is any form of experience or activity in which an individual engages from choice because of the personal enjoyment and satisfaction which it brings directly to him."[4]

All definitions imply that recreation is engaged in without compulsion and that it brings satisfaction to the individual. Hence these two elements are found in all recreation expressions of individuals.

[2]Harold D. Meyer and Charles K. Brightbill, *Recreation Administration* (Englewood Cliffs, N.J.: Prentice-Hall, 1956), p. 1.

[3]Allen V. Sapora and Elmer D. Mitchell, *The Theory of Play and Recreation*, 3d ed. (New York: Ronald, 1961), p. 127.

[4]George D. Butler, *Introduction to Community Recreation* (New York: McGraw-Hill, 1959), p. 10.

## Recreation Versus Community Recreation

Any community recreation activity or service, if it is to have real meaning to individuals, must be purposeful, must provide people with meaningful processes or growth-producing involvements, and must be cloaked in a framework of values. Hence it must generate intellectual, social, and spiritual growth. It is not in the American tradition to use tax moneys or to provide leadership for services that are not goal oriented. Thus one of the greatest limitations to an understanding of recreation is the lack of focus on goals, objectives, and purposes that are consistent with the idea that it helps people approach the democratic ideal.

Therefore it is necessary to differentiate between the word "recreation" and the purposive and socially acceptable "recreations" sponsored by community and public agencies. To define "recreation" as "leisure activity engaged in for its own sake" provides no focus. In fact, it evokes a negative image of fun and frolic that is not a part of our value system. Thus community recreation, or public-sponsored recreation, be it indoor or outdoor, must be, and is, goal oriented and purposive. Indeed, the recreation and park leaders of America are dedicated, one and all, to the arts of living, to the full development of the individual, and to the sponsorship of services that give rise to the excellences of humankind.

The purposes of recreation evolve from the beliefs, attitudes, and value judgments of the people in our society. And while they change slowly, it is apparent that people are beginning to agree that community-sponsored recreation has positive purposes and projections. Its goals are focused toward the improvement and betterment of people through creative, wholesome, and imaginative leisure living patterns. Indeed, unless the purposes of community recreation are directed toward meeting the needs of people through growth-producing experiences, programs will merely act as tranquilizers to amuse the masses and act as palliates for social unrest and unfulfilled wants.

In sum, a clear distinction must be made between the word "recreation," with its great variance of meanings to people, and that which involves people in purposeful activity through "community recreation." On the one hand, there is no focus; on the other, there is direction toward the development of a significant life for all persons, through activity involvement and group associations. John J. Collier, in a recent address, stressed this point clearly when he stated;

> The philosophy of recreation that states that participation in any recreation activity is an end in itself is dying a hard death. Such a belief can only be interpreted by keeping attendance figures and a running total of the number of activities offered. Moreover, it is based on such general terms, it leaves the recreation

leader or the administrator without any valid means of interpreting what he is doing.

Park and recreation needs more money to provide the necessary facilities, the leadership, and the equipment to do the job; but we are not going to get it unless we establish a sound foundation of valid goals and objectives which can be achieved through proper administration and leadership. And our recreation services must be designed to reflect these goals and objectives.

We have labored too long under the assumption that the mere provision of activities will insure good results. No wonder we are placed on the defensive if a vast amount of what we are doing could be done by anyone who can set up a schedule of activities and organize time and people.

We must build our profession on the premise that (1) we as professionals know how to make goals meaningful through recreation experiences that have direction and purpose; and (2) we possess the necessary professional knowledge and skills to make certain that purposes are being achieved.[5]

## AIMS AND OBJECTIVES OF PARKS AND RECREATION

Statement of aims and objectives of community-sponsored recreation and parks programs provides the guidelines for program needs and services. However, one of the problems confronting the profession today is the lack of clear-cut goals and purposes. There is need for a clear delineation of aims—a unification of effort and a pride and dedication in working toward a way of life that gives people a chance for self-fulfillment, creativeness, and a liberation of the human spirit. Differences of opinion only fragment, confuse, and lead to uncertainty. Surely, maturity calls for understanding and striving for unity through acceptance of common goals, if recreation is to be a constructive force in American life. Thus statements of aims and objectives might well embrace some of the general characteristics common to most recreation systems by

1. Being designed for the socialization of the individual through comradeship and group involvement.
2. Being directed toward the growth of human dignity and self-fulfillment.
3. Being consistent with the ideals of education in a democratic society.
4. Being focused toward finding creative and satisfying outlets through a wide range of cultural and wholesome pursuits.
5. Providing means for the individual to maintain good health and physical vitality in order to meet the complex challenges of the day.
6. Being designed to show how the use of leisure can be a major force in the enrichment of personality.

[5]Taken, by permission, from a statement by John J. Collier, July 11, 1979.

7. Being consistent with the idea that recreation contributes to group welfare through greater appreciations, widening interests, and broader understandings.

Recreation has purpose and direction. Indeed, the ends are many and satisfying, but all purposes, either stated or implied, should be focused on strengthening and preserving the American ideal of the best in family and community living.

One of the excellent statements of purpose was that given by Joseph Prendergast, former executive director of the National Recreation Association, when he stated,

> The chief value of recreation, i.e., the creative use of leisure time, lies in its power to enrich the lives of all individuals. It also has many valuable "by-products" in the fields of physical and mental health, safety, crime prevention and citizenship. Because recreation contributes to rich and satisfying living and social cooperation, it should play an important part in the life of everyone, everywhere. It is not merely for those who have suffered misfortune, nor simply to prevent man from encountering misfortune. It is to give all men opportunity for growth, opportunity to be and become themselves.
>
> Children need happy, healthful social play to attain their fullest development; young people require wholesome recreation opportunities to replace questionable amusements which might lead to delinquency; workers need recreation during their off-duty hours in order to keep their spirits and production high; adults and the aged need opportunities to find the most satisfying use of their expanding leisure time. Furthermore, people who play together, sing together, make things together, achieve in its truest sense a community of feeling. Recreation programs also help to preserve local, state and national traditions.
>
> Recreation like other functions found to be essential to the general welfare has become widely recognized as a proper concern of society. Like education, recreation benefits both the individual and society itself. Public recreation makes possible participation on a democratic inclusive basis and insures wholesome recreation opportunities for all.[6]

Some park and recreation organizations have also attempted to identify the special demand characteristics of recreation. The National Outdoor Recreation plan states that recreation yields three basic types of benefits.[7] These are: (1) direct satisfaction to the individual; (2) enhancement of the overall mental and physical quality of the individual—an investment in human capital adding

[6] *The Role of the Federal Government in the Field of Public Recreation* (New York: National Recreation Association, 1953), p. 5.

[7] California Outdoor Recreation Resources Plan, Department of Parks and Recreation, Sacramento, Calif.: State Printing Office, 1976, Chap. 4.

to the productivity of the individual and society; and (3) important third-party benefits such as increased business and property values. Therefore recreation, like education, yields benefits of both a monetary and nonmonetary nature.

## Relation of Aims and Objectives to Administration

If recreation is purposeful and has direction, it is necessary to develop a program to attain it. Thus programs are the medium through which aims are sought and objectives realized. And the means by which people and materials are mobilized and utilized to attain these aims are the goal of administration. Obviously the effectiveness of a recreation system is judged by how well objectives are attained, and if administration is the directing force that propels the organization toward them, a clear understanding of terms is necessary.

When the word "purpose" is used, it connotes the ultimate object to be attained. It is synonymous with "goal," "end," or "aim." Thus, purposes or goals are related to long-range aims, and the terms describing them are inspiring words that are used. For example, the goals of community recreation are improvement and betterment of people through creative, wholesome, and imaginative leisure living patterns. On the other hand, objectives are attainable ends. They can be reached as programs are focused toward the general purpose. They can be likened to progressive stepping stones of achievement leading to the ultimate goals that are sought but hardly ever fully attained.

If the programs of the recreation system are aimed at goals through the achievement of objectives, and if it is through administration that an organization must operate to achieve them, every recreation system must formulate a statement of its general purpose and more specific objectives as based on community needs. The characteristics of desirable statements of recreation aims and objectives as already given can be used as a guide. Each community is unique, however, and must develop its own statement, but a review of objectives of what is to be accomplished through a community's many programs of activities is given here, from a sampling of statements taken from the more progressive systems.

1. *Emotional and Physical Health.* To develop a sound body and mind through wholesome, vigorous, and creative activities.
2. *Character Development.* To build character through rich, satisfying, and creative leisure living patterns focused toward socially desirable attitudes, habits, and values.
3. *Widening Interests.* To open new interests that provide satisfying outlets for individual growth and development.
4. *Citizenship.* To develop through recreational associations of people a respect for the worth and dignity of individuals and faith in democratic action.

5. *Skills*. To develop skills in the arts of leisure-time living that raise the level of the refinement, culture, and happiness of people.
6. *Social Living*. To develop and strengthen social relationships within the family and the community through close group associations and activity participation.
7. *Economic Value*. To strengthen the morale and economic efficiency of the community through expanding leisure-time interests and improved social living conditions.
8. *Community Stability*. To develop community stability by providing an environment that is conducive to wholesome family living and community life.

## THE SCOPE OF PARK AND RECREATION ADMINISTRATION

The concepts and meanings of administration will be thoroughly probed in the second chapter. However, it is well to bear in mind that administration is the process that mobilizes an organization's resources, human and material, toward predetermined goals. Thus the goals must be chartered and a direction given.

An analysis of the approaches to park and recreation administration reveals varying viewpoints as to how this subject should be approached. The very complexity of operations that are embraced under the word "recreation" adds to the difficulty. And the lack of common understanding of what constitutes common goals further confuses the issue. Agencies devoted to the development and use of parks, outdoor recreation areas, camps, community centers, playgrounds, special-use facilities, and the like still have the function of recreation, because these areas and facilities are set aside for the recreational use of people. Likewise, recreation programs and services and special facility uses in industry, institutions, and the armed forces are also a part of the function of recreation. Therefore in approaching the administration of such a diversity of interests, there is a real question of how much consideration should be given to the substantive problems of an area of specialization and how much should be given to the basic processes of administration that are common to all fields.

Regardless of diversity, the ultimate goals of all recreation agencies are similar. There may be some subtle disagreements as to the primary purpose, and some variations may exist in the emphasis given to certain phases of program, but nonetheless all will agree that "recreation" of people is involved and, through it, a better way of life is possible. Thus park and recreation administration is approached on the basis that a similarity of problems runs through all park and recreation agencies, and while technical skill and knowledge are important, a need exists for an understanding of administrative processes and techniques that are common to all. Hence the scope of recreation administration is viewed here in its broadest sense, with emphasis given to the many aspects of concepts, processes, and practices that are common to all agencies, rather than concentrating on the technical skills needed for a specific area.

## THE STRUCTURE AND LEVELS OF PUBLIC RECREATION AND LEISURE SERVICES

In considering the administration of public recreation and parks in this country, one must consider the levels of government and their role in relation to parks and recreation. Each has a contribution to make and a sphere of operating influence; however, one should bear in mind that the nature and concept of administration as discussed here, as well as many of the substantive problems that arise for the administrator, are common in many respects to all levels of governmental operation.

The administrative structure for parks, recreation, and leisure services at all levels of government is affected by the many unique influences that comprise our governmental framework. It would seem desirable in any discussion of the administration of recreation services to start with the federal government; however, it is important to point out that the national government has no direct control in organizing and administering recreation services except through those agencies or organizations that provide services that have implications for recreation use. In other words, while federal agencies exist that make provision for the use of facilities, the national government is largely oriented toward the acquisition of land for park and recreation use and toward policy, planning, research, and coordinating services that deal in some manner with recreation lands or resources.

There are no federal constitutional provisions that make provision for recreation as a national function. Recreation and the delivery of leisure services were left to the states by the Tenth Amendment, which states that, "The powers not delegated to the United States by the Constitution, nor prohibited by it to the States, are reserved to the States respectively, or to the people." This amendment means that states are free to develop laws relating to the administration of recreation services as they feel desirable. And states have accepted the challenge to provide the framework of such services.

Nonetheless, although states have the authority to make provision for state and local recreation services, mention does need to be made of the federal government's role in providing enrichment to the recreation scene.

### The Federal Government

In determining the roles and responsibilities of various levels of government, one should consider the extent of the benefit of the services to the locale and its people. For example, the local government does not invest in facilities and programs beyond its legal area of responsibility. On the other hand, the greater the provision the state and national government can make for recreation services in the surrounding areas for the benefits of all the people regardless of political boundaries, the greater its services should be.

The federal government is involved in three types of recreation services: (1) the operation of areas and facilities for use by the people, (2) the giving of research and other data, and (3) the allocation of funds to state, county, and local agencies. Those agencies that are most involved in recreation can be briefly described as follows:

**The Corps of Engineers.** The Corps of Engineers of the Department of the Army has the primary responsibility of improving and maintaining our rivers and other waterways for the purposes of flood control and navigation. In doing this, it has provided opportunities for fishing, boating, swimming, hiking, camping, and the like.

**National Park Service.** The National Park Service of the Department of the Interior was established in 1916 to administer the national parks, monuments, historical sites, and other areas that would comprise the national park system. Its purpose is to protect and make available to the public the country's outstanding scenic, scientific, and historic areas. The service administers over 175 such areas, totaling more than 27,820,121 acres.

**Children's Bureau.** The Children's Bureau of the Department of Health, Education, and Welfare was established in 1912 to "investigate and report on all matters pertaining to the welfare of children and child life among all classes of people." Under this mandate the bureau has made innumerable studies on aspects relating to the recreation needs of children and youths.

**Sports Fisheries and Wildlife.** This service, under the Department of the Interior, protects the fish and wildlife resources of the nation, through enforcement of federal game laws, and provides use of the facilities for recreation when this is not inconsistent with their primary purpose.

**Forest Service.** The Forest Service of the Department of Agriculture was established in 1897 to protect and regulate the use of the federal forest preserves. Its operational policy has always been to extend the use of such lands for recreation. The Forest Service appropriates money for recreation uses of areas that include individual and organizational camping, picnicking, aquatics, winter sports, fishing, and hunting.

**Extension Service.** The Extension Service of the Department of Agriculture was established in 1914 to give instruction and practical demonstrations in agriculture and home economics, through field demonstrations, publications, and the like, to persons not attending land-grant agricultural colleges. This service provides a means of stimulating and improving recreation in the rural areas and small communities.

**Office of Education.** The Office of Education of the Department of Health, Education, and Welfare was established in 1867 to further the cause of education in the country. However, in doing this it is concerned with the recreational aspects of education and hence disseminates much information on recreation. It also uses its specialists to improve programs and raise the standard of leadership. It is important to note that education has been raised to cabinet level.

**Bureau of Reclamation.** This bureau, under the Department of the Interior, has given special consideration to the recreational use of its reservoir projects. It has not only allowed fishing, boating, swimming, hiking, and the like on its reservoir areas, but it has also allowed the construction of limited recreation facilities such as cabins and concession facilities.

**Bureau of Land Management.** This bureau, under the Department of the Interior, administers over 470,000,000 acres of public lands. Such public lands are open for all types of recreation uses.

**Heritage Conservation and Recreation Service (HCRS).** This service was established by the Secretary of the Interior on January 25, 1978 following a six-month study by a citizen-government task force.[8] This service is designed to coordinate a broad range of conservation and recreation activities by citizens in the public and private sectors. The service forms the nucleus of the National Heritage program, a process of identifying, evaluating, and protecting our cultural and natural resources and serves as the focal point for assuring adequate recreation opportunities. To fulfill its responsibilities in each of these areas, HCRS incorporates the functions of the Office of Archeology and Historic Preservation (OAHP) and the National Natural Landmarks program, both formerly part of the National Park Service, and most of the functions of the former Bureau of Outdoor Recreation.

The new agency's recreation office and OAHP both operate nonland managing programs that emphasize technical and grant assistance and work through state offices. Legislation proposed for the National Heritage program would establish a system for the conservation of natural areas compatible with the existing cultural resource preservation activities of OAHP. This system would include the National Natural Landmarks program. By coordinating the planning, resource identification, grants administration, and state liasion as-

[8] *Heritage Conservation and Recreation Service*, U.S. Department of the Interior, Washington, D.C.

pects of these programs, HCRS can best lead a national effort to safeguard our environment and enhance the quality of life in this country.

The following recreation responsibilities fall under the auspices of HCRS.

1. Development of a Nationwide Outdoor Recreation plan to guide federal, state, local, and private organizations in identifying and meeting recreation needs.
2. Administration of the Land and Water Conservation fund for acquiring land for federally administered recreation areas and for matching grants for states for the planning, acquisition, and development of recreation areas and facilities.
3. Identification of rivers and trails to be studied as potential components of the National Wild and Scenic Rivers system and the National Trails system.
4. Responsibility for the recreational, historic, archeological, and natural science aspects of regional or river basin planning.
5. Evaluation of outdoor recreation and environmental values of specific areas.
6. Transfer of surplus federal property to state and local governments for public parks and recreation use.

Other federal agencies including the Department of Defense, the Bureau of Indian Affairs, the Public Housing Administration, the Bureau of Public Roads, the Public Health Service, the Soil Conservation Service, the Tennessee Valley Authority, and The National Capitol Park and Planning Commission have also in one way or another contributed their resources for the use of recreation.

## The State Governments

The state government is playing an ever increasing role in recreation, but only in recent years has this been true. Before World War II the major concern of this level of government for recreation was in the development of a number of park systems. Today every state is expanding its recreation resources and exploring its responsibility to this basic social service.

At the present time the organization of state services in recreation may be summarized as follows:

1. An independent government agency giving its full time and attention to the advancement of recreation, so defined by law, to render services that the people in a state cannot efficiently do for themselves in the absence of a common authority.
2. A government agency rendering services on a statewide basis of a specialized nature, such as the operation of a state parks system,

control of state forests and fish and game, recreation programs in state institutions, consultation services to the schools, extension services of state colleges, and research and training on the part of state-supported universities and colleges.

3. An interagency committee made up of representatives from the various state and federal agencies interested in public recreation services, designed to further the programs of the agencies, promote cooperation, check duplication, and study needs.[9]

The role of state governments in recreation is similar in many ways to that of the federal government. Not only is this level involved in operating recreation areas and facilities for the use of people, but it also provides the means, through its various agencies, to give advisory services to the political subdivisions of the state. Hence the traditional operation of state parks, forests, wildlife preserves, reservoir areas, beaches, camps, and a variety of other features is one major state recreation function. Likewise, the advisory services of state agencies that assist in planning and promoting the cause of recreation throughout the state by collecting and disseminating data, providing consultative services to political subdivisions, and assisting with problems of the operation of park and recreation programs and services are also a more recent addition to the functions of state governments.

It can be seen from these descriptions of service that one area is described as a responsibility of state park departments, the other, the responsibility of state recreation commissions. Actually, while these two services are provided in some states by separate state agencies, the work of these units has been amalgamated in a number of instances into a single department of parks and recreation. The pros and cons of this issue of unification will not be discussed here, except to point out that it is one design for a state department's operation.

A list of state recreation services should exclude neither the extension programs of state colleges that assist in meeting rural recreation needs nor the services of the state game and fish commissions, conservation and natural resources agencies, and departments of education. All contribute in one way or another to the recreation needs of people. Moreover, many other departments, commissions, and agencies at the state level are indirectly related in one way or another to recreation. Included here would be libraries, hospitals, youth commissions, planning bodies, fair boards, health departments, and the like.

In viewing the responsibilities of the state government for recreation, we should mention the state's role in guiding the organizational and administrative patterns of recreation operation at the local or community level, since it is

[9]Taken from statement made by and used with permission of Sterling S. Winans, former Director of Recreation, State of California, on August 16, 1979.

at the state level that the right to conduct a local recreation service originates. Here is where the basic policies and general organizational framework for community recreation begin. Consequently, while municipalities, counties, and recreation districts have the major responsibility for providing the recreation function, permission to conduct such services requires authorization by the respective state legislatures. Because of the importance of the legal foundation of local recreation, this important phase of administration will be discussed more fully in the latter part of this chapter.

Finally, in looking at the role of recreation at the state level, we see that every state should review its operating practices and procedures and should unify its efforts in this expanding area of social need. In 1974 the governor of California adopted the following policy statement for recreation that could be used as a guide for private and public actions for the years ahead.

1. The resources of the state will be employed to stimulate the active, progressive, and coordinated participation of appropriate federal and local government agencies and of the private sector in providing areas, facilities, and services to meet present and future recreation needs and deficiencies. The state will cooperate in identifying deficiencies and will assist in alleviating those deficiencies according to a system of priorities.
2. Recreational use of lands currently in public ownership will be encouraged. The people of California, acting through their elected representatives, will seek use of suitable lands currently held by all government agencies. Highest priority will be given to prime access routes to beach and coastal lands near urban areas.
3. Local government entities most closely related to the recreation resources and to the sources of recreation demand will be encouraged to provide recreational opportunities.
4. The private sector will be encouraged to develop and operate appropriate recreation resources and recreational opportunities on both public and private lands, while giving full consideration to the quality of the environment.
5. All state public development and public works programs will be conducted in such a way as to preserve and whenever possible to enhance the environmental quality of California for its citizens.
6. The state will encourage at all levels of government and within the private sector the use of natural, historic, and archeological resources for outdoor education interpretation so that the citizens of this state may be able more adequately to enjoy, appreciate, and understand the state's ecology.
7. The state recognizes the Pacific Ocean and its estuaries as a resource that has not been thoroughly understood. Its many benefits have

therefore not been fully realized, including its potential for underwater recreation opportunities. Working with the private sector, the state will encourage preservation, enhancement, and development of these important coastal and estuarine areas.[10]

This plan was based on the premise that "the preservation, enhancement, and enjoyment of our environment continue to be of major concern to all. The search for an improved quality of life includes increasing the opportunities for various forms of recreational use of mountains, shorelines, deserts, and valleys and the enjoyment of open space, wilderness areas, and unspoiled waters."[11]

### Local Government

Recreation touches the lives of most people at the local level of government. Here individuals interact in a close relationship with their government officials and families feel the need for day-to-day recreation programs.

The major types of local government units involved in providing recreation service are the city and county. In some parts of the country, the borough, township, and village may have varying degrees of responsibility for this function. Also, the special district is frequently used as a unit of government to provide a recreation service. The use of this governmental pattern, be it a recreation, park, or school district, will be probed more thoroughly in Chapter 5.

Because the major responsibility for year-round recreation service involving people in their day-to-day-living environment is borne by municipalities or other units of local government, nearly all aspects of organization and administration as presented in this book are focused primarily at this level of operation.

Municipalities are responsible for the provision of recreation opportunities within their boundaries. The programs provided are extremely varied in scope, as are the type of recreation areas and facilities. Actually, it is at this level that many activities are provided under the supervision of professionally trained leaders, and the number and variety of organized programs is the feature that distinguishes this service from those of the county, state, and federal levels of recreation operation. As recreation moves along continuously from the neighborhood to the community, city, district, county, state, and federal levels, it can be clearly seen that its role varies in scope and purpose. At the municipal level, programs are focused on the needs of people living in an urban environment. At the county and regional level, there is greater involve-

[10]*Recreation Policy of the State of California*, California Outdoor Recreation Resources Plan, State Printing Office, Sacramento, Calif., 1974, p. 20.

[11]Ibid.

ment in the use of areas and facilities and less in highly organized programs. A thorough discussion of the role of the county in recreation will be given in Chapter 10.

### Roles and Responsibilities

Clearly, the recreation and park problem at the various levels of government is compounded by exploding populations, expanding leisure, increased mobility, higher standards of living, early retirement, and greater opportunities to learn, appreciate, and use the services and facilities that are dedicated to leisure. Furthermore, the competition for recreational lands and the confusion as to governmental responsibilities at a time when the recreation demand is increasing faster than the population accentuate a need that can only be met through coordinated recreation and park planning.

This problem is not one that is only faced by each state. Indeed, it reaches out to every state and other political subdivision in the nation. One of the solutions to the problem may be the establishment of spheres of government responsibility for various phases of recreation and park needs. This unique proposal was suggested and graphically presented in a 1962 statement of Recreation Policy of the State of California and illustrated in Figs. 1-1 and 1-2,[12] and is still apropos today.

## LEGAL FOUNDATIONS OF RECREATION

It is necessary to have legal authority for a public recreation system to acquire, develop, and maintain recreation areas, to construct and operate buildings and facilities, and to carry on the duties necessary to develop a program of activities. Therefore a local recreation system can only function through authority granted to it by state law. In short, recreation agencies must operate within a legal framework specified by state statute

The legal powers under which recreation and park systems may function vary considerably from state to state. However, all such legislation consists of four basic types: generally, general enabling acts, special laws, regulatory laws, and provisions for home rule.

### Recreation Enabling Acts

A recreation enabling act is a general state statute that authorizes the establishment of a local recreation service. It usually specifies (1) the authorization for

[12]*Recreation Policy of the State of California*, California Outdoor Recreation Resources Plan, State Printing Office, Sacramento, Calif., 1962, p. 14–15.

SOLUTION *to the problem*

ZONES OF RECREATION NEED INDICATE PRIMARY GOVERNMENT RESPONSIBILITY

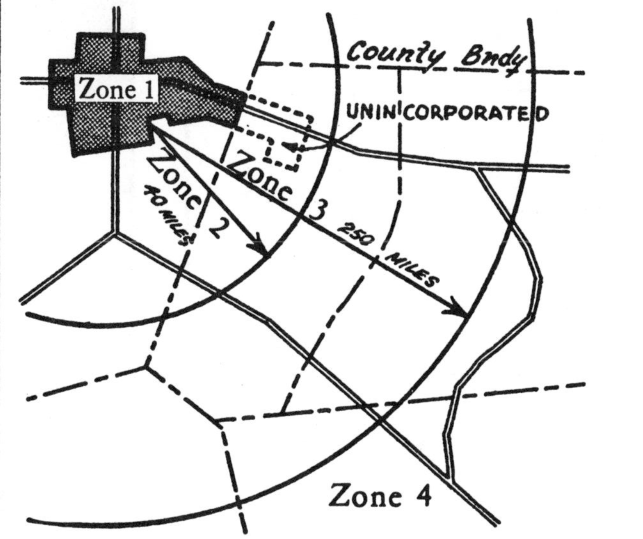


| RECREATIONAL NEEDS | PRIMARY RESPONSIBILITY |
|---|---|
| **ZONE 1**<br>COMMUNITY RECREATION<br>Incorporated<br>Unincorporated<br>When Needed | <br><br>City<br>County<br>District |
| **ZONE 2**<br>Regional Day Use | County<br>Multicounty |
| **ZONES 3 and 4**<br>Overnight, Weekend and Vacation | State and Federal |
| Leadership for Coordination | State |

***A solution to the problem can be achieved through co-ordinating agency responsibilities by zones-of-recreation-need, thus reducing overlaps and gaps in services.***

**Figure 1-1.** Zones of recreation need.

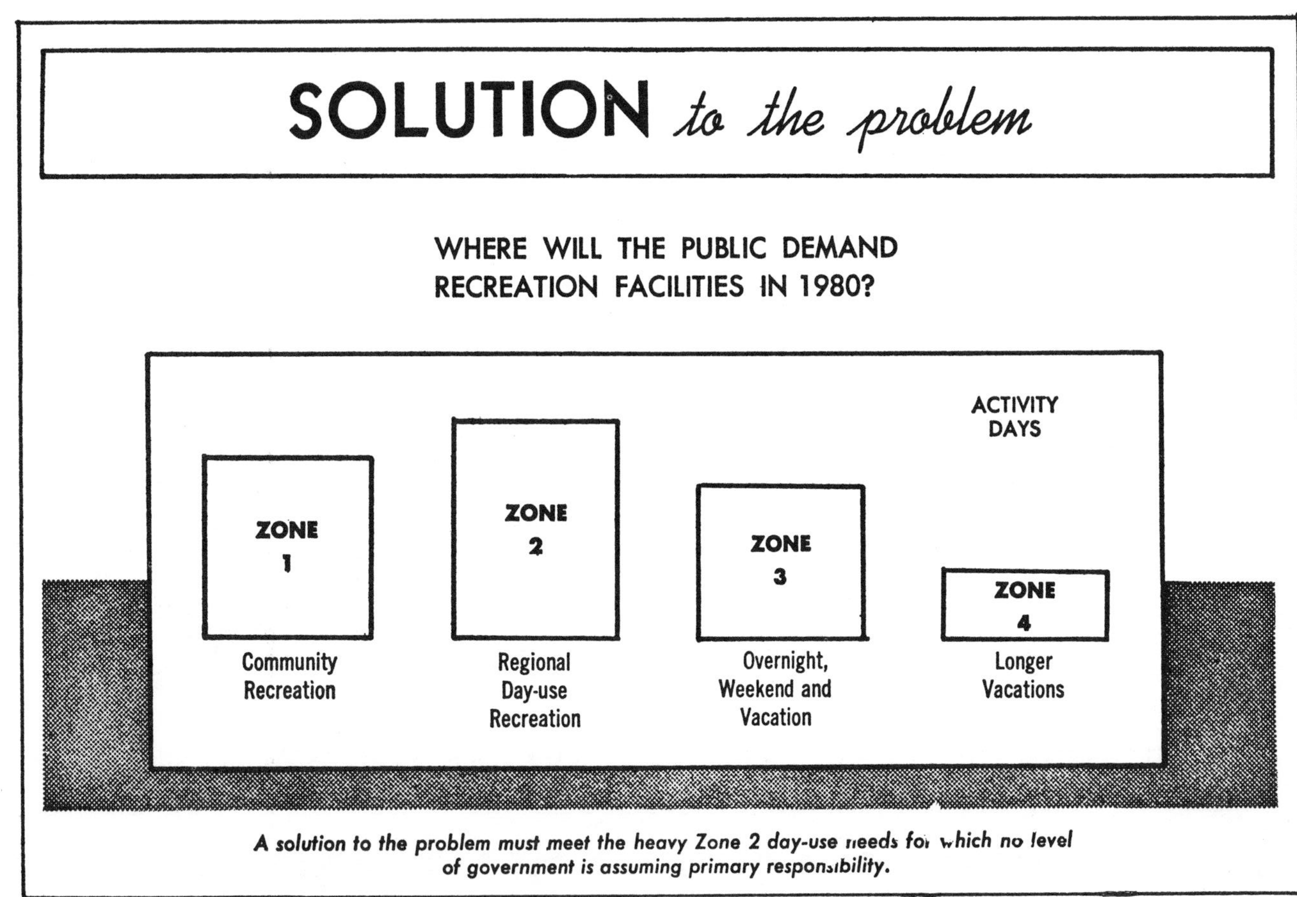


**Figure 1-2.** Demand for recreation facilities.

the local governing unit to exercise certain powers, (2) the provisions for allowing a local unit to establish a board or department to administer the powers given, (3) the authorization of particular powers that may be exercised, as well as the authorization of powers given under specified limits or under certain conditions, (4) the provision for joint exercise of powers by two or more public bodies to establish and operate a recreation program, and (5) the provision for financing the powers granted. In other words, it provides a blanket approval, as well as a framework, for the establishment of local public recreation programs and therefore eliminates the need for a special state statute to be passed each time a local unit of government wishes to establish a recreation service.

The Michigan Enabling Act, passed in 1917, is a good example of the authority granted to local government units to conduct a recreation service.

> An act authorizing cities, villages, counties, townships and school districts to operate systems of public recreation and playgrounds.
>
> *123.51 Public Recreation System*
>
> Section 1. Any city, village, county or township may operate a system of public recreation and playgrounds; acquire, equip, and maintain land, buildings or other recreational facilities; employ a superintendent of recreation and assistants; vote and expend funds for the operation of such a system.
>
> *123.52 Same; powers of school district*
>
> Section 2. Any school district may operate a system of public recreation and playgrounds, may vote a tax to provide funds for operating same, and may exercise all other powers enumerated in section 1.
>
> *123.53 Same; operation*
>
> Section 3. Any city, village, county, township or school district may operate such a system independently or they may cooperate in its conduct in any manner in which they may mutually agree; or they may delegate the operation of the system to a recreation board created by any or all of them, and appropriate money, voted for this purpose to such a board.
>
> *123.54 Same; location*
>
> Section 4. Any municipal corporation or board given charge of the recreation system is authorized to conduct its activities on (1) property under its custody and management; (2) other public property, under the custody of other municipal corporations or boards, with the consent of such corporations or boards; (3) private property, with the consent of the owners.[13]

[13]Compiled Laws of Michigan, Sections 123.51–123.54, Act 156, 1919, p. 284.

The general functions specified or implied in the Michigan act are (1) the development and organization of recreation activities and programs, (2) the acquisition, development, and maintenance of areas and facilities for recreation, and (3) the use of school, village, county, or other public properties for recreation purposes (with the consent of such bodies). These stipulations also make it clear that a recreation board could be established and would have the authority to carry out the recreation function.

The Kansas Enabling Act ( Appendix A, pages 375–379) is more specific than the Michigan act, in that it makes no mention of a county operation. In other words, it only provides for public recreation under the auspices of a city, a school district, or both. For a county to provide this service, another act is provided, authorizing the establishment of such programs.

Recreation enabling acts sometimes specify that the law is only applicable to cities of a certain size, in which case it is essentially class legislation. Others provide the optional features shown in the Michigan act that allow school districts, counties, townships, or villages, as well as cities, to operate a system of public recreation.

## Special Laws

Some state recreation laws are enacted for the purpose of giving a specific locality the power to provide recreation facilities or certain forms of recreation. For example, such statutes may give permission to a city to conduct Sunday band concerts, use state lands for a day camp, or provide for a municipal auditorium. In these instances it can be seen that special laws are passed for specific types of recreation that are considered unique to a particular locality and are not of concern to other communities within the state. In Massachusetts a special act gives the city of Boston the right to "provide for the closing to vehicular traffic, during periods of time, of any public way or any part thereof . . . except a state highway, or a parkway, or boulevard . . ." for the purpose of promoting recreation, play, or sport. New York City is another large city that closes its parks to traffic on weekends to allow cyclists and skaters to pursue their recreation in safety.

## Regulatory Laws

All states have enacted regulatory or controlling laws in the best interest of the public as it uses or participates in certain forms of recreation. Examples of these laws can be found in regulations governing sanitary conditions of municipal pools and camps, regulations concerning maximum speeds for motor boats, or regulations governing hunting and fishing.

| REG$ | OT$ | DT$ | T |
|---|---|---|---|
| $ 70.66 | $ 0.00 | $ 0.00 | $ 7 |
| $ 70.66 | $ 0.00 | $ 0.00 | $ 7 |

## Home Rule

Many states provide for home rule by allowing cities and county governments to adopt their own local charter.[14] In localities where charters are in effect, provisions for recreation may or may not appear. When they do not, a charter amendment may be instituted that gives the local governing unit the power to establish and maintain a recreation service. The two very excellent examples of such charter amendments included in Appendix A (392–398) are those for the cities of Phoenix, Arizona, and Detroit, Michigan, which are fairly representative of this kind of legislation.

## Local Legislation

Any local unit of government may establish a recreation program or service, if the legal authorization exists for it to do so. Although the privilege to perform a recreation function originates at the state level, it is important to emphasize that the state enabling acts are not mandatory in the sense that a local unit of government must establish a recreation system. Instead, they are permissive, and a locality may or may not wish to use the powers provided. In any event, the legal authorization is there to be used as needed.

If however, a locality does wish to establish a recreation program, it can do so under authority of the state enabling acts, special state laws, or provisions of local charters. The procedure involved is for the local governing body to pass an ordinance or resolution that sets forth in detail the organization, powers, and duties of the proposed recreation system. Of course, it must conform to the provisions of state law or local charters.

In many instances the managing authority for recreation is given to a board or commission. Here again, whether an ordinance provides for an administrative and policy-making authority or one that acts merely in an advisory capacity depends on the powers provided by state law or local charter. The state of California, for example, does not allow general-law cities or counties (as distinguished from charter cities and counties) to establish administrative or policy-making boards. The differences between these two types of boards will be probed more thoroughly in Chapter 6.

[14]A *charter* is a written instrument, duly executed as to form, that specifies the powers, duties, and responsibilities of a local governing unit to conduct its own business affairs for the benefit of the people governed. It is a grant from the state that, under the limitations imposed, general home-rule powers can be established by the people to be affected by the instrument. Usually, the state statute specifies the procedure to be followed in instituting this kind of rule. In general, it will specify that there will be an election of a charter board by the voters of the city or county—this board to prepare a charter that in turn is to be submitted to the voters for their approval. If approved, it becomes the law of the said city or county.

### Laws with Referendum Features

In a number of instances the state recreation enabling acts provide for a referendum feature, which means that the issue of whether a locality is to have a recreation service is referred to a vote of the people at a local election. Such an election not only determines whether the people want a recreation system but also usually specifies the tax levy that is to be imposed to finance the proposed program.

### Special Districts

A number of states have passed legislation authorizing the establishment of recreation districts or park and recreation districts. The general purposes of such districts as well as their advantages and disadvantages will be discussed fully in Chapter 5. However, an example of a state statute that permits people of an area to establish such a special taxing district is that for the state of Oregon, shown in Appendix A.

## HISTORICAL PERSPECTIVES IN ESTABLISHING A PARK AND RECREATION SYSTEM

In the development of public recreation, parks, and leisure services the administrative patterns for making provision of these services was being developed during the period of the 1920s–1940s. Public recreation services at the local and state levels were not developing at the same time, but the general outline of their struggle for existence was quite similar. Actually, it was during this period that the recreation and park movement was struggling for existence. However, a number of factors emerged to make possible an orderly evolution that led to the administrative structure as we know it today.

During the 1920s, and even before, park commissions and boards were becoming well established in American communities. In most instances they were independent of the city government's administrative structure for a number of reasons, which were also the case in the establishment of an administrative structure for recreation services. Both were struggling for recognition.

On the other hand, one must consider the political and social milieu of the times and must realize that prior to 1900 the strong mayor–council form of government was the primary form of administrative structure for most American cities. It was during this period that many of these strong mayor–council forms of government were highly political and the "spoils" or patronage system was commonplace. Political involvement meant jobs, and those interested in the park movement, and later the recreation movement, felt very strongly that these human services should be non-political and that jobs should be provided on the basis of merit. Therefore state park and/or recreation enabling laws were passed to make possible the establishment of policy-making boards or commissions in which the funding source was identified through a special millage or other sources. Thus these bodies were given the freedom to operate in a climate that was free from political control and apart from city government. As the recreation movement evolved, the same concerns were felt that led the early leaders of this movement to use the park organization as a model.

Shortly after the turn of the century the commission form of city government evolved and was promoted as the savior for city government. However, this form of government also had built-in weaknesses, which gave little hope for a move to encompass recreation and/or parks as a unit in the family of city services. Needless to say, some cities of both types did incorporate recreation and/or parks in their family of services, but this was not the rule. One therefore needs to envision a city governing body with two independent appendages, one a park board and the other a recreation commission— both of which were policy-making bodies.

The National Recreation Association actively promoted the establishment of independent recreation commissions for the preceding reasons and both bodies—parks and recreation—had none or a slight concept of merging, inasmuch as each had its own philosophical identity. Such a merger was to come later.

During this time little attention was given by local governing boards to embrace either human or social services, and "recreation" was still suspect. Some thought of it as a "frill," unimportant and unacceptable as a needed community service. Therefore the early leaders of the movement battled for support, for recognition, for elimination of political patronage, and for the establishment of independent boards or commissions.

The social changes that were evolving during the 1930s and 1940s gave credence to the need for recreation services, and as they gained general acceptance, the council–manager form of city government was also growing in popularity, until a point was reached where it became the most popular form of city government for the smaller, middle, and even large cities. Only the extremely large metropolitan cities continued to operate under the strong mayor–council form of government because of their political traditions.

The council–manager form of city government had great influence on park and recreation organizations, because it was established for efficiency and founded on nonpartisanship. The political emphasis was gone, and merit, professionalism, and accountability were the bases for its organization and operation. Proponents of this form of government did not believe in separate and independent commissions and boards with policy-making authority. These early city administrators believed that the arguments for having such bodies were now nonexistent, and as this form of government gained momentum, their charters showed parks and/or recreation as a department within the structure of city government with advisory boards and commissions established to assist the administration.

As a result, independent policy-making park and recreation boards and commissions were losing ground or were eliminated, except in some cities where a long tradition of such organizational structures were entrenched. City managers were now looking for professionally trained supervisors and leaders. It can be said that recreation and park services finally became acceptable and respectable, and the struggle for a funding source in large measure became more secure. While the era of policy-making park and recreation boards or commissions was in decline and were being replaced by advisory bodies, there was still much apprehension on the part of recreation professionals that they were still vulnerable without an independent source of funds and final decision-making authority. However, services of all kinds were being recognized and needed at the municipal level of government, thus allaying many of the fears of the professionals in the field.

The cities faced problems of acceptability and funding for recreation, as did the counties. Recreation and parks within this unit of government also gained respectability, as did special recreation taxing districts. Special taxing districts having their own independent policy-making boards will be discussed in Chapter 5, and the county will be discussed in Chapter 10.

As the spoils or patronage system was beginning to disappear in American cities, except in the very large metropolitan areas, city managers, county administrators, and state governmental officials were looking toward professionally trained recreation and/or park administrators to provide this service and to be held accountable for it. In turn, this provided the colleges and universities with the impetus to develop the necessary curricula to train such professional leadership.

It should be noted, however, that it was not easy for the recreation and park leadership to give up their independent status in which they were only responsible to their own policy-making boards or commissions. Many feared that losing their earmarked recreation funding source would place their organizations in jeopardy during periods of recession or budget cutbacks. But this did not prove to be true when the services provided were meeting the needs of the community and were of such quality that they could compete with other governmental services.

## THE CHALLENGE OF ACCOUNTABILITY

Recreation, park, and leisure service organizations should be held accountable and able to validate their programs of service during periods of financial exigency. If they cannot, perhaps cutbacks are necessary and a new look given to the needs of people in our present and changing society. It should be the goal of every professional in the field to take a position in favor of and work:[15]

1. For humane, beautiful, livable cities and countryside and for conservation, land-use, and environment planning and coordination of policy aimed to organize and make available the gifts of nature to satisfy the needs of a growing population.
2. For the recovery of leisure, the cultivation of the unharried, the leisurely, the voluntary, the spontaneous, for satisfaction, fulfillment, creativity, and genuine recreation.
3. For a healthy and creative way of life, including the cultivation of the joys and exercise of walking, hiking, working physically, and participation in the cultural arts.
4. For community and neighborhood social and civic life, valuing diversity as the richness of the metropolis, and strengthening voluntary association as an American genius and need.
5. For greater opportunities for older youth: for work, for greater place in the adult community activities, and especially for inclusion to replace alienation in the more deprived areas.
6. For equality of recreational opportunities for those who have the least opportunities.

It is for these goals that we must excite public imagination and be held accountable, because recreation and parks will be a mirror of our beliefs and values as a people. Therefore each of us should dedicate his or her powers and add his or her individual will to the undertaking. Marshall Lyantrey, when he was in Africa, asked his gardener to get him a certain tree. The gardener said, "this tree will not reach maturity for two hundred years." The marshall replied, "In that case, there's no time to lose. Plant it this afternoon."[16]

## Questions for Discussion

1. Define the word "recreation." Why do you feel the word has such different meanings to different people? Or does it?
2. Why must "community recreation" be goal oriented? Or should it?
3. Do you feel the word "recreation" is restricted in its use? Does it include the use and development of parks, camps, seashore areas, and outdoor wilderness resources? Why or why not?
4. Why must public recreation and parks have a legal foundation?

[15]Statement made and used by permission of Roy Sorenson, February 11, 1962.

[16]Hutchins, Robert M., "*The Nature of Human Life*" *Bulletin*, Center for the Study of Democratic Institutions (March 1961): 2.

5. What is meant by a state recreation enabling act? How is it different from an ordinance that establishes a local park and recreation service? Explain this relationship.
6. Do you believe a park and recreation enabling act should provide for a referendum feature? Explain.
7. Explain the role of recreation at the city level. The county level. The state level. The federal level.
8. What influence does the federal government have on state and local park and recreation operations?
9. Does a state have an influence at the local level of recreation operation? If so, in what way?
10. Why is there need for a state policy on recreation? What will it accomplish?

## SELECTED REFERENCES

Butler, George D., *Introduction to Community Recreation*. McGraw-Hill, New York, 1967, Chap. 1.

Carlson, Reynold E.; Deppe, Theodore R.; MacLean, Jan, *Recreation in American Life*. Wadsworth, Belmont, Calif., 1972, Chap. 1.

Edginton, Christopher, and Williams, John G., *Productive Management of Leisure Service Organizations: A Behavorial Approach*. Wiley, New York: 1978, Chap. 1.

Hjelke, George, and Shivers, Jay S., *Public Administration of Recreation Services*. Lea and Febiger, Philadelphia, 1972. Chap. 1.

Kraus, Richard G., *Recreation Today*. Goodyear, Santa Monica, Calif., 1977, Chaps. 1, 2.

Kraus, Richard G., *Recreation and Leisure in Modern Society*. Prentice-Hall, Englewood Cliffs, N.J., 1971, Chaps. 1, 2.

Kraus, Richard G, and Curtis, Joseph E., *Creative Administration in Recreation and Parks*. Mosby, St. Louis, Mo., 1977., Chap. 1.

Sessoms, H. Douglas; Mayer, Harold D.; and Brightbill, Charles K., *Leisure Services*. Prentice-Hall, Englewood Cliffs, N.J., 1975. Chap. 1.

# CHAPTER 2

# Administration: Basic Concepts

The administration of recreation and park services is concerned with those relationships and processes that make possible the most efficient departmental operation in accomplishing organizational objectives. In essence, then, administration is the process that mobilizes an organization's resources, human and material, to attain predetermined goals. Newman states that "administration is the guidance, leadership, and control of the efforts of a group of individuals toward some common goal."[1] Pfiffner and Presthus state that it "is an activity or process mainly concerned with the means for carrying out prescribed ends."[2] Actually, all definitions imply that the purpose of administration is to get the job done in the most efficient manner. Therefore any discussion of recreation and parks administration must be concerned with this problem in mind: How can goals be reached with minimum effort and maximum efficiency?

Administration implies that goals and objectives must be predetermined, for only as energies are directed toward ends can any accomplishments be forthcoming. Hence the goals discussed in the previous chapter are those toward which recreation and park administration is focused, but since these

[1]William H. Newman, *Administrative Action* (Englewood Cliffs, N.J.: Prentice-Hall, 1950), p. 1.

[2]John M. Pfiffner and Robert V. Presthus, *Public Administration*, 4th ed. (New York: Ronald, 1960), p. 3.

goals can only be reached through programs, it is important to point out that the machinery of administration is synchronized for the development of programs and services.

## THE ADMINISTRATIVE PROCESS

Every recreation and park system needs executives with administrative skill.[3] This skill is related to the ability to develop a smoothly operating organization to get the job done. But how does an administrator achieve this? What techniques or processes are involved in molding an organization into a smoothly operating unit? The answers to these questions involve administrative action.

In examination of the process of administration it is clear that much more is involved in getting a job done than merely performing a myriad of substantive tasks. Indeed, basic processes or techniques seem to be common to all administrative endeavors. Hence identification of these management skills or techniques will provide clues to the successful achievement of objectives. Stated more succinctly, it is possible to abstract those elements that are consistent with, or similar to, all administrative effort.

One way of exploring the processes involved in administrative action is to analyze the problems that confront every administrator and to relate them to the process involved in solving them.

1. *Planning*. What are the objectives of the department? What program and activities should be provided to meet the objectives? What policies should be formulated? What should be the scope of operation?

The answers to these questions center around the need for planning. In other words, every administrator is concerned with the process of planning, or deciding what is to be done.

2. *Organizing*. How should the plans be carried out? How will the work be allocated or divided? What organizational units should be established to carry out basic functions? What relationships should exist between units of operation?

These questions relate to the problems involved in organizing the department, or identifying the various tasks within the organization. Hence the process of organizing is a common element in all administration.

3. *Staffing and Resourcing*. Who is to perform the many and varied tasks? What human and material resources are available? How are they to be allocated?

[3]The terms *executive* and *administrator* will be used interchangeably in this discussion.

These problems relate to the assembling and allocating of resources, human and material, to perform the tasks at hand. The questions are concerned with the processes of staffing and providing the needed resources to do the job.

4. *Directing*. Who is to oversee how the work is being carried out? How are orders to be issued to get the organization operating and carrying out its function? Who will direct general operations? How will this direction be given?

These questions are related to the process of directing, or the issuing of orders and instructions as to how the various tasks are to be performed.

5. *Coordinating*. How will the various units of work be fused together in a team effort? What means should be used to ensure that divisions of work are functioning in harmony and are synchronized in effort?

These questions are concerned with the process of coordinating, or unifying the various segments of operation into a smoothly functioning team effort. It attempts to synchronize or fuse all the complex elements of the organization into a functioning whole.

6. *Controlling*. How are the assigned tasks being carried out? Are they conforming to agreed-upon plans? Are they meeting standards?

These questions relate to the administrative process known as controlling. In other words, this process involves the controlling of the organization's many operations to see that objectives are being met and adjustments made when needed. Related to it is a seventh process—the evaluation of organization tasks in relation to predetermined goals.

7. *Evaluating*. Were the objectives accomplished? Could the services be improved?

These questions relate to an appraisal of the recreation organization.

In viewing the elements that constitute the responsibilities of administration, we see that much of what is considered a universal consensus of what is involved here relates to the solutions of the preceding questions. In essence, then, the work of administration consists of techniques or processes that can be classified as planning, organizing, staffing and resourcing, directing, coordinating, controlling, and evaluating.

Various writers have attempted to classify the nature of administration. All use similar descriptions. And while some modify their terms, the meanings are much the same. Shortly after the turn of the century Henri Fayol described common administrative functions as "planning, organization, command, coordination, and control."[4] Gulick and Urwick's monumental work on the nature of administration in the 1930s did much the same.[5] Their famous "posdcorb" was used as a memory device by many students in public administration during this period. Each letter of this term is merely a cue to a word that is descriptive of a function of the administrator: "planning," "organizing," "staffing," "directing," "coordinating," "reporting," and "budgeting." Newman uses the terms *planning, organizing, assembling resources, directing,* and *controlling.*[6] Sears lists these processes as *planning, organizing, directing, coordinating,* and *controlling.*[7]

Some writers have approached the function of administration in other ways. Pfiffner and Presthus, for example, list the tools of administration as *leadership, decision making, communication, planning,* and *research.*[8] On the other hand, Culbertson, Jacobson, and Reller classify these processes as *communicating, building morale, administering change,* and *decision making.*[9]

Tead, in his analysis of the administrative process, gives the responsibilities of administration as follows:

1. To define and set forth the purposes, aims, objectives, or ends of the organization.
2. To lay down the broad plan for the structuring of the organization.
3. To recruit and organize the executive staff as defined in the plan.
4. To provide a clear delegation and allocation of authority and responsibility.
5. To direct and oversee the general carrying forward of the activities as delegated.
6. To assure that a sufficient definition and standardization of all positions have taken place so that quantity and quality of performance are specifically established and are assuredly being maintained.
7. To make provisions for the necessary committees and conferences and for their conduct in order to achieve good coordination among major and lesser functional workers.
8. To assure stimulation and the necessary energizing of the entire personnel.

[4]Henri Fayol, "Administration Industrielle et Générale," in Constance Starrs, *General and Industrial Management* (London: Sir Isaac Petman, 1949).

[5]Luther Gulick and L. Urwick, eds., *Papers on the Science of Administration* (New York: Institute of Public Administration, 1937).

[6]William H. Newman, *Administrative Action* (Englewood Cliffs, N.J.: Prentice-Hall, 1950), p. 4.

[7]Jess B. Sears, *The Nature of the Administrative Process* (New York: McGraw-Hill, 1950), p. ix.

[8]John M. Pfiffner and Robert V. Presthus, op. cit., p. 6.

[9]Jack Culbertson, Paul Jacobson, and Theodore Reller, *Administrative Relationships* (Englewood Cliffs, N.J.: Prentice-Hall, 1960), chaps. 4–8.

9. To provide an accurate evaluation of the total outcome in relation to established purposes.
10. To look ahead and forecast as to the organization's aims as well as the ways and means toward realizing them, in order to keep both ends and means adjusted to all kinds of inside and outside influences and requirements.[10]

It is clear from these descriptions that all relate to the point that administrative processes are merely techniques or means used to effectively mobilize and direct men and materials toward the achievement of goals. And while these processes involve a motivation of personnel, we should stress the importance of leadership, for, indeed, skill in energizing a staff or group in carrying out the processes as outlined is basic to sound administrative action. Leadership is implied in the preceding directing process. Since leadership is not a specific process, it is not classified as such here. On the other hand, it is a catalyzing factor that acts as a stimulant to successful performance and the attainment of goals.

### Substance Versus Process

Some students of administration may argue that, after all, it is the substance, skill, or knowledge of an organization's operation that is basic to effective management of a recreation and park system. However, while skill and technical knowledge are of great value to an administrator, they alone are not enough to guarantee administrative success. Something additional is needed, and most authorities in administration agree that this extra ingredient is administrative skill related to the process of management. An examination of these processes may clarify this point.

## PLANNING

There is no question about the need for planning in providing recreation and park services. To plan is to look ahead, but, more important, it is a goal-oriented process—a means by which an administrator explores in advance a course of action to accomplish his or her mission. In short, to plan is charting the way to reach a goal. It can readily be seen that planning involves the deliberative examination of a myriad of ideas and data before decision making takes place. It is a continuous process, for actually, the work of a department is constantly changing, but it is essential in administrative action to look ahead and prepare a plan of action through study of known facts and deliberation before proceeding.

[10]Ordway Tead, *The Art of Administration* (New York: McGraw-Hill, 1951), p. 105.

Moreover, while planning is essentially looking ahead, assembling facts, preparing for contingencies, and evaluating decisions, it should not be overlooked that a recreation and park system is goal oriented. Identification of these goals and planning for their attainment are basic to the planning process. Indeed, not to establish objectives is like charting a course without a destination. Furthermore, once goals are established, all later management processes are expedited.

## ORGANIZING

Organizing is the process of dividing the function of a recreation and park system into related units of work, assigning responsibilities, establishing cooperative relationships, and determining lines of responsibility and authority. In short, organizing a department or agency is merely a means of grouping work into units of specialization and then defining and integrating them into patterns of work relationships.

### Departmentalization

The first major phase of the organizing process and one of the first tasks facing a recreation and park administrator after objectives have been determined is that of establishing a plan by which the work of the organization will be divided into its component parts. In other words, this is the task of classifying the various positions into groups on the basis of their similarity. How this will be done depends upon the kinds of services rendered, the areas of specialization, and the organizational philosophy of the executive. Actually, there is no hard and fast rule for departmentalization or for division of work into administrative units. However, such divisions or classifications should be as consistent as possible throughout, and, whenever possible, they should be mutually exclusive.

First, let us consider the general guides for grouping of similar tasks into a departmental organization. Most recreation and park systems divide their activities according to function or similarity of work. This is known as "grouping by functions." A function is merely a body of related duties or activities that are grouped together for purposes of execution. Known also as "organization by purpose," this approach concentrates on single phases of operation and makes possible specialization of effort. All activities of a related nature within the department are combined into general purpose groupings that may be called divisions, bureaus, sections, units, offices, or any other name.

While functions may vary in degree or significance, it is generally agreed that, within a recreation and park system, the following major functions can in general carry out the basis tasks of the organization.

1. *Program.* This function is related to the operation, or the work the organization is undertaking, namely, recreation. This function is generally a major division in the organization and may be known as "recreation," "operations," "activities," "program services," or some other identification.
2. *Construction and Maintenance.* This function is concerned with the planning, development, construction, and maintenance of recreation lands, buildings, and structures as used in the program of services by the department. Identified as a major division within the department because of its importance in making land and buildings usable for recreation, this function is usually identified as "parks," "planning and development," or "construction and maintenance."
3. *Special facilities.* This function may be a single unit or division in which all special services are classified under this category, or it may be divided into a number of functions. For example, under this grouping would be included camps, marinas, golf courses, aquatic facilities, stadiums, zoological parks, museums, forestry, and the like.
4. *Finance and accounting.* This function is concerned with the financial record keeping of the organization and the receiving and disbursement of moneys for operation and improvements. It may be identified as a division of administration, business and accounting, or business and finance. Common titles for the person heading up this division are office manager, administrative officer or assistant, administrative analyst, or business manager.
5. *Personnel.* This function is designed to increase the overall efficiency of the organization through the development of personnel policies, procedures, and operational practices. Concerned with recruiting, soliciting, classifying, training, and other personal relationships that assure the finest degree of development of the human resource in the organization, this function is vital to successful operation.

It can readily be seen from examination of the preceding functions that two important operations are not listed as separate functions. These are (1) administration and (2) public relations. A case could be made for adding administration as a function, since it encompasses a grouping of similar responsibilities related to the integrating and coordinating of all these operations into a smooth working relationship. However, in our view, this process relates to each of the functions and is not a single identity. Likewise, public relations could be elevated to functional status in large departments or organizations, since the gaining of acceptance of the services as provided is essential. Again, this operation is not singled out as a separate function, for the reason that the body of duties involved here is usually assigned to all members of the administrative staff, with final authority resting with the executive. Actually, this becomes an important task of the administrator and is one of the chores he or she has in planning and mapping the procedures for reaching departmental objectives. On the other hand, in large systems, the public relations operations may be elevated to functional status if their importance to the organization merits this.

It is important to point out that it is difficult to differentiate between functions. Many overlap with other categories, especially with that of process. Also, it is not uncommon to see functions listed as administration, supervision, leadership, and research when actually they are processes, for broad usage of the term is prevalent.

Since government services are usually grouped by purpose or function, only brief attention will be given to a discussion of other organizational patterns, except to clarify them. For example, tasks may also be grouped by process, area or place, clientele, or time. Gulick's division of tasks, given over 25 years ago, is still pertinent today.

> In building the organization from the bottom up we are confronted by the task of analyzing everything that has to be done and determining in what grouping it can be placed without violating the principle of homogeneity. This is not a simple matter, either practically or theoretically. It will be found that each worker in each position must be characterized by:
>
> **1.** The major purpose he is serving, such as furnishing water, controlling crime, or conducting education.
> **2.** The process he is using, such as engineering, medicine, carpentry, stenography, statistics, accounting.
> **3.** The persons or things dealt with or served, such as immigrants, veterans, Indians, forests, mines, parks, orphans, farmers, automobiles, or the poor.
> **4.** The places where he renders his service, such as Hawaii, Boston, Washington, the Dust Bowl, Alabama, or Central High School.[11]

By dividing the work of a department by "process" is meant identifying tasks according to how a thing is done rather than what is done. In other words, this is related to the process involved in performing the job, such as engineering, accounting, purchasing, stenography, or the like. Frequently, departmentalization by process overlaps that by function. Few recreation and park systems use a pure functional pattern; most use combinations. "Area" classification relates to the grouping of tasks on the basis of the district or

[11]Luther Gulick, "Notes on the Theory of Organization," in Luther Gulick and L. Urwick, *Papers on the Science of Administration* (New York: Institute of Public Administration, 1937), p. 15.

other area served. It is used in large cities where decentralization is necessary. By "clientele" grouping is meant the division of work on basis of the individuals served. For example, some agencies may divide their services into programs for children, youths, and adults. Not infrequently, physical education departments are separated into two units, in particular, men and women or boys and girls. Finally, by "time classification" is meant division of work tasks on the basis of the time when they are performed. Examples would be a division between day and evening services or in terms of other time shifts.

## Administrative Integration

The second phase of the process of organizing is that of integrating the work of the various divisions of a department into a functioning and coordinated unit focused toward common goals. This is accomplished through (1) the establishment of an integrated organizational structure, (2) the identification of a unity of command, (3) the development of an adequate span of control, (4) the establishment of compatible relationships in the delegation of authority and responsibility, and (5) the use of staff units for research and specialization.

**Integrated Structure.** Integration of the organizational structure refers to the ways by which an organization is pulled together or coordinated through a hierarchy of responsibility and authority. This is accomplished through what is known as the "scalar principle."[12] Also called the "hierarchical principle," this concept is based on the integration of functions through established lines of authority. In other words, this principle rests on the premise that a department will have levels of authority and responsibility that start with the top executive and end with individual workers performing the specialized tasks of the organization. Such a structure would resemble a pyramid with the administrator as its apex. Between the apex and the base would be the several units organized by function or purpose and headed by operatives having degrees of job responsibility. This hierarchical arrangement would be coordinated by lines of authority running from the top to the various operating levels. For example, in a recreation and park department, the line of authority over recreation programs would extend from the top executive to the superintendent of recreation (division), to the supervisor of community centers (bureau), to the recreation center director (unit), and finally to the recreation or play leader (worker). An example of such a hierarchical triangle is shown in Fig. 2–1.

**Unity of Command.** The number of operating levels within the hierarchy will depend on the size of the organization and the complexity of its services. The

[12]James D. Mooney, *Principles of Organization*, rev. ed. (New York: Harper & Row, 1947).

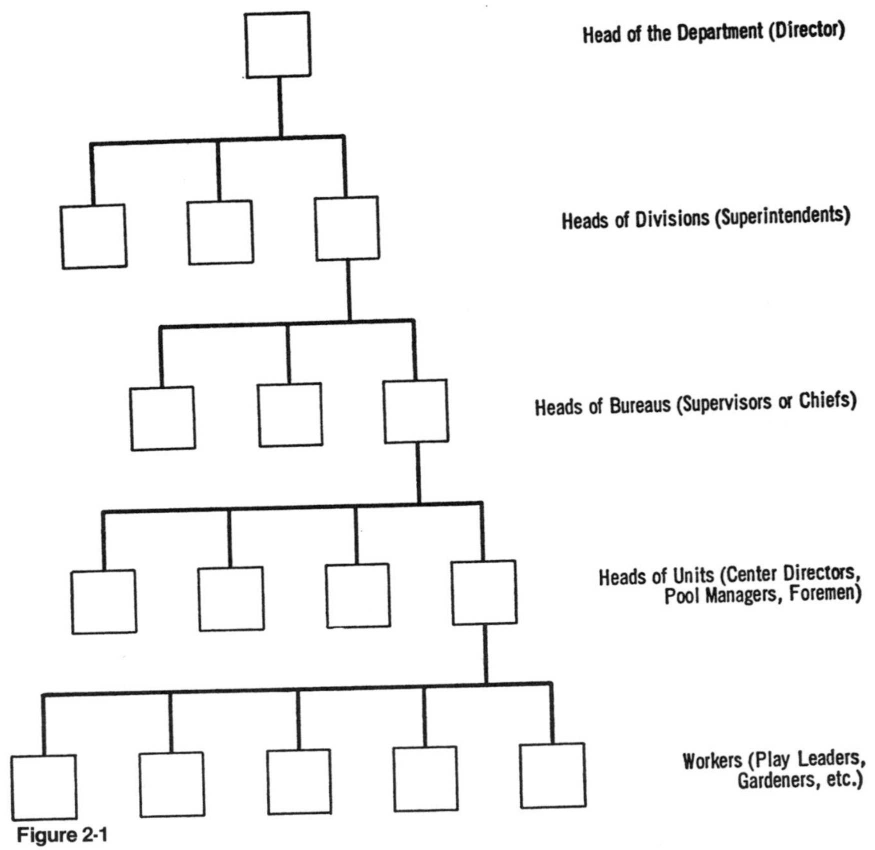


Figure 2-1

small department may have only an executive at the top of the hierarchy and recreation leaders and maintenance personnel at the base. In the larger and more complex organizations additional levels of authority emerge. The chief administrator, or, in this instance, the director of recreation and parks, exercises his or her authority over all other positions in the organization, through what is known as the "chain of command," which, in essence, means that responsibility is distributed through a pyramid of authority from those holding top administrative posts to those at the lower levels. Thus the occupant of every position has but a single superior to whom he or she is responsible. The principle of this unity of authority, in which every individual in the system has but a single "boss" is known as the "unity of command" principle. Enunciated by Luther Gulick and others, the principle recognizes that employees can only work effectively and efficiently if they have but a single superior. It is important to point out that nothing but confusion arises from work situations in which this principle is violated. Indeed, workers should not have more than one person from whom they receive orders and reports. Furthermore, the duties of their position and to whom they are responsible should be absolutely clear to them. Fuzzy relationships only lead to duplication of effort, disruption of work, and loss of control.

The "unity of command" principle works in both principle and practice, but some modification is needed when a director finds it necessary to make use of functional supervisors to assist in technical areas and in special program features. For example, a director of a community center may be responsible in the line of command to the superintendent of recreation. On the other hand, he or she may be subject to the orders of a supervisor of personnel and in-service training or a supervisor of athletics. It would appear that the unity-of-command principle would be violated in this instance, but actually, the director in this case is subject to different kinds of supervision. The superintendent is his or her "line" superior, while the supervisor is a specialist in a technical area of program who is seeking improvement in an area of specialization. When separate areas of authority are defined, and if it is clear that an employee is responsible to a single "line" superior if questions of conflict occur and that a functional supervisor only provides specialized direction in tasks that are focused on the improvement of service of the operating unit within the organization, this dual responsibility should not be an impossible situation for an employee, nor does it destroy the principle of the unity of command.

**Span of Control.** *Span of control* means the number of subordinates a person can direct or supervise in an effective manner. Every administrator has a limit on his or her time and energy, and to go beyond this makes additional supervisory work ineffectual. Hence to establish guidelines over the number of persons an executive can effectively manage is important to administrative action.

Much of the early writing on span of control implied that the number of persons an executive could supervise should be small, as few as five or six if their work is interrelated.[13] Others have tried to fix a figure at less than this number, while some have suggested 12 or more. Today, however, attention is not focused on a specific figure as much as on other factors or variables that influence the number of subordinates an administrator can effectively control. Such factors include:

1. *Diversity of services.* The need for supervision is greater when division of work is complex or diversified than when the work is repetitive and requires few decisions or personal conferences.
2. *Executive Skill.* The energy, adaptability, and skill of some executives vary from those of others, and hence the span of supervisory efforts of these people will vary.

[13]L. Urwick, op. cit., pp. 52–54.

3. *Skill of employees*. The time and effort required for untrained and unskilled employees is greater than for those who are more highly competent.
4. *Nonsupervisory Relationships*. The duties of the employee other than those related to a supervisory function will seriously cut down on time spent in supervision.
5. *Stability of Operation*.The coordination and control is greater in a highly dynamic work situation in which turnover of employees is high, hence limiting time for supervision.
6. *In-service Training*. The requirement of considerable on-the-job training in some units means less time can be spent on supervising others.
7. *Type of Activities*. The supervision of other executives or administrative activities requires different responsibilities from that of supervision of operating employees. Thus each process requires a different amount of time, and hence there is variance in span of control.

It can be seen from the preceding discussion that organizational theory on span of control has not centered on naming a fixed number that will apply in all circumstances. Instead, the number will vary with the factors presented. Nonetheless, it has been suggested that the span for executives should be three to eight persons, while an operative span could be as many as 30.[14] It is true that many supervisors in recreation and park systems are not closely supervised by the executive, although they do have access to him or her at all times. Actually, the delegation of responsibility to subordinates combined with an open-door policy gives the administrator more time for planning and permits a wider span of supervisory control. Certainly, however, there are many factors making for variance of spans in individual situations.

Span of control has a direct bearing on organizational structure and the number of supervisory levels. For example, an organization structured with a single level and a maximum span of control of nine could be visualized as shown in Fig. 2–2. In other words, *A,* as a symbol representing the head of a unit, would be supervising nine individuals. No administrative levels exist between this head and the operating staff. In short, this is the simplest organizational structure for an agency having ten employees (one executive and nine employees). In this instance each individual performing a single task or multiple tasks reports directly to the executive.

Using the same number of operating employees, an executive may decide to cut the span of control to three but, will need to add three additional super-

[14]R. C. Davis, *The Fundamentals of Top Management* (New York: Harper & Row, 1951), pp. 269–276.

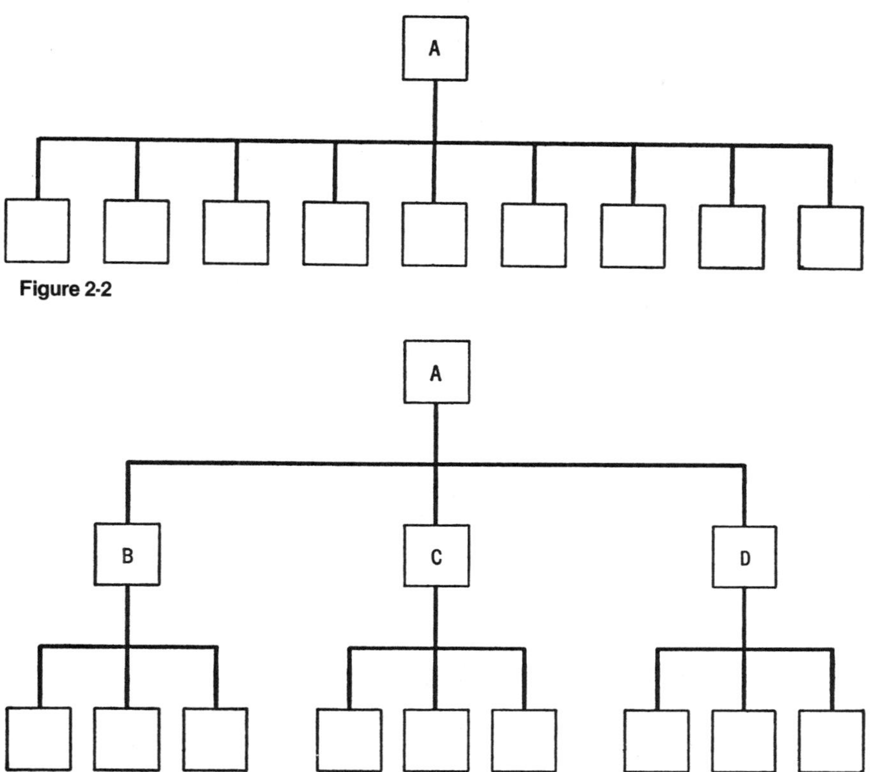


Figure 2-2

Figure 2-3

visors, *B, C,* and *D* to do this. Here we have two levels of supervision, as shown in Fig. 2–3.

If the executive desired a span of control of two, there would be need for three levels of supervision and six additional supervisors, *B, C, D, E, F,* and *G*. The organization would look like that shown in Fig. 2–4.

In viewing the foregoing structures, we can readily see how span of control affects the levels of supervision. One level makes for a "flat," or horizontal, structure, while the addition of a layering of supervisory levels makes for a "tall," or vertical, structure. This can further be illustrated by showing a horizontal, or "flat," structure with two levels of supervision and a span of control of eighteen (Fig. 2–5, top) and a tall or vertical structure showing four levels of supervision and a span of control of three (Fig. 2–5, bottom).

If there are a number of levels of administration, the expense of operation is obviously increased because of additional supervisory personnel. However, it is not implied here that this is undesirable. Actually, the complexity of an or-

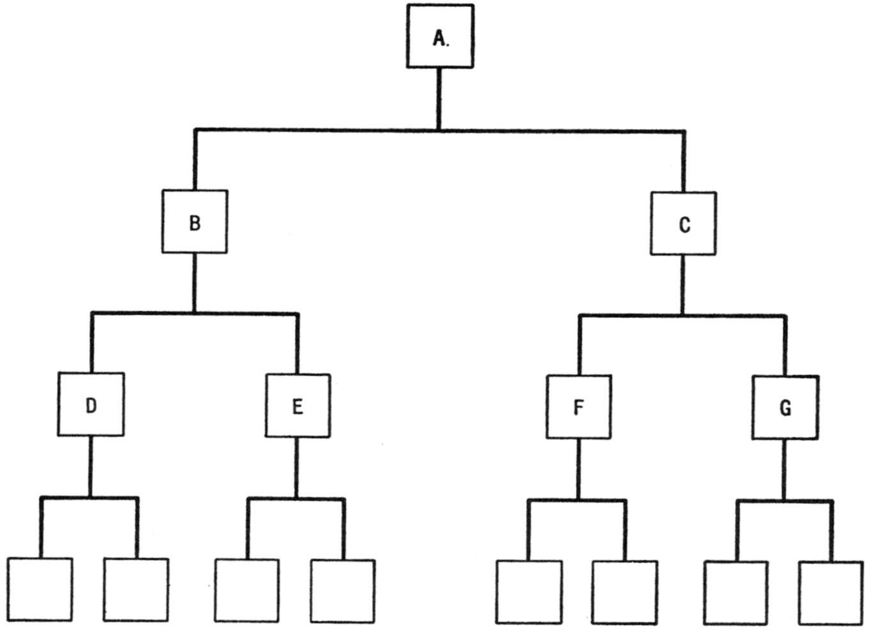


Figure 2-4

ganization and other factors already mentioned affect "layering," or the number of administrative levels needed for efficient operation. And every administrator should give considerable thought to this problem of organization. Some general considerations to bear in mind are:

1. *Effect on Communication.* The establishment of clear-cut lines of communication is more difficult whenever a number of levels exist between the top and the lower levels of the organization. Hence within the chain there is more opportunity for delay and the "garbling" of information and instructions.
2. *Effect on Delegation.* The horizontal, or "flat," organization forces delegation of responsibility and decision making on the part of subordinates. The executive or supervisor has less time to give to a greater number of unit heads and consequently, fewer controls exist. Staff members have more time to themselves and can be more independent in developing their own ideas. This can be good or bad, depending upon the employee. Those who wish to go their own way have greater opportunity to do so. On the other hand, a "tall," or vertical, organization encourages close supervision and frequent interaction. This can lead to an executive or unit head making minute

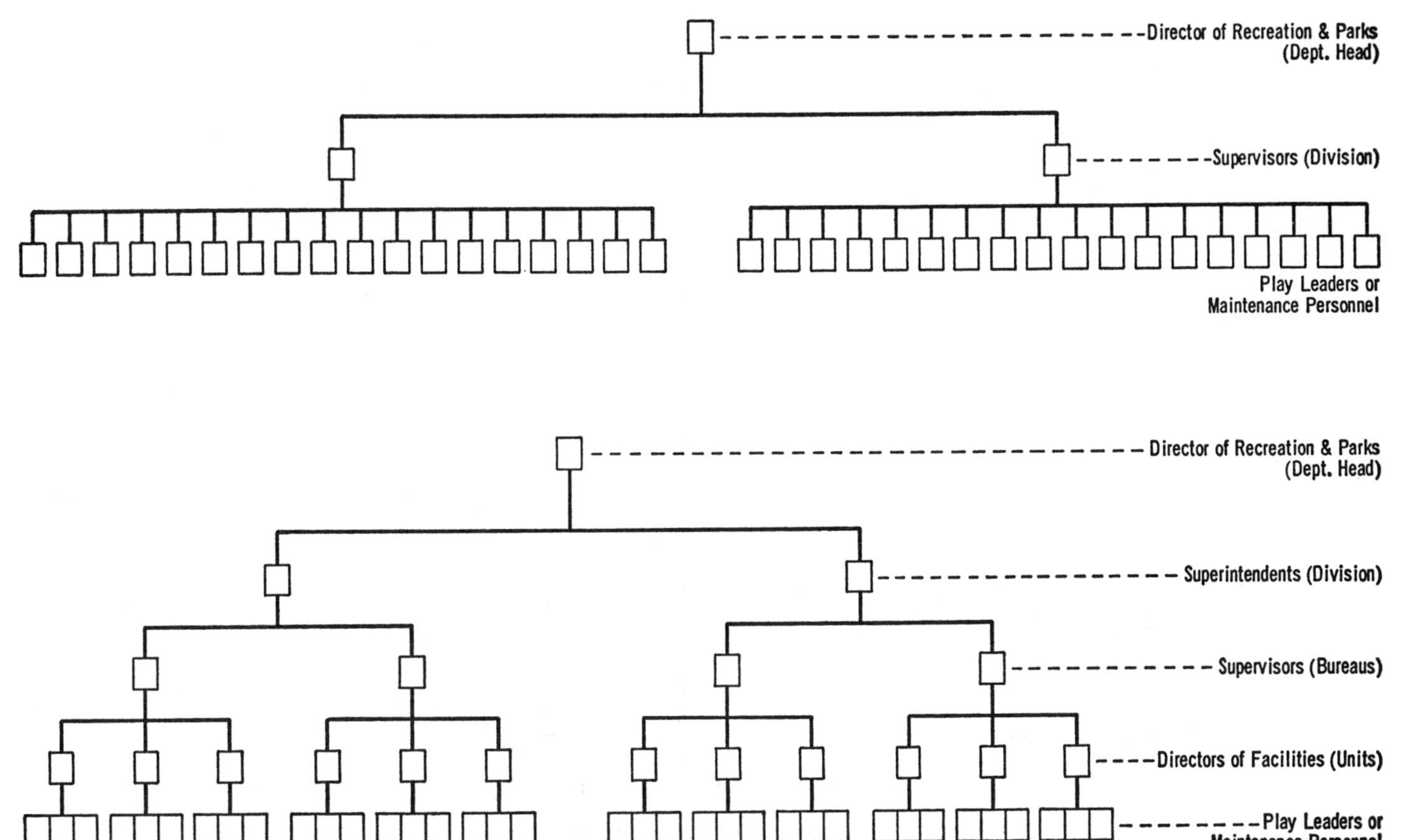
Director of Recreation & Parks
(Dept. Head)
Supervisors (Division)
Play Leaders or
Maintenance Personnel
Director of Recreation & Parks
(Dept. Head)
Superintendents (Division)
Supervisors (Bureaus)
Directors of Facilities (Units)
Play Leaders or
Maintenance Personnel

Figure 2-5

decisions and meddling in the work of his or her subordinates. Some employees having day by day interaction with their superiors either become oriented in a relationship in which the "boss" does all the decision making or become resentful of this domination. Not all superiors, however, dominate; some are democratic and encourage an atmosphere of cooperation and teamwork. But the environment for autocratic control and stifling of initiative and decision making is more apt to exist in the "tall" organization, where the span of control is relatively small.

3. *Effect on Control.* The "tall" organization, without doubt, provides for greater control of the work of an organization. Furthermore, a "tall" organization encourages the development of "staff" divisions or specialization of function. However, "staff" units do add to further controls and make for greater complexity of interpersonal relationships.

In sum, it can be seen that a balance must be maintained in the operational relationships of an organization and that decisions on whether to have a "tall" or "flat" structure affects this balance. Many supervisory levels and a small span of control create problems; a "flat" organization, on the other hand, gives less control and may also create problems by reducing efficiency. Perhaps the answer to this problem is related to the development of in-service training programs in administrative management. As greater emphasis is given to decision making, perhaps "flatter" organizational structures will develop.

**Authority and Responsibility.** In integration of the work efforts of an organization it is essential for a clear relationship to exist among all heads of divisional units in terms of their respective spheres of responsibility and authority. Nothing stifles initiative as much as being held responsible for a given task but not being given discretion and authority for its performance. Hence if work is divided into hierarchical units with basic authority resting at the top, unit heads in the chain must have delegated to them by their immediate superior their task responsibilities, along with a degree of authority in performing the tasks, although such delegation does not rid the superior who delegated the task from final responsibility for it. Therefore unit heads representing a level of operation in the hierarchical chain should be given an opportunity for decision making in performing the responsibilities vested in them. It can be seen that, in a large recreation and park system, a series of redelegations are made from top to bottom units, each of which has its area of responsibility and authority. In the smaller department there are fewer redelegation points. The justification of this authority pattern is obvious. No one wants to be held responsible for a job and then not given a sphere of discretion in discharging its responsibilities. Actually, nothing is as damaging to morale and the self-respect of a person heading a divisional unit than to be given a task to perform and then to have his or her superior make all the decisions as to how it is to be accomplished. Still, this fundamental principle is constantly being violated by executives who are fearful of losing their decision-making function in matters of trivia and detail that should be left far down the hierarchical ladder. In fact, the fear of delegation is one of the most common executive faults. To bring outstanding staff into an organization and then handcuff their abilities by not letting them perform their jobs with a degree of discretion is the very epitome of poor administration. There is no implication here that an executive gives away his or her final responsibility. Indeed, no executive can do this through delegation. However, he or she can assign tasks and give a sphere of authority in having them accomplished. Recreation and park executives making decisions on every issue of minutia that is passed on from the lower divisions because of the fear of subordinates making decisions that may be the "boss's" prerogative provide one of the sad commentaries on contemporary administration. Yet there will be no change, unless a pattern of delegation is instituted. In some medium to large departments there are no redelegation points. All decision making rests with the executive, be it decisions on cleaning solvents for the janitors or the planning of large recreational structures. Surely, here is a symptom that calls for action.

Another aspect of administration poses a problem of some magnitude, in discussing this topic—that is, placing workers in the hierarchial structure at a level commensurate with their responsibilities. Titles confuse this issue, whether a position of "line" or "staff." Does the assistant director of parks and recreation have a higher rank and receive more pay than the superintendent of parks? Does a functional supervisor of a special acivity such as arts and crafts have a higher rank than a general supervisor of recreation and/or parks, or vice versa?

More and more, a move is taking place to provide titles that carry status even if a change in title provides no additional responsibility or authority. It can almost be looked on as a move toward giving an employee "status pay." Furthermore, pressure is often put on a manager to raise the level of an operating unit from a subordinate level to a coordinate one. In other words, raise a "bureau" to a "division" level, which is one step higher in the hierarchy. It is natural for a supervisor or unit head to be as close as possible to the executive at the top of the organizational hierarchy. This implies more authority, rank, and prestige. And it should be done if the facts show the units have equal responsibility and authority, even though of a different kind.

A way out of this dilemma is to identify clearly regardless of title, duties, functions, and responsibilities as well as the authority that rests with each of the positions in question and to establish a standard that shows their place in the organization hierarchy. New titles should be encouraged; it improves morale and provides a level to which employees may aspire. Raising units to higher levels in the hierarchy needs to rest on their own merits and not to be based on superficial judgments or prestige factors.

**Staff Versus Line.** In developing an integral organizational structure, one should make a clear-cut distinction between staff and line activities. A "line" activity relates to the basic purpose or function of the organization; it is program oriented. Those working in a line activity are engaged in those general-purpose programs or services that directly meet the recreational needs of the public. On the other hand, those activities that are purely advisory in serving the general-purpose units of the organization are known as "staff" functions. Therefore as used here, "staff" functions are those duties and responsibilities that supplement the primary purpose of the recreation and park system. Staff workers do not operate program units. Instead, they perform general, technical, and auxiliary services essential to the carrying out of program functions by line workers[15] and are as follows.

1. *"General Staff."* General staff activities means the work of a general nature that an administrator will usually delegate to an assistant. In fact, the person holding a position as assistant to the director, administrative assistant, or administrative analyst will usually do whatever is needed by the executive. It may consist of planning or be a mass of detail and routine functions. However, whatever is done is performed under the authority of the executive.
2. *"Specialist Staff."* Staff activities usually refers to the category of specialist staff, in which the work is of a technical or specialized nature. Much work in an organization is of this kind. Individuals with specialized and technical skills are needed for effective decision making. They give advice, counseling, and consultation to line officials and make studies, investigations, and reports that add to the over-all efficiency of the organization. Their know-how in specialized areas makes them authorities who command respect. And although they lack a command authority, they do provide the answers in their area of specialization. Planning, personnel, finance, and legal services call for specialized knowledge and skill. Likewise, program areas call for specialization. Individuals in music, dance, drama, arts and crafts, and the like may have functional authority only. In otherwords, they may only assist the line officials, be they recreation center directors, recreation leaders, or the superintend-

[15]An excellent discussion of line and staff activities is given in John M. Pfiffner and Robert V. Presthus, op. cit., pp. 243–252.

ent. Strictly speaking, staff specialists only operate through line officials. If they perform their service directly for the public, they actually are performing a line function.

3. *"Auxiliary Staff."* Auxiliary staff activities are of a general nature, are common to all units within the organization, and exist in every recreation and park system. In fact, these activities are related to the process involved in performing them. Sometimes called housekeeping activities, they include stenographic processes, purchasing, central mailing, equipment maintenance, transportation control, record keeping, and maintenance of buildings, communication facilities, and the like. While grouping of such functions into a single unit of specialization takes place only in those organizations that have enough personnel to warrant it, one can see that it is economical to group together those work processes that cut through all units. For example, some recreation and park systems combine printing, mimeographing, mailing, and distribution of office supplies into one unit. Others combine automotive and equipment maintenance into one group. Obviously, such pooling should only be established if it furthers the efficiency of the operating units. Some authorities include planning, personnel, finance, legal services, and the like under auxiliary services. However, since these are areas of specialization, they are included in the specialist category discussed in (2).

**Staff-Line Relationships.** It should be stressed again that staff work is used to increase the overall efficiency of the organization in providing the best service possible, and such efficiency can only be achieved if a clear authority relationship is established between line operations and staff work. Line and staff can only work cooperatively if the objectives of each is clear to the other. That is, the line official must realize that staff is only seeking ways to improve operations and is in no way trying to undermine operations.

## ASSEMBLING STAFF AND RESOURCES

Once objectives have been determined and a departmental framework developed, the administrator's next task is to assemble an able staff and to provide the necessary resources to carry out the basic functions of the organization. Indeed, this task of bringing together the personnel needed to fill the positions, providing the facilities to operate, obtaining the needed equipment and materials, and securing a sound financial base to pay for both human and material resources calls for great managerial ability and administrative capacity.

Although every recreation and park system must mobilize and allocate its resources to do the job, the most important single element in this process is the assembling of personnel to perform the leadership and other tasks that are needed to achieve organizational objectives. To be able to pick an able staff and provide it with an environment conducive to its development is basic to effective recreation and park administration. Also, while recreation and park executives leave much of the actual selection of employees to the supervisors or unit heads under whom the new staff members will be working, final approval still rests in their hands. Subsequent chapters will deal with the nature of leadership, departmental staff, and personnel practices and policies. This will be followed in still later chapters by a thorough discussion of finance practices and budgeting, as well as of the physical resources of the recreation and park organizations.

## DIRECTING

Directing is the energizing force that activates the organization, or, strictly speaking, it is the process involved in the planning of work to be done, the issuing of orders, the giving of instructions, and the providing of leadership in motivating action toward goals. Directing cannot be separated from leadership. Indeed, the effective administrator must be an able leader if he or she is to direct his or her organization effectively. Actually, the recreation and park executive must motivate his staff to contribute their best efforts in performing their tasks, and the techniques used are only methods embraced within the general framework of leadership. Hence only the techniques involved in the "command" aspects of directing will be discussed here. The next chapter will discuss more thoroughly the aspects of administrative leadership.

The first consideration in directing the organization is to develop an effective plan of what needs to be done and to communicate decisions for implementing this plan to the various units within the organization. In short, the administrator must translate policy into work units. In fact, this step is known as "managerial planning" and is closely allied to another important facet of directing, decision making. Indeed, managerial planning and decision making are the very foundation of the directing process. Some writers feel that decision making is so important that it should be an administrative process in its own right. However, decision making is related to directing. It is the job of the administrator to make decisions that affect the entire focus or direction of the organization. At the same time, planning and decision making must be interwoven into the cloth of the organization so that the entire staff knows clearly and concisely what needs to be done. To do this, clear lines of communication must be established through the issuance of instructions or orders, either oral or written, and the preparation of administrative codes or manuals. In sum, directing as a process must consider planning the work, decision making, and communicating instructions for performance of the work.

## Managerial Planning

It is the job of every recreation and park administrator to decide the types and quality of recreation service to be provided for the public. To plan how this is to be done through the resources available is, in essence, the basis for managerial planning. The executive must keep a perspective and balance among the constituent units of the organization and must provide a continuity of program services. Planning of the work, of course, depends in large measure on the amount of funds available to do the job. To translate this money into jobs and physical resources and to allocate it wisely to accomplish the goals of the organization through approved work programs is basic to effective directing.

It is the job of the administrator to be constantly alert to change, to expedite areas of need, to change direction as required, and to exercise leadership as needed. Only the administrator can see the overall pattern and chart the direction for the organization. Too often, staff members and unit heads see only their own problems; it is the job of the administrator to provide a guide that shows relationships and the reason for expediting one phase of program more than another.

## Decision Making

Directing an organization requires making innumerable decisions. Decision making involves the selection of one course of action over another and involves deciding on an action to be taken in preference to alternatives. It is the job of the recreation and park administrator to make choices or to have the conceptual skill to make rational decisions on the basis of facts, recommendations, and judgment.

Decision making does not rest exclusively with the chief executive, unless he or she so wills. Actually, decision making should also be assigned to the unit heads in the hierarchical structure. Thus each unit level would be concerned with its own operational problems and its head would have decision-making authority to meet them. Delegation of decision making to the level that can handle it best relieves the top administrator of an additional task and involves subordinates in an important phase of management.

Since a decision once made commits the organization to a definite course of action, this action cannot be taken lightly, especially when competing alternatives are also compelling and attractive. For this reason, decision making is extremely difficult for some administrators. If the wrong decision is made, it can have adverse effects on the operation of the organization. Therefore it is imperative that all evidence and facts supporting each alternative action be available and weighed before a final decision is made.

Many influences have a bearing on rational decision making. The organization itself and its historical patterns for the making of decisions can act as a constraint in some instances. Is the organizational climate such that the making of decisions is encouraged? Do outside pressures come to bear upon the choice to be made? Many times, there is great pressure from both inside and outside the organization for the decision maker to come up with a choice that favors one group over another or with an answer that is controversial. Regardless, as previously stated, decisions need to be made on the basis of facts, recommendations, and judgment. Not to do so means that an executive is shirking his or her responsibilities.

### Communication

It is not unusual to think of directing as only communicating decisions on what needs to be done. Much of this communication takes place through the issuance of instructions or orders and through the preparation of administrative codes and manuals.

The giving of instructions is not so simple as it seems. Too often, it is taken for granted and often assumed that subordinates know what to do and how to do it, when, in effect, orders or instructions may be ambiguous, vague, or unreasonable.

Orders are merely communications on what needs to be done. They can be either written or oral. The kinds of orders will depend on the skill of the administrator in organizing his or her department and in managerial planning and delegation of authority. However, he or she can never be relieved completely from giving orders or instructions.

Generally speaking, all communication in the organization is related to supervision. In fact, they are inseparable, for all "supervision" means is assistance in improvement. However, it should not be implied here that communication only takes place through orders, either written or oral, although they are important means of modifying of changing human action. Actually, communication occurs through any sense perception with resulting behavior modified as desired. Effective supervision is based on leadership skill, but, since direction is provided in no small amount by the giving of orders, certain basic principles can very well be used as guides. These are:

1. An order should be complete and concise. It should give details on what needs to be done and when.
2. An order should only be given to recipients who have the authority and the means of carrying it out.
3. Orders should be clear and reasonable. No one should be expected to do the impossible.
4. Orders or instructions should be in writing if this is at all possible.
5. All orders should have a follow-up. A time should be specified when a report is due, an action is to be taken, or a task is to be accomplished, and it should be carefully adhered to.

6. The purpose of an order should be explained. Compliance is easier when the recipient knows why the order has been given.
7. If an order is complex and several staff members are involved, a consultation with the recipients of the order may encourage greater involvement and cooperation.

Although the issuance of orders is often necessary, orders should be used with discretion. The use of leadership skill in initiating changes, interpreting policy, and bringing about desired ends is by far the superior method. To rule by authority rather than through leadership is a poor technique for directing.

It is common practice to standardize many orders and issue them through handbooks or administrative manuals and codes by which they are available for ready reference when points of issue are involved. Much can be said for this practice. An administrative code is "legal"-oriented; it is the sum of laws, ordinances, resolutions, and other legal enactments that provide guidelines for action. On the other hand, an administrative manual is a compilation of the administrative rules, regulations, and guides for the use of employees in better understanding their responsibilities. It covers procedures and rules concerning how tasks are to be accomplished. It provides a general framework of the details of operation and can eliminate much order giving or instruction that is standardized procedure.

As previously stated it should not be implied here that all communication only takes place through orders, either written or oral. An entire recreation and park organization comprises a network of interrelationships that necessitates an effective communication system if the goals and objectives of the organization are to be achieved.

All that is involved in communication is the process of transmitting ideas, images, or thoughts from one person to another. However, this is not easily done in a managerial setting. The giving of orders as just described is a means of communication that flows from higher to lower authority. One of the pitfalls of this downward type of communication is that passage of messages or orders from the top of the organization down the hierarchy are diluted as they pass through the many intermediate levels and that misunderstandings may occur at the point of execution. This is especially true in the larger park and recreation enterprises.

Therefore an upward system of communication should also be developed in which ideas and messages are passed up the organizational hierarchy to management. This is not an easy task. Here again, a filtering process takes place; and in many instances, either deliberate or otherwise, few messages get through to top management. Problems are disturbing and there is a tendency on the part of staff at the various levels to protect the "boss" from bad news. Good news reaches the top much more easily, however. This communication block should be overcome, and it can be if the organizational climate is positive and subordinates are encouraged to express their views.

Managers and supervisors should not overlook lateral communications—the kind that takes place when no superior–subordinate relationship exists since workers are at the same level in the organizational hierarchy. Also, this method of communications exists with those in other organizations, departments, or agencies. Informal organizational patterns encourage some lateral communication flow, and the desire to shortcut the communication process by not following the traditional lines of authority is also a factor. By discussing problems with co-workers and by coming to conclusions without action being taken by the executive is not an uncommon operational procedure. It can strengthen employee cooperation; but, at the same time, it can also create problems if the executive is not aware of what is happening. This communication process bears watching and is a method that has both positive and negative overtones. If lateral communication is used extensively, a second look needs to be given to the organizational structures of the enterprise to see if responsibilities are well defined and no excessive "paper flow" exists.

Mention also needs to be made about informal communication or in the parlance of many, is called the "grapevine." All formal communication systems coexist with an informal one. The term "grapevine" originated in Civil War days when intelligence telegraph wires were strung "hodge-podge" from tree to tree resembling a grapevine and many messages as a result became garbled and confusing.

The term "grapevine" has become synonomous with informal communication or the desire of workers to communicate or socially to interact on matters, whether they be fact or fiction. It fosters rumors, informational errors, and ugly gossip but can also play a positive role. Information of importance to management can rapidly reach the decision makers without going through channels and in most instances the information is correct. One study on the "grapevine" showed that over three-fourths of the information relayed in this fashion was correct.[16] It also is and has been used to leak information that can be used as a sounding board or "trial balloon" for reactions on important matters before action is taken.

A park and recreation executive should know what kind of information is being communicated through the "grapevine." It should be used by management only in the most skillful manner if desired results are to be obtained, but the danger always exists that the desired information or messages will become distorted as it grinds through the "gossip mill." If most of the information or communication flow of importance is transmitted in this way, it should clearly indicate to managers that the formal communication system needs repair. Regardless, there is no way to stop the "grapevine." It is well to know that it ex-

[16]Eugene Walton, "How Efficient is the Grapevine?" *Personnel* 38(2) (March–April 1961):48.

ists and is not just a gossip mill; it is influential and can be guided. Many managers or administrators listen to it, study it, and influence it in a positive fashion. Rumor and maliciousness must be stopped and it can be if the worker is provided the facts on situations that cause it.

## COORDINATING

The process of coordinating is closely related to that of directing. In fact, it is many times discussed as a part of the directing process. But it could also be considered an integral part of organizing and all other phases of administration. By "coordination" is meant the harmonizing or drawing together of the diverse elements of an organization toward its central purpose. Coordination seeks teamwork. It is the force that welds the working team into a unified and cooperative unit. As was stated in a government publication, "coordinating means to arrange our work in proper order and relation to those we work with; to balance and harmonize actions with those concerned both within and outside of our organization."[17]

Coordination is a complex task for the recreation and park administrator. To integrate the work activities of his or her organization toward common objectives is a major undertaking and one that requires the highest leadership skill.

Every recreation and park system has a form and structure. However, how it functions will depend on how the staff identifies with and relates to organizational objectives. Thus the director is involved in human organization and must inspire his or her staff toward team efforts, even when members may have conflicting interests and separatist tendencies. Indeed, as units become preoccupied with their own immediate tasks, they often magnify their role and wish to splinter off into more identifiable units. This is particularly true with those areas of work that call for technical and specialized skill. Clearly, then, it is the job of the administrator to weld his or her staff into a harmoniously integrated unit directed toward organizational objectives, notwithstanding the personal, immediate, and sometimes selfish interests of some of the staff.

Coordination can best be achieved when staff is involved in the decision-making process. Adherence to this principle will instill in individuals a personal identification of their role with the organization's larger values and greatly enhance team effort. This point will be probed more fully in a subsequent chapter, on leadership.

[17]United States Air Force, "How To Direct and Coordinate Work," *Management Course for Air Force Supervisors AFP 50-2-11* (Washington, D.C.: U.S. Government Printing Office, 1955), p. 10.

## Methods of Coordination

Much of coordination takes place through the development of an integrated structure as explained in the description of the process of organizing. Actually, adherence to sound principles of organization provides a "built-in" coordinated structure. Likewise, the effective implementation of other administrative processes will also yield desirable outcomes. This is succinctly stated in a government publication as follows:

> In management, coordination deals with synchronizing and unifying the actions of a group of people. Basically, coordination should flow or result from effective planning, organizing, directing and controlling. Continued need for special coordinating devices may, in fact, be an indication of poor planning and organization. It is certain that far too much time will be consumed in coordinating efforts if the other functions of management are not well performed.[18]

Some of the coordinating devices that could be used by recreation and park systems are described below.

**Effective Communication.** A basic means of effecting coordination is through the establishment of clear lines of communication. To have current information of what is going on is vital to effective teamwork.

**Voluntary Cooperation.** Much coordination takes place through the cooperative efforts of the staff. To encourage members to meet together, socialize, and visit the work areas of the various units of the organization does much to make for better understanding. Establishment of lounge rooms for informal discussion during work breaks also does much to break down barriers.

**Personal Contacts.** Much can be accomplished in coordination if opportunity is given for face-to-face contacts with the individuals concerned when adjustments are needed.

**Use of Committees.** An excellent means of coordination is effected through use of committees. Committees provide a structure for voluntary coordination. If used effectively, they can provide the setting for common understandings and team efforts that are essential in the coordinating process.

Some of the specific committees that have been structured in recreation and park systems have been an administrative council consisting of division heads and those responsible for certain phases of the department's program, a supervisor's council, a foreman's council, a recreation leaders' committee, and, in some instances, a coordinating committee.

[18]U.S. Department of Agriculture, *Essentials of Good Management* (Washington, D.C.: U.S. Government Printing Office, not dated), p. 25.

Actually, the establishment of committees representing the various levels or ranks of the organization that perform specific functions is fairly commonplace. Basically, their purposes are to share thinking on mutual problems and to propose solutions that are in the best interest of the organization. However, not to be overlooked is a suggestion by Tead and others that each committee or other body should have representation from each of the other committees or bodies that have close relationships with it.[19] In other words, groups or committees should not consist only of "insiders" but should also have members from other representative groups of the organization.[20] While use of a committee structure to effect coordination has limitations, it does have advantages that outweigh its shortcomings. Most of all it provides a means for sharing in deliberations and gives employees a sense of belonging.

**Staff Meetings.** Not to be overlooked as a device for coordination is the use of staff meetings. To call all members of the department together for a general meeting provides an excellent means for sharing of ideas and interpretation of policies and needs of the organization. However, in many instances, the staff meeting is not used effectively. Rather than using it to unify the actions of the staff, stimulate ideas, or provide a structure for problem solving, administrators too often call such meetings to announce decisions or report action taken. Except for tradition, such meetings serve a doubtful purpose and perhaps should be abolished. Unless a staff meeting is open to participation by its members and has a purpose for being other than tradition, it is better not to have one. Without sharing and the given and take of discussion, decisions or reports can just as well be give through a written communication.

## CONTROLLING

The controlling process seeks to determine the effectiveness with which orders and policies are being executed. In other words: Is the recreation and park system working according to plan? Are its objectives being met? Are duties and responsibilities being properly carried out? Are programs and services to the public being maintained up to quality standards? The answers to these and similar questions are of concern to the administrator. Indeed, it is his or her job to see that objectives are being met and programs conform to plan.

Actually, controlling cannot be separated completely from directing, planning, or organizing. Once planning takes place and the procedures are provided to set the organization in motion, the amount of control will depend on the thoroughness with which the foregoing processes have been put in operation. Nonetheless, control mechanisms are needed if the administrator is to

[19] Tead, op. cit., pp. 188–189.
[20] Ibid.

be fully aware of what is happening in the organization. Only insofar as he or she has information as to progress and achievements can he or she provide effective supervision, chart new directions, or correct mistakes. In the final analysis, then, this process is concerned with the checking of performance through control devices such as (1) reports, (2) measurements, (3) direct observations, (4) staff conferences, and (5) independent surveys.

### Reports

To control, an administrator must have information on how the organization is progressing, and one of the chief means of securing this information is through a system of reporting. Every executive needs information at regular intervals on his or her spending program and work achievements. This usually takes the form of monthly progress reports that are presented as statistical summaries. For example, for the operational division, figures are presented that show a comparison of attendance at playgrounds, centers, golf courses, museums, and the like against similar figures from previous weeks, months, or years. Accomplishments of the maintenance units are also shown with comparisons with previous periods of time.

Also, final reports of revenue and expenditures with comparisons with previous years and checks on how the various units are staying within their budget are an absolute must if control is to be maintained. Indeed, control of the budget is essential to control of the organization. To work without being regularly informed on how the budget is being spent, the amount remaining in each appropriation, and the overspending or underspending of budget items is truly to work in the dark.

In sum, each important operating unit within the department should prepare reports that summarize the basic data needed by the administrator for purposes of supervisory control. And the recreation and park administrator should have regular reports on the state of his budget.

### Measurements

Another control medium is that of securing information that measures performance against an acceptable standard. While such standards have not been developed to any great extent by recreation and park bodies, they should not be overlooked as a means of securing objective data on work output.

### Direct Observation

Another control device that should not be overlooked is that of first-hand observation of the programs being provided or inspection of the services being rendered. Actual perception of a situation provides an excellent opportunity to measure quality of performance. On the other hand, such inspections should not provide exclusive judgments. Too many other factors enter the picture for a person to be able to judge a situation by perception alone. Also, caution and good judgment should be used in not using visitations and inspections as a device to "shortcircuit" first-line supervisors who are responsible for the work being observed.

### Staff Conferences

Every administrator delegates authority and responsibility to subordinates, and a common means of knowing how goals are being achieved is through face-to-face conferences with key staff members. The administrator is confident that problems and adjustments will be brought to his or her attention at these meetings. In fact, many executives use this method as their principal source of information. They have respect for, and confidence in, the judgment of their staffs. An executive's leadership in cooperating with subordinates who are just as keenly interested in the efficiency of the organization as he or she is makes for a cooperative venture in reaching goals. Indeed, development of a leadership climate requires a continuous two-way flow of information that alerts the executive to areas of need.

### Independent Surveys

As departments become involved in their own systems of reports and control devices, it is sometimes wise to secure an appraisal or survey of management methods. An objective overview of the organization and its operating procedures through an administrative survey provides the staff with a new look at their department. It is best to call in an outside firm or research team to do this work. In most instances, surveys or appraisals are made only for judging programs, finance practices, or facility needs. Overlooked are the details of reporting, office organization, and management procedures that can be of great assistance to the administrator in directing and controlling the organization.

## EVALUATING

The last administrative process to be considered here is evaluating. Some writers consider this process a part of controlling, and actually it is. As used here, however, it represents the final appraisal of the department's program and services and of the extent to which objectives were accomplished. Evaluating is a continuous process, and as such is a controlling process, but once the program is expended, only by a critical look at what was accomplished in terms of what was planned can some of the weaknesses and shortcomings of operations be revealed. Effective controlling can take the place of evaluating, if "postmortems" are also provided. Every recreation and park system should examine each of its end products (services, in this instance) and see if they could be improved in depth and scope, and if they could be provided more efficiently and economically.

## PRINCIPLES OF ADMINISTRATION

To sum up the elements of the administrative processes and concepts, we can use the following broad principles as reliable guidelines. These principles were explained in greater depth in previous discussions of this chapter and are applicable to all aspects of the operation of a park and recreation system.

### Purpose and Objectives

*The aims, objectives, and goals of the organization must be clearly identified and understood by the employees of the enterprise.* A park and recreation organization is a complex unit and all employees should know its mission. Not to know its purposes and objectives leads to misunderstandings and conflict and a lack of unity. Furthermore, this leads to confusion on the part of the public being served which in turn can create funding problems when departmental budgets are analyzed.

### Single Head

*Every park and recreation enterprise should have a single head of the unit.* No organization can operate effectively by dividing responsibility and authority at the top level. When such division occurs someone must take over a coordinating role and this person soon assumes the cloak of being the single executive. Examples of departments that have tried to separate parks and recreation, each head in turn being responsible to a single board, have not been successful. Then the board becomes the coordinating body and concerns itself with administrative detail that is not a part of its function. The services of an organization must be coordinated, and this can only be done through a single executive head.

### Administrative Hierarchy

*No department structure should be so highly organized that decision making is lost or diluted in passing through the various levels of the hierarchy.* As a department becomes highly organized with various levels such as divisions, bureaus, sections, units, and the like the lines of communication can become so long that it becomes difficult for decisions, information, or messages to go up the "chain" or down without becoming "garbled," delayed, or diluted. When this happens, it is important for the executive to take a careful look at the organization structure to see if it is "top heavy" with supervisory personnel. In large organizations such "layering," or additions of supervisory levels, may be necessary, and if so, a change in communication patterns may be necessary if problems of this nature occur.

## Planning

*Every park and recreation organization should establish a program of long range planning in which future needs are identified and a plan of action is established to meet them.* To plan is to look ahead and every organization must explore in advance how its mission can be accomplished. Not to look ahead leads to stagnation. Therefore, a timetable of need priorities should be charted and vigorously followed.

## Span of Control

*Every administrative official should be given a span of control that makes his or her supervision and leadership effective.* An executive or supervisor can only supervise or direct in an effective manner a limited number of subordinates. This number is dependent on the complexity of the duties and responsibilities of the positions being supervised and the limit of time and energy the executive can give to this phase of his or her responsibilities.

## Position Description

*Every employee in a recreation, park, or leisure service organization should have a clear cut statement of the duties, responsibilities, and authority of his or her position.* To do an assigned job, an employee must know what is expected of him or her. Not to have such a description leads to frustrations, poor morale, and lack of incentive. It is an interesting commentary that when employees find themselves in such nebulous situations, there is a tendency on their part to perform or carve out a job that fits their own strengths and training. More often, however, confusion results, and the employees so affected start looking for other positions.

## Unity of Command

*Every employee should have only one immediate supervisor who has direct authority over his or her work.* No individual in an organization should take direct orders from more than one person. To have more than one "boss" makes for an intolerable situation for an employee. It creates stress and anxiety for him or her and is destructive to his or her productivity. Conflicts inevitably arise when this principle is violated.

### Authority and Responsibility

*All unit heads within the parks and recreation enterprise should be delegated authority commensurate with their responsibility in carrying out the tasks assigned.* Nothing stifles initiative as much as being held responsible for an assigned task and not being given the authority to carry it out. It is ironic that some executives take great pride in hiring outstanding individuals to perform specific tasks in which they have proven ability and then do not let them use this ability by not delegating authority to them to carry out these responsibilities. This principle is constantly being violated by supervisors and administrators. Whether it be ego or fear that they will lose control of the organization, that the person will undermine their position by doing too good a job, or that they will lose status by giving away some of their power, the facts are clear that few executives find it easy to delegate authority.

If it could be made clear to executives that their success is dependent on the abilities of their employees and selecting the most competent of such employees will only enhance their image, perhaps this fear of delegation could be diminished.

### Standing Operating Procedures

*Every organization should prepare a policy manual that establishes standard operating procedures for the many routine tasks that confront the administrator.* Standardization of record keeping, reporting, accounting, time schedules, evaluations, and many other routine procedures saves management time and energy.

### Controlling and Evaluating

*Controlling the effectiveness with which orders and policies are being executed is essential to the effective operation of any organization.* Controlling is related to evaluating and supervising. Any park and recreation program would be seriously hampered without a system of evaluating or controlling its effectiveness. In other words, are duties and responsibilities being carried out? Are services up to standard? Are the programs conforming to plan? The principle of evaluation and control applies to all aspects of the organization and is essential if the administrator is to be fully aware of what is happening in the execution of policy and the general operation of the enterprise.

## RECREATION AND PARK ADMINISTRATION

A discussion of administrative processes should not ignore the need for skills in the actual operations of the department. Not to have knowledge of, or proficiency in, operational phases of recreation and parks would be a fatal shortcoming, for, indeed, the administrative processes involved cannot work in a vacuum. They are inextricably enmeshed with functions, whether programs, finance, personnel, areas and facilities, coordination, or public relations.

To simplify this point, one can point out that individuals involved in programs, personnel, finance, and the like must plan, organize, direct, and coordinate, just like any other administrator, if their responsibilities call for it. Administration, then, is not just a responsibility of those in the top echelons of the organization. It is also a part of the work of the operating heads, and the amount of time spent on any or all of the processes listed previously will depend on the character of the work and the responsibilities assigned.[21] However, it should be emphasized that administration is skill—skill in working with human beings to bring out their best. And the highest compliment that can be given an administrator is that he or she is a leader who likes people, respects their dignity, and is able to weld his or her staff, with all their differences in aspirations and other aspects of individuality, into a unified and cooperative working unit.

Administration, then, is personal. In fact, "Administration which is impersonal is condemned to mediocrity."[22] Administration is concerned with human beings and the means of directing their behavior and motivations toward the expeditious accomplishment of ends. Basically, the administrative process works through people; thus the human relations element is an important aspect of all administrative action. Indeed, human relations and skill in getting the job done through personal leadership are the essence of superior administration.

In sum, administrative skill is directly related to the ability to get people to work together toward goals that they come to feel are desirable. Call it "executive leadership" or "skill in leadership" if you like, but the complexities of administration call for this ability. The recreation and park field can only develop to its maximum if this phase of administration is fully developed, and the recreation and park field will only develop to its maximum as this element is forthcoming to meet the needs of an expanding economy and an expanding world of leisure.

[21]This point is discussed more fully in Tead, op. cit., pp. 5–6.

[22]C. H. Garland, "Some Reflections on Administration," *Public Management* (February 1932):56–57.

## Questions for Discussion

1. Define *administration.*
2. What is meant by the "basic processes of administration"?
3. How is the process of directing related to the leadership function?
4. What are the elements of the organizing process?
5. Define *unity of command.* Define *span of control.*
6. Compare the pros and cons of a "flat" organization to those of a "tall" organization.
7. Differentiate between "staff" and "line" activities. Give examples of each in a recreation and park department. Which do you feel is the most important? Why?
8. What are some of the problems involved in giving orders? Can formal orders be eliminated?
9. How may the coordinating process be effected in a recreation agency?
10. State several basic principles of organizing.
11. Why must authority be delegated along with responsibility?
12. Should an employee "go around" his supervisor to the chief executive if a controversy arises? Why or why not?

## SELECTED REFERENCES

Edginton, Christopher R., and Williams, John G., *Productive Management of Leisure Service Organizations: A Behavioral Approach*. Wiley, New York, 1978, Chap. 5.

Gibson, James L.; Ivancevich, John M.; and Donnelly, James H., Jr., *Organizations: Structure, Processes, Behavior*. Business Publications, Dallas Tex. 1973. Chaps. 2–5.

Gross, Bertram M., *Organizations and Their Managing*. Free Press, New York, 1968, Chap. 2.

Hjelte, George, and Shivers, Jay S. *Public Administration of Recreation Services*. Lea and Febiger, Philadelphia, 1972, Chap. 3.

Lutzin, Sidney J., Ed., and Storey, Edward H., Assoc. Ed., *Managing Municipal Leisure Services*. International City Managers Association, Washington, D.C., 1973, Chaps. 3, 4.

Massie, Joseph L., and Douglas, John, *Managing: A Contemporary Introduction*. Prentice-Hall, Englewood Cliffs, N.J., 1973, Chap. 6.

Pfiffner, John M., and Fels, Marshall, *The Supervision of Personnel*. Prentice-Hall, Englewood Cliffs, N.J., 1964, Chaps. 5–7.

Robbins, Stephen P., *The Administrative Process*. Prentice-Hall, Englewood Cliffs, N.J., 1976, Chaps. 2–4.

Stoner, James E., *Management*. Prentice-Hall, Englewood Cliffs, N.J., 1978. Chaps. 1, 2.

# CHAPTER 3

# Human Behavior in Organization

One cannot discuss effective administration without considering the human element and the part that people play in making an organization effective. It can be concluded that people make a park and recreation organization productive. And the question can be raised, "What makes people behave as they do?" "What can be done to make people work more effectively?"

There is no doubt today that a park and recreation administrator, manager, or supervisor can only be a success as long as he or she can work effectively with his or her subordinates or staff. This is self-evident, but it took some years for this fact to be accepted.

## EARLY BEGINNINGS

Early organization models established by Weber[1] and followed by Taylor and Fayol were mechanistic in their approach and were based on a classical model for a bureaucratic organization with divisions of labor, functional specialization, hierarchy of authority, and a system of work procedures. Impersonality prevailed and operations were mechanized. Weber's model minimized the human element and emphasized institutionalized authority through carefully prepared procedures developed around form and structure.

[1]Max Weber, *The Theory of Social and Economic Organization*, A. M. Henderson and Talcott Parsons, trans., Talcon Parsons, ed., New York: (Free Press, 1947.)

Frederick Taylor, who is noted as the founder of the scientific movement, also stressed this mechanistic approach as opposed to the human factors in organizations; and likewise, Henry Fayol paid scant attention to this important element.

## HUMAN RELATIONS' CONCEPT IN MANAGEMENT

Mary Parker Follett was one of the first writers to recognize the human element in organization and to point out the importance of these factors in enterprise.[2]

### The Hawthorne Experiments

Few experiments have been quoted more extensively than the Hawthorne study as to its impact on human relations and the productivity of an enterprise. Actually, the human relations' movement began as a result of this study that was undertaken at the Western Electric Company's Hawthorne plant in Chicago between 1927–1932.

During this period, research was being focused upon means of increasing productivity through a study of the impact of fatigue and other physiological factors that exist in the work environment. The amount of light, temperatures, and other working conditions were selected as elements of the physical environment.

In the first series of the experiments the researchers expected individual output to be increased as light levels increased. While output did increase, it still continued to increase when the light level was drastically decreased. This was not what the researchers expected. Nor was it what they expected in their second and third experiments when changes in working conditions and the effect of group piecework showed little change on productivity. It had to be assumed therefore that psychological rather than physiological factors were at work. It was concluded that the improved morale as a result of participating in the experiment had an influence on productivity, as well as the special recognition given the group and the development of an environment that fostered group interaction.

The significance of the Hawthorne experiments was its impact on administrative thought. It stimulated behaviorist models of administrative organization and interest in the human factors in the motivation of employees and brought into focus the importance of informal group relations in the work en-

[2]H. C. Metcalf and L. Urwick, eds., *Dynamic Administration: The Collected Papers of Mary Parker Follet*, (New York: Harper and Row, 1942.)

vironment. Most of all, it recognized the importance of employee behavior in the success of an organizational enterprise.

One could discuss many of the theories proposed by the early leaders of administrative thought, as well as those now experimenting in this field. None should be overlooked. Max Weber, Frederick Taylor, Oliver Sheldon, Mary P. Follett, James D. Mooney, Lyndall Urwick, Chester Barnard, Herbert Simon, Frederick Herzberg, Douglas McGregor, Victor Vroom, A. H. Maslow, and many others made significant contributions to modern thinking on the human side of enterprise. The author has chosen only a few for examination, although all are in one way or another related.

## McGregor's Theory Y

McGregor's Theory *X* and Theory *Y* are widely quoted and have received considerable attention as to how people behave.[3] He proposed two opposing points of view of human beings in organizations, one being largely negative and called Theory *X* and one being basically positive, labeled Theory *Y*.

Theory *X* makes the assumptions that human beings dislike work and will avoid it if they can, are innately lazy and unmotivated, and must be coerced and controlled if any work is to be accomplished. Theory *X* further postulates that employees place security and stability over all other factors of their work and will display little ambition or initiative.

On the other hand, Theory *Y* presents the point of view that employees will accept responsibility, will view work as natural as rest or play, will exercise self-direction and self-control, and will strive for excellence. Theory *Y* places employees in a positive light in contrast to Theory *X* with its negative overtones. If one subscribes to Theory *X*, it means that employees work best under the usual bureaucratic pattern of organization. Further, the administrator who holds these views will establish rigid organizational structures with authoritarian guidelines, strict practices of supervision and compliance, and an atmosphere of fear through threats of firing or economic reprisals.

Needless to say, Theory *Y* as an organizational model has been supported by management and behavioral researchers who believe in the human relations' concept of management. Unfortunately, neither of these models work in their pure form. Few administrators subscribe to all the assumptions of one or the other, but the administrator who holds the posture that employees will work best in a positive environment and establishes an organizational structure that places a premium on reliance, self-control, recognition for achievement,

[3]Douglas McGregor, *The Human Side of Enterprise*, (New York: McGraw-Hill, 1960).

imagination, creativity, and the like will be better able to solve organizational problems and to establish a climate for greater work output.

## Informal Groups

Behavioral research on organizational management and structure has found that an organizational enterprise comprises a system of people who work together in an integrated manner and develop close personal attachment to others. People are naturally inclined to form informal social groups, and these groups can significantly affect the operation of an organization.

Formal organization provides structure and a design focused toward a predetermined plan. It clearly defines lines of responsibility and authority and charts the means of accomplishing organizational objectives.

In the previous discussion of organization in Chapter 2 attention was given to organizational designs and specific relationships that are structured by the very nature of park and recreation enterprises. It is true that formal organizational structure with assigned duties and relationships is needed and does provide degrees of motivation to which an employee will respond.

On the other hand, an informal organization is a network of social relations that arises through the interaction of people within the structure of an organization. It is created without formal planning, is structureless, and only develops as people associate with each other. Actually, one of the outcomes of the Hawthorne experiments was the emergence of informal organizations.

When men and women work together and frequently see each other on the job, at lunch or after work, they develop close informal social relationships and establish working ties that give them feelings of satisfaction and of belonging. They air grievances, seek advice, and get reinforcement concerning their own feelings about change or working conditions.

Further, these informal groups give the worker a sense of security, provide companionship, provide strength for bargaining, open channels of communication, and provide a means for sharing common problems.

A person may belong to several of these groups, each of which has its influence on him or her. He or she soon comes to rely on the consensus of feeling by the group and eventually gives way to the pressure of the group toward actions that may be contrary to his or her own beliefs.

Because these groups establish patterns of behavior most favorable to themselves, any modification of practices that upsets their social relationships will be resisted. Change in job assignments, reorganization of the physical plant, changing work routines, or any other work modification may disrupt pleasant social relationships and cause anxiety that has no foundation in fact. In addition, if morale is low and work output is not up to standard, it could be the result of group interaction. Therefore park and recreation administrators need to know the impact of these informal social practices and their effect on the organization.

**Social Behavior Versus Organizational Objectives.** Informal social groups may be practicing or performing tasks that are not congruent with the formal structure of the enterprise but nonetheless are more efficient than those in operation. The wise executive may very well change job practices to bring them in line with current practices. However, a thorough investigation should be made of such practices to see if they are in line with operational policy and are in the best interests of the organization. If they are not, every effort should be made to change the practices to meet the organization's objectives.

In sum, informal groups satisfy certain wants or desires of the worker on the job. They perpetuate values that are held important to them; they receive social satisfactions through the development of status, recognition, and opportunity to relate to others; they open channels of communication; and they provide restraints or social controls on their members. However, these positive influences can also work in reverse, which would require management control. In perpetuating values they may want to maintain the status quo and resist positive change; they may relate to others in a way that is detrimental to the organization; they could use their communication network to establish a grapevine of rumors, and they could develop social controls that lead to conformity and attitudes not consistent with the norms of the organization.

## MOTIVATION IN ORGANIZATIONS

Motivation of employees is one of the most difficult tasks facing an administrator. It is an element that is almost impossible to measure because it deals with people, all of whom are different. But since motivation has such an important effect on the quality of services of a park and recreation enterprise, one should examine human motivation as to whether it can be influenced or whether it can be managed toward positive outcomes.

Every manager and supervisor searches for understanding of the extrinsic and intrinsic factors that drive employees to do what they do. Furthermore, strategies developed to drive workers to greater accomplishments depend largely on the leadership style of the supervisor or manager and his or her fundamental beliefs in the nature of human beings. If the supervisor or manager believes that an employee is intrinsically good in the sense of being goal directed, self-activated, and self-controlled, the results make for cooperative social relations and a propensity by the employee to become more productive and more involved in the activity of the enterprise. Moreover, a worker or employee reacts in a positive or negative fashion to the way he or she is treated. Thus it is extremely important that the administrator look for the best in his or her employees and has a positive feeling about them. McGregor's[4] Theory *X* and Theory *Y* are good examples of opposing points of view about the behavior of people in an enterprise and the way a supervisor or executive may view them.

A further tool for motivating employees is the establishment of a system of management by objectives (MBO) in which employees and administrators work together in a cooperative manner in reaching objectives and goals.

## MANAGEMENT BY OBJECTIVES (MBO)

It is a commonly accepted principle of management that organizations must direct themselves toward objectives in order to function effectively.[5] To do otherwise is to function without aim or purpose. This notion of working toward common objectives is certainly clear and easily understood. Some managers do it instinctively; that is, they effectively direct their operations toward sound objectives without utilizing any formal system. Unfortunately, however, a large number of managers and their organizations just function from day to day, drifting along without aim or purpose.

Essentially, management by objectives (MBO) is a formal system that attempts to establish a sense of purpose, focus, and concentration for individuals, organizational units, and the total operation. In principle, the greater the focus on the desired end results, the greater the likelihood of achieving them.

In addition, a second essential principle of MBO relates to the process of establishing those objectives; the greater the participation in setting objectives, the more meaningful they will be and the greater will be the motivation for achieving them. MBO cannot exist without an open system of communications in general and a free flow of information from subordinates to supervisors in particular.

The purpose of MBO, then, is to establish a formal mechanism as well as an agency philosophy that requires that specific goals be established, timetables be set up, and certain persons be designated as accountable for achieving the desired end results. All of this takes place through widespread participation.

Numerous writers have discussed and expanded on these basic elements since the first efforts of Peter Drucker and Douglas McGregor in the early 1950s and therefore a multitude of definitions and key elements have evolved. Management theorists talk about MBO as a process, a philosophy, a system, a

[4]McGregor, op. cit.

[5]Section on "Management by Objectives" taken from statement made by and used with permission of David Culkin, Instructor, Department of Recreation Resources Administration, North Carolina State University at Raleigh, October 1, 1979.

planning tool, a decision-making model, a method of performance appraisal, an integrated approach to management, a method of improving employee skills, a system of equitably determining wages and salaries, a way of allocating resources, and a system for keeping control of all the various aspects of an operation.

The original basic ideas of MBO have been expanded to such a degree that elaborate and highly refined MBO systems now exist. Unfortunately, the multitude of definitions, the complicated nature of many of these systems, and their proliferation in the management arena have caused a lot of confusion, especially among managers who are involved with or considering its implementation.

Until recently, management by objectives was being implemented almost exclusively in private business industry. A number of federal agencies, such as the Heritage Conservation Recreation Service and the National Park Service, have implemented MBO with varying degrees of success, but for the most part, park and recreation departments have not been quick to implement the system. This is probably due more to the fact that confusion about MBO has left many managers skeptical rather than the fact that no need exists.

Certainly, a number of problems exist regarding implementation. First of all, it requires a strong commitment on the part of the leadership. The system will not function by itself; leadership has to make it function. Second, it is imperative that an atmosphere of participative management exists in the organization. As stated previously, widespread involvement of all key personnel in the formulation of objectives is an intimate part of MBO; the system will not work in an authoritarian, centralized operation.

Third, it is easy for employees to succumb to a commitment to the day-to-day operations and fail to consider the processes necessary for accomplishing those tasks. In other words, very busy, task-oriented employees often dislike taking the time to focus on the steps needed for implementation. For example, it is hard to concentrate on a discussion of the mission of your organization when you are not yet ready for a meeting scheduled in several hours.

Establishing specific, measurable objectives is another aspect of MBO that creates problems because it is frequently difficult to transform qualitative goals into specific, obtainable targets. For example, a goal might be to improve agency relations with the semipublic recreation sector. Specific objectives might be to (1) have an informal meeting with one of the administrative heads at least once per month, (2) offer to establish a joint program at least once every two months, and (3) host a gathering of all key semipublic personnel prior to July 1. Even if these targets are achieved, there is still a question as to what degree, if any, the goal has been achieved.

Also, there is a common tendency on the part of management to want immediate results from MBO. It is a mistake to attempt the implementation process quickly or to expect quick results once implemented. A certain amount of time is necessary before employees can break old habits and feel comfortable with a new system. Finally, certain MBO systems involve a considerable amount of paperwork, and this understandably annoys personnel. Many effective MBO systems utilize very little paperwork, however, and stand as good examples that large quantities of paperwork are not necessary.

These problems should not overshadow the fact that many organizations do function haphazardly, without goals and objectives, and a management by objectives system would probably do a great deal to improve their efficiency. One frequently mentioned benefit of a smooth functioning MBO system is improved communications. When supervisors and subordinates begin talking about responsibilities, tasks to be accomplished, allocation of resources, and timetables; everyone gets a clearer understanding of what is happening and what needs to happen.

A second commonly cited benefit relates to improved planning. If the plans of various departments and divisions are integrated, better coordination should result. The same applies for individuals working within the various organizational units.

It should be understood that many options are open to the manager who is considering MBO. In some cases it may be better to implement only a modified system or only at the upper organization level or only in certain organizational units. Implementation by phases is another alternative. Some excellent resources have recently been published which will assist the park and recreation manager who is considering MBO. Without a doubt, MBO has a great potential for many park and recreation operations and will get more attention in the years ahead.

## HUMAN NEEDS AND ORGANIZATION

There have been countless discussions of human needs and the means by which they can be satisfied on the job. If these need satisfications provide triggers for motivating an employee to perform more effectively, they are, indeed, important to management. The question to be answered is what needs are satisfied by employees on the job and how do they rank in importance. Therefore one means of understanding human behavior and motivation is to look at the basic needs of people and to relate them to on-the-job satisfactions. It is a truism that a person behaves in a manner that best satisfies his or her needs.

It will be the approach in this chapter to look at psychologist Abraham H. Maslow's satisfaction-of-needs theory[6] inasmuch as it is widely accepted and

[6]Maslow, Abraham H., *Motivation and Personality*, 2d ed. (New York; Harper & Row, 1970).

quoted because of its ease of understanding. He hypothesized a need priority of five levels.

1. *Physiological*: Need for shelter, air, food, and sex, among others.
2. *Safety*: Need for security and protection against danger, deprivation, and protection from emotional harm.
3. *Love*: Need for affection, acceptance, social activity, belonging and friendship.
4. *Ego Esteem*: Need for self-respect, achievement, status, recognition, attention, and self-esteem.
5. *Self-Actualization*: Need for self-fulfillment, self-realization, growth, accomplishment, and achieving one's potential.

The basic physiological need or the need for survival is placed on the bottom of the hierarchy in order of priority, but as it is satisfied, or partially satisfied, need number two becomes more dominant. In other words, as a need is satisfied, a person then seeks the next higher need. These needs, therefore, operate in a sequence of domination.

Actually, Maslow's hierarchy of five needs can be categorized into higher and lower levels, with physiological and safety needs placed at the lower level and love, esteem, and self-actualization at the higher level. This was done on the premise that people's lower order needs are predominantly satisfied through economic effort, which relates to salaries, wages, tenure, working conditions, and work environment. On the other hand, higher level needs are more intrinsic in the sense that they are satisfied internally by the employee through his or her efforts toward gaining a sense of belonging, acceptance, status, recognition, esteem, growth, self-respect, and achieving one's potential.

It is these higher order needs that provide a challenge to administrators, managers, and supervisors, since most employees in a park and recreation organization are past the physiological and safety need level. This suggests that in the future the satisfaction of these higher order needs may dominate future negotiations between the employee and the employer. Although labor negotiators have not placed pressure on the employers to satisfy these higher level needs on the job, we may expect pressure for them to do so in the future.

Many classifications of needs have been written and explored. While Maslow poses the satisfaction-of-needs theory, Frederick Herzberg proposes a motivation-maintenance theory,[7] David McClelland developed research around people's need to achieve,[8] and Victor Vroom presents his expectancy theory.[9]

[7]Frederick Herzberg, *Work and the Nature of Man* (New York: World, 1966).

[8]David C. McClelland, *The Achieving Society* (Princeton, N.J.: Van Nostrand, Reinhold, 1961).

[9]Victor van Vroom, *Work and Motivation* (New York: Wiley, 1964).

Vroom states that a person's desire to produce at any given time depends on his or her particular goals and perception of the relative worth of performance as a path to the attainment of these goals.

As we examine Maslow's needs theory, it is apparent that managers are most concerned with the higher order levels of employee needs or those above the physiological and safety categories. These basic needs are essentially finite whereas needs of the third, fourth, and fifth categories are essentially infinite. Actually, these five orders of needs are somewhat artificial in that the whole person cannot be separated; all needs interact, as can be seen when examined.

## Physiological Needs

All people have needs that pertain to survival and body maintenance. Indeed, this is a basic survival need. Its satisfaction relates to food, drink, shelter, rest, and the like. Furthermore, it relates to management in that the work environment should provide for adequate heat, light, ventilation, and attractive physical work environment.

## Safety Needs

These needs focus on peoples' desire for security, certainty, stability, and safety in life. Not only is it applied to a desire for economic security but also embraces emotional and psychological security. Although most attention is focused on economic security, emotional insecurity is deeper. Most park and recreation enterprises provide economic security to their employees through steady employment, pensions, and unemployment and disability insurance,

A more subtle problem is the need for emotional security. Some employees have doubts as to whether they can cope with present and future job requirements. An employee likes routine, established social relationships, regular schedules, and knowing the rules of the game. Every employee wants to know whether economic or technological changes will affect established and routinized work relationships. Employees need the assurances that they can cope with new conditions of work and change.

Employees who feel insecure in their jobs will many times express their tensions by criticizing and making caustic remarks about the organization and its management. These soured employees present problems that face nearly every recreation organization. Needless to say, these employees are extremely insecure, and any remark they overhear is distorted and misinterpreted, even though it was not directed or applicable to their job situations. Such employees need constant approval, and most supervisors cater to them with the hope that the problem can be resolved. Unfortunately, some of these employees never change.

## Love and Belonging Needs

All people have an inner desire for love, affection, and social activity. A person must receive the affection of someone to feel appreciated, important, and valuable, if he or she is to feel a part of things. While it might be stated that this need should be met off the job, it nevertheless is important to management, because a worker spends one-third to one-half of his or her waking hours on the job. Furthermore, employees develop strong friendships that make their jobs more bearable. They belong to informal social groups, which were discussed earlier, which provide companionship and give the employees a sense of belonging. Such socialization makes the job more rewarding. Indeed, just working together as a team helps build morale.

It is true that most employees want to establish a positive relationship with their supervisors. They want to be treated fairly and like praise when it is deserved. Management can create an atmosphere of openness and freedom from stress and tension by establishing relationships between supervisors and workers that are friendly and compatible within the framework of the purpose of the organization. When an employee works in a vacuum with no opportunity to develop friendships, affection, social acceptance, or a sense of belonging, he or she feels lonely, isolated and left out, and needless to say, work output declines.

## Ego-Esteem Needs

Ego-esteem needs can be described as (1) a desire of people for achievement, mastery, and competence and (2) a desire of people for prestige, status, importance, and recognition.

These needs can be met on the job by managers establishing a work environment that provides for challenging work assignments, recognition of excellent performance, open communication and performance feedback, and involvement in goal setting.

## Self-Actualization Needs

This need relates to a person's ability to accomplish and achieve, to fulfill one's potentialities. An employee tries to find meaning in his or her work and to reach his or her highest level of personal growth.

This need varies with different individuals. Managers and supervisors should be aware that one employee may achieve self-actualization through a high quality of performance, whereas another will find it through creative effort or being given new responsibilities. By being sensitive to the needs of employees, management can provide different and varying approaches to achieve this need. Many employees never reach this fifth level of need. Some never have the opportunity of reaching it; others choose not to for one reason or another.

## Implications of Need Satisfactions

An understanding by management of the needs of employees can have lasting and positive results on the effectiveness of an organization. Although most of the needs as previously stated are self-evident, they nonetheless challenge management to provide on-the-job satisfactions for employees. But this is not easy, because the fulfillment of these needs calls for a deep understanding of human behavior and its relation to motivation. However, managers are aware that needs satisfaction must be a basic component of their operational policy and that psychological realities must be integrated into operational practices.

Every recreation enterprise is influenced by a manager or supervisor's perception of those people with whom he works, be they leaders, maintenance workers, specialists, board members, and the like. Therefore the human component is the major asset through which the organization can function effectively. Thus management can benefit by investing in its human resources through the development of programs that enhance skill and competence of both its professional and nonprofessional staff. Recognition of the needs of employees leads to a better understanding of the conditions by which growth-enhancing programs can be developed.

## Operational Generalizations on Motivation

It has been the experience of the author that certain organizational practices and policies will greatly improve morale and inspire motivation leading to increased efficiency of a park and recreation enterprise. These are:

1. *Employee Trust.* Assurance that management has a feeling of trust in its employees is basic for good operational practices. Since an organization can only accomplish its goals on the basis of the quality of its employees, management must have explicit trust that they will live up to work expectations.
2. *Salary and Merit Pay.* Good salaries, promotional opportunities, and regular salary increments on a merit basis for those who are deserving should be a part of the personnel policy of the organization.
3. *Opportunity for Expression.* Employees should have the opportunity to express themselves concerning good or bad operational practices without a feeling of reprisal or fear of position security.
4. *Praise for Work Well Done.* Employees should be praised for work well done, but on the other hand, work that is not up to expectation should be called to the attention of the employee through a means of

communication that provides an opportunity to discuss how the work performance can be improved.

5. *Decision Making.* Problems should be approached in a positive manner, and employees should be given explanations of why and how decisions were made, if such decisions affect their operation or position.
6. *Honesty.* Management should be completely honest with employees and not misrepresent any information that would be of concern to them.
7. *Career Opportunities.* New opportunities should be provided workers who reach the top of their job classification, thus extending the career ladders of such employees. None should be "dead-ended" in a position that has no promotional opportunities without a clear understanding between management and the employee that such a position leads nowhere without transfer to another position.
8. *Work Environment.* All employees' work environments should be planned to provide adequate lighting, ventilation, cleanliness, and work space that is adequate.
9. *Use of Titles.* Attractive titles give a worker an opportunity to gain status and recognition in an organization.
10. *Place in the Hierarchy.* Most employees wish to be in a position as close as possible to a high level administrator. Adding new levels in an organizational structure can create anxiety and loss of status for workers if they feel they are down graded in the hierarchy, although the position remains the same. A good example of this has been when park superintendents with years of experience feel they are demoted when they are required to report to a director of parks and recreation instead of to the park board, as was the previous practice. Even with no change of duties, title, or pay, their feeling of being demoted with loss of status could not be quelled.
11. *Opportunity for Growth.* All employees want the opportunity to grow and develop in the organization. This can be accomplished through in-service training programs and developmental plans sponsored by the organization.

## BASIC ASSUMPTIONS OF MOTIVATION

The following basic assumptions underlie an effective program of motivating employees.[10]

[10]Prepared with some modifications by Miles C. Romney, Vice-chancellor for Academic Affairs, State System of Higher Education, Eugene, Oregon, October 1, 1979.

1. Park and recreation professionals, no less than workers in other occupations or professions, are guided in what they do by a consuming desire to satisfy their own "felt" needs. They will give their fullest attention to the improvement of their program when it is evident to them that this is the surest route to the fulfillment of the needs they feel.
2. Need satisfaction for the recreation professional is rooted, on the one hand, in the indwelling sense of fulfillment that comes from work superlatively done and, on the other, in his or her desire for prestige, security, or authority.
3. Extrinsic motivation can be expressed in either a positive or negative way. Positive motivation is represented by factors such as promotion, salary increments, awards, praise, granting of authority over others. These enhance the professional worker's need satisfaction. Negative motivation—which reduces the professional worker's need satisfaction—is represented by factors such as unwanted or low-esteem assignments, withholding of salary increments or promotion, criticism from the supervisor or department head who is in a position to control salaries, promotion and the other emoluments at the disposal of the organization.
4. Positive motivation, as simple and easy as it may appear, unless it is properly used, may generate misunderstanding, hostility, or even rejection by the persons it was intended to benefit, for example, in industry, where on some occasions efforts of management to improve the workers' lot have sometimes been seen by workers as a sly way to gain further ascendancy over the worker by disarming him or her with favors granted in a paternalistic manner.
5. Experienced professionals can easily read and understand an organization's ethos. Its sum is more than the total of its parts. Thus when the ethos says clearly to the employee that the institution's commitment is to effective program planning, the message is not easily lost on him or her.

   The professional draws his or her clues of the organization's devotion to and interest in an improved program of services by the things the organization does to express that interest and to reward those who demonstrate a capacity for a truly dedicated and professional level of work. He or she seeks these clues through an examination of such matters as:

   **(a)** Is there an emphasis on the amount of time given for in-service training and employee improvement programs in recruitment of employees?

**(b)** What solid evidence is there of the manager's concern that the organization be known for the quality of its program? What evidence exists that the manager seeks in any systematic fashion to keep informed of the state of the program in the organization or of the efforts being made to encourage continuing interest in a good program of services?

**(c)** Is there any evidence that the department heads and supervisors are seeking in any continuing, systematic fashion to keep informed as to the state of the program in the organization or the efforts being made to encourage continuing interest in a good program of services?

**(d)** Is there any kind of organizationwide agency having special responsibility for promoting leadership training?

**(e)** What kinds of special provisions are there to encourage individual employees interested in the improvement of their job? For example, what funds, facilities, and time are allocated to improvement of skills and the like?

**(f)** What efforts are there to be made in a systematic fashion to consider leadership abilities and achievements in the making of decisions as to promotion and salary increases?

6. Park and recreation professionals, like others, find that change is not easy and is likely to be even more difficult if it is felt that change is being imposed from above, particularly if it appears that the "administration" (managers, department heads, supervisors) is pressing for changes on the basis of uncritical acceptance of roseate promises of increased efficiency and economy. If the employees sense that the administration is building up false hopes as to what may be achieved by new instructional devices or methods, for example, their resistance may be more a matter of resistance to administrative incursions into the employee's domain than it is to the changes themselves. When this occurs, it is difficult, if not impossible, to secure from the employees an impartial, open-minded assessment of a potentially useful method or device.

   As a practical matter, it is the employees who ultimately must live with, and make operative, any programs or other changes looking to the improvement of services. No effective or lasting changes in services can be achieved without their involvement, cooperation and support.

7. Many changes in park and recreation services, as evidence suggests, have been prompted by outside pressures, or have been aided by such pressures. Organizations tend to be imitative. Program and fa-

cility innovations elsewhere are more likely to be considered to have relevance for the local organizations under the following circumstances:

**(a)** When the innovation is found in a department considered by the employees to resemble closely the local operation. Found in a setting appreciably different from the employees' own departmental setting, the innovation may well be considered by the employee to be untested and unproven. Thereby it loses much of its appeal for the local employees.

**(b)** When the worker can observe, firsthand, the innovation in action. No amount of descriptive material, however well prepared, whether oral or written, can equal in persuasiveness employees visits to an institution not unlike their own to observe firsthand that an innovation works.

**8.** Employees are stimulated to program and facility innovating when their work is the object of overt interest by others—notably their fellow workers and administrative colleagues, particularly from within the organization. The Hawthorne effect apears to be as potent in park and recreation management as it was shown to be in industry and in the army.

In short, simply paying attention to what employees are doing apparently can improve their work; the advantage of this kind of 'improved service' is that it tends to ensure better results (at least temporarily) even from approaches that are inherently no better than those they replace.

**9.** Any systematic plan for the improvement of service in a park and recreation organization that aims to serve diverse community interests must recognize that employees differ in their leadership abilities, in their interest in efforts to improve their work, and in their readiness to participate in movements aimed at improving their service output. This has two important implications for any organizational plan for seeking to stimulate interest in the improvement of their work:

**(a)** No organization can move on a solid front involving all of its staff, for all are not ready. Organizations must, of necessity move on a broken front involving those employees or leaders who are in a state of readiness, meanwhile seeking ways to bring others to that state as soon as possible. This commonsense principle is supported by psychological theory that suggests that: (1) when an individual is ready for action, action is satisfying, (2) when an individual is ready, and he or she is not permitted to act, he or she experiences a sense of frustration, and (3) when an individual is not ready to act and is nonetheless obliged to by an outside force, he or she may feel both threatened and frustrated.

**(b)** The organizational plan must have numerous, different aspects to it, in order that all the diverse interests represented by those employees who are psychologically ready to act can find some aspect of the plan of interest to them, at the level at which they are prepared to act.

Employees must begin at the point where they are in their thinking. Some being further advanced in their thinking than others are prepared to operate at a different level from those less well prepared. The important thing in employee improvement is not that everyone begin at the same level in their efforts to improve but, rather, that each begin at the level of his or her present understanding and need and that he or she progress from where he or she is to some higher plane. The author's experience with park and recreation leaders over a period of many years bears out in dramatic fashion the importance of this principle.

## ORGANIZATIONAL CHANGE

In any discussion of human behavior in organization, it is important to look at the ramification of change and its effect on operational practices. We live in a world of change. In fact, more changes have taken place in all facets of life during the past 50 years than in the previous history of human existence. If park, recreation, and leisure service organizations are to survive in today's world, they must face the challenge of change and react to it in a positive fashion.

It is not easy to bring about change in an organization for the reason that employees like the traditional, orderly, and the established way of doing things. A change can be a threat to familiar patterns of work and can also pose a threat to job security. Therefore patterns of resistance develop among employees to thwart or avoid change when possible.

### Resistance to Change

Change is related to being up-to-date, to being modern, to being innovative in reaching organizational goals. But the employee looks on change as to its effect on him or her. One of the most threatening changes in an enterprise relates to technological advances or automation that may or may not eliminate jobs. Others negatively relate to reassignment of duties, the uncertainty of coping with new equipment, new operating procedures, and new supervisory personnel. Further, any change that poses a threat to an employee's status will meet with strong resistance as will any changes that affect ongoing informal social relationships.

In general, nearly every aspect of change in an organization will meet with degrees of resistance if it affects in one way or another the work of an employee. Knowing this, management is challenged to initiate needed changes in a manner that produces the least resistance.

### Acceptance of Change

How can a manager bring about change with a minimum of resistance. One way, of course, would be through economic incentives. This means that an employee would be assured of job security and an opportunity for increased earnings. Knowing that management will guarantee no loss in pay to the employee and knowing that a proposed change will provide an opportunity for a possible increase in pay can allay suspicion of loss of job security. Even with job security the employee has fear of the unknown concerning other aspects of his or her job. Therefore management can diffuse much of this fear through a system of communication that lets employees know what is going to happen, why it is going to happen, and what the results will be. A two-way communication system between supervisors and employees concerning a proposed plan involving change can do much in eliminating resistance to it. In addition, a procedure that brings the two groups together can bring forth new ideas and enrichment of the proposal by those that will be involved in implementing it.

Another means by which resistance to change can be reduced is through a system of involving employees in the decision-making process. Many recreation enterprises use this method effectively. However, a note of caution should be brought to the attention of those involved in the process. Decision making by consensus is not necessarily a good management tool. But if used properly in getting ideas from employees, it gives the worker a feeling of having some control over his or her work environment and job practices. Too many times this practice has been used as an "out" for supervisors or executives when tough decisions must be made.

Group decisionmaking should only be used to get the feeling of the group concerning change and a commitment by its members that they will accept the ultimate decision when made by the supervisor or executive. Further, group decisions or acceptance should only be made on matters or issues within the scope of the experience of the group. To bring together staff or employees to discuss matters foreign to them would only be an exercise in futility.

Change can also be made with a minimum of resistance by placing it on a "trial basis." This is not always possible, of course, in those instances when changes must be introduced quickly. But changing operational practices, introducing new procedures, reorganizing areas of responsibility, and the like can be introduced with the understanding that the change is tentative and will only be accepted or rejected after a period of trial and evaluation. This procedure has built in weaknesses in that employees who have fixed attitudes against change may undermine the program, and those for it may overreact to its potential. Needless to say, it does provide a testing period and a climate that generates less resistance by employees when they see firsthand the result of the action taken.

It is important that supervisors or managers do everything possible to let their employees know what is happening. The lack of knowledge about what is going on provides a threat to employees, which in turn leads to resistance. However, there are times when changes must be made and resistance to them will be evident by some employees. Unfortunately, it is not possible to satisfy everyone in all instances. A decision must be made, and the supervisor or manager can only counter negative reactions by being honest as to why the action was taken.

Some believe that the best way to avoid violent resistance is to develop an overall plan, communicate this plan to employees, and then piece by piece make the changes needed until the plan is operational. This is a slow plan method based on the premise that a small change is less disruptive than a big one. Without doubt, such a procedure does work. However, some outstanding executives believe that judgment is the determinate of what plan, if any, should be used. If the executive or supervisor has the trust of his or her employees who know that fairness will be practiced, very slight resistance evolves. Actually, both quick and slow changes can be made on the basis of this same premise.

In sum, if changes are made in an organization and consideration is given to the human element of the effect such changes have on employees, the problem of resistance is slight. Employees have a valid interest in their jobs. They do not like uncertainties; they do not like a threat to the status quo; and they do not like a threat to their job security. Management needs to know that problems involving change need to be brought out in the open and that open channels of communication are basic to sound operational practices.

## Caution for Change

A word of caution needs to be brought to the attention of those employees, supervisors, or managers who plan to introduce change in an organization before they are thoroughly aware of the implications of a proposed change. This is especially true of new employees who assume they need to prove themselves as capable, and knowledgeable in their new position. A new employee does not need to feel he or she must make immediate changes in the operation of an organization unless an emergency situation exists. It is far better for the new employee, supervisor, or manager to become thoroughly acquainted with his or her work environment; to know the strengths and weaknesses of the organizational structure; to know employee responsibilities; to know the power structure in the community, as well as in the enterprise itself; to know the background and historical reasons for certain practices that may be unique to the organization; and to feel secure in knowing the general operation of the position they are filling.

### Questions For Discussion

1. What is the significance of the Hawthorne experiments? What did the experiments prove or disprove?
2. Explain fully McGregor's Theory *X* and Theory *Y*. Which do you prefer? Why?
3. How do informal groups become established in a park and recreation enterprise? Do they help or hinder a park and recreation executive in the operation of the department?
4. How would you motivate some members of your staff to work more effectively if such motivation is necessary?
5. What is meant by Management by Objectives (MBO)? Do you feel it is an effective tool for management? Why or why not?
6. Why should "human needs" of workers be of importance to executives and supervisors in a park and recreation enterprise? How do you determine these needs? What are the five levels of need priorities Maslow discusses?
7. Even though organizational change is necessary, why are some employees against implementing such changes? What are some ways of bringing about such changes with a minimum of resistance?
8. What characteristics distinguish informal from formal organization?
9. How can management help employees achieve their need for self-esteem?
10. How important do you think security is to the employee of today? Are other needs more important? How would you categorize your needs?

## SELECTED REFERENCES

Drucker, Peter., *Management*. Harper & Row, New York, 1974., pp. 312–366.

Edginton, Christopher R., and Williams, John G., *Productive Management of Leisure Service Organizations: A Behavioral Approach*. Wiley, New York, 1978, Chaps. 3, 4.

Gibson, James L.; Ivancevich, John M.; and Donnelly, James H., Jr., *Organizations: Structure, Processes, Behavior*. Business Publications, Dallas, Tex., 1973, Chap. 8.

Gross, Bertram M., *Organizations and Their Managing*. Free Press, New York, 1968, Chaps. 9, 10.

Harris, Ben M., *Supervisory Behavior in Education*. Prentice-Hall, Englewood Cliffs, N.J., 1963, Chap. 10.

Massie, Joseph L., and Douglas, John, *Managing: A Contemporary Introduction*. Prentice-Hall, Englewood Cliffs, N.J., 1973, Chaps. 3, 5, 8.

Newman, H. William; Sumner, Charles E.; and Warren, E. Kirby, *The Process of Management*. Prentice-Hall, Englewood Cliffs, N.J., 1972, Chaps. 7–10.

Robbins, Stephen P., *The Administrative Process*. Prentice-Hall, Englewood Cliffs, N.J., 1976, Chaps. 16–18.

Sayles, Leonard R., and Strauss, George, *Human Behavior in Organizations*. Prentice-Hall, Englewood Cliffs, N.J., 1966, Chaps. 1, 4, 6, 9.

Stoner, James E., *Management*. Prentice-Hall, Englewood Cliffs, N.J., 1978, Chaps. 2, 12, 16.

# CHAPTER 4

# Leadership

Leadership is a popular word, which is surrounded by many myths and misconceptions of what it is, what it purports to do, and how one establishes a leadership identification. Indeed, the literature constantly makes references to its importance.

A park, recreation, or leisure service organization cannot function with any degree of efficiency without dynamic leadership. It is the oil that lubricates the operation; it is the unifying force that gives vitality to the system; it is the mortar that holds the organization together.

Because the word "leadership" has become so popular, it is understandable that it has become cloaked in widespread misunderstandings and misconceptions. Too often, the term has been used to identify a leadership type who possesses social, physical, and personality traits that endow him or her as the "born leader." In some recreation systems this fiction is believed. Because of the confusion and conflicting points of view with respect to leadership, it is well to define the term, present some generalizations on its nature, and translate the role of leadership into operational practices. Therefore this chapter will define leadership, explain leadership roles, and point out current practices used by professional recreation and park personnel in achieving leadership roles. Indeed, it is the wise administrator or manager who searches for the real meaning of leadership and how it can be harnessed.

## DEFINITION OF LEADERSHIP

Leadership is the inspiration, the energy, and the motivating force that transforms a group into a conscious and purposeful action body. In other words, leadership is the motivating force that triggers action toward the achievement of organizational goals.[1] Tead states that leadership "is the activity of influencing people to cooperate toward some goal which they come to find desirable."[2] Pfiffner and Presthus describe it as "the art of coordinating and motivating individuals and groups to achieve desired ends."[3] Morphet, Johns, and Reller state that "any person provides leadership for a group when he (1) helps a group to define tasks, goals, and purposes, (2) helps a group to achieve its tasks, goals, and purposes, (3) helps to maintain the group by assisting in providing for group and individual needs."[4]

Clearly, all definitions of leadership imply that it is a motivating force that moves a group toward an accomplishment that they believe is desirable. People must be involved; a person cannot be a leader unless he or she is interacting with another person or persons. There must be communications media through which thoughts and ideas are transmitted from person to person, and action or movement toward achieving goals or purposes. Unless the group is motivated positively toward achieving the organizational or group goals, and unless there is positive movement toward their achievement, the leadership process breaks down.

Using these definitions, we see that leadership is associated with leader behavior, and thus a "status" or "titular" leader may or may not be a leader in the true sense of the word. To explain further, a park and recreation director is not *ipso facto* a leader. Actually, he or she maintains a command position or has a leadership role that is recognized as carrying authority within the formal organization. He or she may be a leader as well as a holder of a "headship." If so, he or she is displaying evidence of effective administrative behavior in working with individuals and groups in bringing them into congruence with respect to recreation purposes and goals. This will be explained further under "formal leadership."

## LEADERSHIP ROLES

The question is asked to how an individual becomes a leader. Separating fact from fiction that surrounds this term is not an easy task, but the means by

[1]Ordway Tead, *The Art of Leadership* (New York: McGraw-Hill, 1935), p. 20.

[2]Ibid.

[3]John M. Pfiffner and Robert V. Presthus, *Public Administration* (New York: Ronald Press, 1967), p. 88.

[4]Edgar L. Morphet, R. L. Johns, and Theodore L. Reller, *Educational Organization Administration* (Englewood Cliffs, N.J.: Prentice-Hall, 1967), Chap. 5, p. 127.

which a person is conceived as a leader can be pointed out. Broadly speaking, a person may acquire a leadership role or be identified as a leader through:

1. Inheritance or symbolic status.
2. "Halo" effect on preeminence in an identifiable field.
3. Formal appointment, election, or designation by a higher authority.
4. Group emergence or functional identification.

In park and recreation management, little concern needs to be given to categories (1) and (2), except in an academic way. The latter two categories are of greater importance, although any listing without qualification can be questioned. Actually, a study of research findings on leadership leaves much to be desired. Nonetheless, an understanding of the leadership roles an individual acquires in an organized park and recreation system can be of value to administrators and supervisors alike. Leadership is a human phenomenen; it is a force that operates among people; it guides activities in a given direction toward specific goals. And since a park and recreation system is largely made up of people, it demands understanding.

## Inheritance or Symbolic Status

Many examples exists of a person receiving a leadership role by virtue of inheritance or birth, whether through a royal family or a religious sect. Whereas such a person is a symbolic leader and may have real power, he or she does exert leadership of his or her many subjects through establishment of a leadership symbol. Far removed from his or her people, except in a prepared role, the leader can create an image as one with divine power or superhuman abilities. The value of this image as a rallying force in times of crisis cannot be overestimated.

The technique of developing a leader image in the minds of masses of people is well known by ruling groups as well as by contemporary politicians. There is no magic in this. Actually, a manager of a park and recreation system in a city of sufficient size where remote relationships with the public exist can create an image of dynamic leadership through a planned program of publicity and public relations. Indeed, many executives have even created an image of infallibility with their professional peers through speeches and innumerable involvements in district and national professional conferences, workshops, institutes, and committees.

## Halo Effect or Preeminence

Some individuals have reached such high levels of achievement in areas of specialization that a "halo" image has developed around them, and they are identified as leaders in their fields. Their opinions are highly respected and they hold positions of great influence. Indeed, they are the top men in their fields of specialization. But such expertness carries certain hazards with it. Too often, people look on these individuals as omniscient and capable of carrying a leadership role in areas outside their fields of specialization. Such eminent figures are sometimes led to believe that their influence gives them license to speak with authority and to influence others in fields completely unrelated to their own.

Examples of this can be found among prominent athletic figures who, because of their athletic prowess, assume the role of leaders of youth, scientists who take the platform to influence directions in various social services, and entertainment personalities who act as counseling authorities. Actually, it is only in those instances when competencies are related that a person can transfer his or her expertness to other areas of specialization. Of course, we are not implying here that an individual does not have the right of free speech or oral expression. We observe, however, that a prestige figure by the very nature of his or her eminence does not have unlimited powers of leadership and infallibility in every other aspect of human endeavor. Such transfer does not take place notwithstanding the halo image that is presented.

## Formal Leadership or Designation

By "formal leadership" is meant the leadership role that is given an individual by reason of his or her holding an official position or office in the hierarchy of an organization. Specifically, this leadership status is associated with the position rather than the individual. By designation through appointment or election, he or she is officially the head of the organization and his or her leadership status ceases when the position or office is vacated. Actually, such leadership is the kind found in many formal organizations; in fact, it could be called "official leadership," or "titular leadership."

The park and recreation executive and many members of the supervisory staff hold formal leadership posts. Many aspire to these positions by reason of the authority, status, and remuneration given them. There is no question but that these institutional positions carry a leadership image. At least, the authority vested in them causes people to pause and perhaps obey. However, to see leadership only as a power or command tool is to overlook the real nature of leadership as a cooperative and shared experience. The mere holding of a position as a recreation and park superintendent is no indication that the incumbent will live up to what is expected of his or her leadership. Nor will it guarantee that he or she can sustain and stimulate the interest of the staff in striving to reach recreational goals. Nonetheless, there is no question that some positions in a recreation and park system provide better opportunities to demonstrate qualities of leadership than others. Recreation and park managers, supervisors, and unit heads in a system have excellent platforms to project their programs and to enhance their image by the very structure that exists.

## Emerging or Functional Leadership

Most authorities agree that leadership is group oriented and involves motivation of group effort. Therefore the person who helps a group achieve meaningful goals or directs activities toward ends that are worthwhile to the group is the one endowed with leadership ability. In other words, this emerging or functional leader is one who works as a guiding, coordinating, motivating, and unifying force within the group, in helping it get what it wants with a minimum of friction.

On the other hand, some students feel that leadership is a team effort, that it is the group working as a unit toward goals, with each individual contributing to the leadership structure. Regardless of whether leadership is an individualistic or a group function, the persons constituting a group give it form, dimension, and motivation. It is true that a leader within a group must be guided by group values and accepted as one who establishes guidelines toward meaningful goals. Furthermore, it is true that an individual cannot project his or her leadership on the group but, instead, must work within a complex of reciprocal social relationships through which he or she achieves status as a satisfier of the needs of the group. Indeed, a leader must be identified with the group; must be looked on as one who gets things done and satisfies needs; must have a positive relation to his followers; and must be able to translate goals into action.

In looking at actual leadership behavior within the structure of the group, research and empirical studies have found that a leader gives a group cohesiveness and unity of purpose through use of a "human relations" approach and sensitivity to individual needs. Cartwright and Zander express the view that "leadership consists of such acts by group members as those which aid in setting group goals, moving the group toward its goals, improving the quality of the interactions among the members, building the cohesiveness of the group, or making resources available to the group."[5]

## What is a Group?

What is a group? Since executives and/or supervisors work in a group setting and must be sensitive to the dynamics of group interaction and since it is

[5]Dorwin Cartwright and Alvin Zander, eds., *Group Dynamics: Research and Theory* (New York: Harper & Row, 1953), p. 538.

agreed that leadership is group oriented, it seems apropos to examine the nature of a group.

A group must consist of at least two or more persons who share and interact in working toward common goals. A group or mass of people do not constitute a group until goals are established and interaction takes place. Hemphill gives 10 categories that describe a group:[6]

1. *Size.* The number of members in the group.
2. *Viscidity.* The degree to which the group functions as a unit.
3. *Homogeneity.* The degree to which group members are similar with respect to age, sex, background, and so on.
4. *Flexibility.* The degree to which the group has established rules, regulations, and procedures.
5. *Stability.* The frequency with which the group undergoes major changes or reorganizations.
6. *Permeability.* The degree to which the group resists admission of new members.
7. *Polarization.* The degree to which the group works toward a single definite goal.
8. *Autonomy.* The degree to which the group operates independently of direction by other or larger groups.
9. *Intimacy.* The degree to which group members are acquainted with one another.
10. *Control.* The degree to which the group restricts the freedom of members' behavior.

From this list we see that many intangible factors are involved in the social phenomenon known as "leadership." Furthermore, individual needs are based in large measure on social values and consideration of human relationships working within the dynamics of group interaction. The implication for recreation leadership seems obvious. The leader must (1) work within the group to make activity meaningful, (2) establish a climate that motivates cooperative effort toward group goals, (3) emphasize employee interests, (4) give fair treatment to all, (5) provide for participation in decision making, (6) encourage personal development, (7) recognize employees as important to the function of the organization, and (8) give the staff a sense of belonging, a feeling of security, and a sense of achievement. In short, the park and recreation executive and the leadership staff must always work to create an atmosphere of mutual trust and respect.

[6]John K. Hemphill, "The Leader and His Group," in C. G. Browne and T. S. Cohn, eds., *The Study of Leadership* (Danville, Ill.: The Interstate, 1958), p. 369.

## NATURE OF LEADERSHIP

It is generally agreed by most writers that leadership is a group phenomenon—the product of human interaction; it is not the product of status or position. Thus symbolic status, "halo" effect, formal designation (formal leader), and group emergence (functional leader) indicate the setting of leadership. The major focus remains in the direction of group effort and group values as found in the functional leader.

Before looking closely at the interactional group approach to leadership, we will view other avenues to the study of this phenomenon. These approaches have been identified by some writers as theories of leadership. However, they will be considered here as (1) the trait approach, (2) the situational approach, and (3) the group approach, which was previously discussed.

### The Trait Approach

The trait approach of leadership is based on the premise that an individual must possess certain personal and social characteristics, or traits, in order to achieve leadership. In other words, this approach stresses the point that individuals achieve their leadership role and status as a result of personality. Such beliefs led to the "great man" theory that an individual is destined to greatness and leadership because of unusual traits or uncommon abilities. And many people still view this explanation as completely plausible and realistic.

Many studies have been made to determine the validity of the trait approach.[7] Some attempted to find a consistent pattern of the characteristics of leaders. Others looked to a single core of traits related to leadership achievement. However, from this research, no clear evidence has emerged that leadership exists primarily in the individual; rather, personal characteristics are only a part of the total relationships and interaction between the leader and members of his or her group. Furthermore, no individual has a cluster of leadership traits or a leadership personality that he or she brings into every type of situation to motivate people toward their objectives. Indeed, a potential leader must bring to a group a combination of traits, or characteristics, that integrate in a positive way with others in reaching goal situations. A person having a leadership role in one situation may not have one in a different situation. This has given rise to the view that leadership is related to the function or the situation.

[7]Good discussions are presented in Cecil E. Goode, "Significant Research on Leadership," *Personnel* (March 1951) 342–350; R. M. Stogdill, "Personal Factors Associated with Leadership: A Survey of the Literature," in G. G. Browne and T. S. Cohn, eds., *The Study of Leadership* (Danville, Ill.: The Interstate, 1958), p. 58; and Charles Bird, *Social Psychology* (New York: Appleton, 1941).

However, before leaving the "trait approach," it is important to point out that there is some evidence to suggest that personal characteristics may play an important part in affecting group behavior, even though research has had only moderate success in relating them to leadership. Therefore a discussion of the "trait" approach should not be dismissed completely, in view of the fact that some authorities do believe that personal traits are a factor.[8]

Stogdill brings out the point that the personal factors associated with leadership can be classified as follows:

1. *Capacity.* Intelligence, alertness, verbal facility, originality, judgment.
2. *Achievement.* Scholarship, knowledge, athletic accomplishments.
3. *Responsibility.* Dependability, initiative, persistence, aggressiveness, self-confidence, desire to excel.
4. *Participation.* Activities, sociability, cooperation, adaptability, humor.
5. *Status.* Socio-economic position, popularity.[9]

Stogdill is quick to point out, however, that such a classification is not surprising, inasmuch as a person must work within the group setting to achieve his or her leadership status, but "a person does not become a leader by virtue of the possession of some combination of traits, but the pattern of personal characteristics of the leader must bear some relevant relationship to the characteristics, activities, and goals of the followers."[10] In other words, although traits may be in evidence, they are only important insofar as they relate to the activities and goals of the group.

Other researchers have been more positive in singling out certain traits, or personality qualities, that may be related to leadership. A few of these are energy or drive, intelligence, emotional maturity, ability to work with people, ability to communicate, and enjoyment in a leadership role.

**Energy or Drive.** With few exceptions, promising leaders have a great amount of energy and drive. They have initiative, motivation, and a keen desire to succeed. They are bundles of nervous and mental energy. In fact, they are action oriented and give push and direction in motivating groups toward solution of goals. Although they have high energy output and are aggressive in goal seeking, leaders usually maintain composure, and rarely do they show indications of tension, anxiety, or extreme nervousness.

[8]Cecil A. Gibb, "Leadership," in Gardner Lindzey, ed., *Handbook of Social Psychology* (Reading, Mass.: Addison-Wesley, 1954), Vol. II, p. 879.

[9]Stogdill, op. cit. (note 7), p. 58.

[10]Ibid., p. 58.

**Intelligence.** It would be hard to conceive of a leader without at least normal intelligence. On the other hand, it is not necessary for a leader to have exceptionally superior intelligence. In fact, studies have shown that if there is too great a discrepancy between the leader and his or her followers, the leadership function is hampered. Actually, it is not high intelligence itself that makes a leader, but the need for intellectual flexibility in working with people or groups. Also, the very fact that a leader is called on to deal with abstract relationships and complicated problems indicates the need for intellectual capacity.

**Emotional Maturity.** Leaders tend to be emotionally mature. And this is essential if a person is to lead effectively. Actually, it would be practically impossible for an immature, selfish, egotistical, and maladjusted individual to earn leadership status in a group situation. Instead, leaders in general have high frustration tolerance, are well adjusted, and have well-rounded interests.

## Ability To Work with People

Of all the general characteristics needed for effective leadership, none show up more often than that of skill in handling and working with people. Sometimes described as skill in human relations or given a host of other names, this characteristic relates to one's ability in getting along with others. After all, leadership is concerned with human beings; hence, the leader must be "people-minded." He or she must like them and respect their dignity; must deal with their individuality and be skilled in interpersonal relationships. Needless to say, such a leader is generally "well liked." He or she inspires confidence among his or her followers and receives their loyalty. The leader is sensitive to people's needs and knows how to get the most out of them. In short, he or she respects his or her followers, challenges their spirit, and understands their psychological needs.

## Ability To Communicate

Perhaps this characteristic of leadership should be listed under leadership skills. Nonetheless, leaders must be able to communicate with members of their group. Words are the tools of communication; they carry the ideas. The inability to communicate seriously hampers or even cripples a potential leader. Indeed, the person who can communicate effectively in either writing or speech has a tremendous advantage over a person who cannot. This ability should not be confused with the mere mechanics of articulateness, vocal ability, or verbal expression, for the ability to communicate extends beyond speech skill. Ideas must be shared and appreciated. There must be substance in any method of communication. Too often, one hears an expressive but dogmatic address presented by a person who is facile in speech but shallow in ideas. However, it should be pointed out that the ability to handle words and to be expressive is a great advantage to a person seeking a leadership post. Such a skill calls attention to the individual and tends to generate the confidence of the group. Furthermore, if the speaker is convincing, forceful, self-confident, and enthusiastic in presenting his or her ideas, the group will be more likely to give their support than if he or she were contemplative, withdrawn, uncertain, and lacking in strength of conviction. It is unfortunate that many recreation systems place such a high premium on mere articulateness on the part of their leadership staff when actually it is only important when combined with other capacities and abilities. On the other hand, speech skill and other abilities working in unison provide a means of motivating people toward goals, and this, in essence, is leadership.

### Enjoyment in a Leadership Role

Leaders tend to enjoy their leadership role. They like the responsibility of seeing that things are accomplished. Taking risks and making important decisions come easy to them; they are happiest when "things are humming" and goals are being accomplished. Confident of their ability, they do little soul searching about decisions; they have deep convictions about the job to be done and the means of accomplishing it. Usually, in fact, they not only enjoy their leadership role but are happiest when performing it.

H. A. Overstreet further stresses the point of view that there seem to be a number of common attributes or essential qualities needed for recreation leadership. His listing of ten indispensable qualities is summarized as follows:

1. *Wisdom with People.* To project and work with others; a sensitivity to the needs of people.
2. *Community Intelligence.* To know one's community and its people; to know its structure and organization, its institutions, and the interactions of people.
3. *Ingenuity with Material.* To make the most of meagre materials but give people rich experiences from them.
4. *Long Patience.* To be realistic concerning the shortcomings of people; to have a wisdom of insight and not to expect too much but aim at goals of high purpose.
5. *Sincere Tolerance.* To respect the mind and actions of others; to develop a mature philosophy of tolerance toward one and all.
6. *A Sense of Humor.* To see oneself in proportion and see the humor of acknowledging one's shortcomings.
7. *Democratic Attitude and Procedure.* To initiate a climate of participation, develop freedom of action, and provide an atmosphere for growth of human dignity and democratic attitude.

8. *Skill in a Particular Field and Several Avocations.* To be able to do something fairly well; to develop as many skills as possible.
9. *Emotional Maturity.* To be emotionally mature and free from childish regressions and fixations.
10. *A Deep Happiness in the Work.* To find joy and happiness in one's work; to feel deeply and happily dedicated to one's profession and feel that it is worthwhile and purposeful.[11]

## Summary of Trait Approach

Although many interesting points of view are held on the trait approach, it is generally agreed that traits cannot work in a vacuum. Instead, it is the motivation resulting from the interaction of these traits within the group that is important. Traits may function differently in different situations, but viewing them in their relationship to group behavior is all that can be expected. Furthermore, it is apparent that the concept of "traits" is all too broad as given in the preceding list. Physical, psychological, social, and personality traits are used interchangeably; as are learned responses grouped here. Actually, it has been suggested that a "character trait" is "persisting motivation—to some degree unconscious—which may produce variable behavior, but variable within limits."[12] This is not the case with characteristics or learned responses such as communication ability and other skills. Thus many of the personality traits as listed are actually acquired traits that can be learned by training or experience.

Furthermore, Stogdill concludes that his findings "provide 'devastating evidence' against the concept of the operation of measurable traits in determining social interactions."[13] Thus the study of traits would indicate that leadership is not passive nor is it determined by the possession of personality traits. Instead, it is related to group interaction and the manner in which a leader helps a group reach its goals. Furthermore, the trait approach is too much concerned with description and not enough with analysis of the arrangement of traits in a functioning situation. All in all, the present analysis of the trait approach opens up new avenues for further study in this important area of human interaction, and the place of traits in the leadership process should not be lightly discarded until new and sharper instruments for studying them further have been perfected.

[11]Address given before the Society of Recreation Workers of America, Boston, 1939, by Dr. H. A. Overstreet, Professor of Philosophy at the College of the City of New York; also published by *Recreation* (December 1939).

[12]Alvin W. Gouldner, ed., *Studies in Leadership* (New York: Harper & Row, 1950), pp. 40–41.

[13]Stogdill (1958), op. cit.

## The Situational Approach

While the "trait" approach concentrates largely on the possession of inherited capacity or other innate genetic psychological traits, the situational approach stresses the point that leadership is more related to the environment or the situation in which it occurs. In other words, it is related to the social milieu. This would indicate that manifestations of leadership will vary greatly depending on time, place, and circumstances. From this point of view therefore leadership only emerges if the environment is right for the interaction of the leader and the group. Thus different attitudes and responses will vary in time and space, and hence leadership qualities in one situation will not necessarily meet the needs in a different situation.

The conclusion for recreation executives would be that a person should lead best if the specific situation was right for use of his or her skills. However, the situational theory leaves much to be desired. It emphasizes too strongly the importance of environment, or external forces over which the leader has little control. Actually, it is difficult to divorce personality factors from situations in which such traits would be highly valued. Furthermore, it is impossible to separate the leader from the group with which he or she interacts. While specific situations call for certain leadership behavior, it is difficult to deny that there are leadership attributes or qualities that, given the right situation and group, would place a person higher in leadership status than he would otherwise be without them.

## The Group-Dynamics Approach

This approach is based on the premise that leadership is a group phenomenon in which the leader and the group work together in a unified and coordinated relationship to reach group goals. In this approach, leadership is determined by group performance. It implies that a leader interacts with the group and only in a sense leads them, if he or she plays the role to facilitate the achievement of group ends. Thus leadership situations are a cooperative venture in which both the leader and the led actually work together in satisfying their needs.

Furthermore, authorities feel that leadership is a team effort; it is the group working as a unit toward its goals. Thus each individual contributes to the leadership structure. Regardless of whether leadership is an individualistic or a group function, the persons comprising a group give it form, dimension, and motivation. Truly, a leader within a group must be guided by group values and accepted as one who establishes guidelines toward meaningful objectives. Also an individual cannot project his or her leadership on the group, but instead must work within a complex of reciprocal social relationships through which he or she achievs status as an implementer of the needs of the group. Actually, a leader must be identified with group; must be looked upon as one who gets things done and satisfies needs; have a positive relation to his or her followers; and be able to translate goals into action.

In looking at actual leadership behavior within the structure of the group, research and empirical studies have found that a leader gives a group cohesiveness and unity of purpose through use of a "human relations" approach and sensitivity to individual needs. Cartwright and Zander express the view that "leadership consists of such acts by group members as those that aid in setting group goals, moving the group toward its goals, improving the quality of the interactions among the members, building the cohesivenss of the group, or making resources available to the group."[14]

## LEADERSHIP THEORY AND THE PARK AND RECREATION LEADER

Having summed up some of the basic thinking on the conception of leadership, it is now possible to look at its relation to recreation administration. Leadership contributes to group enhancement; therefore common goals that the group feels to be meaningful and worthwhile are basic in the leadership process. Without them, activity is nothing. A recreation leader must interpret group goals and implement them through purposeful work activity. Furthermore, he or she should create an environment in which every group member sees meaning in what he or she is doing, receives satisfaction, and feels a part of the team. It is obvious from the discussion of leadership theory that the leader does not lead by controlling others, that he or she does not lead by reason of having exclusive personal qualities or attributes, and that he or she does not lead just because the environment is right, but that he or she does lead through interaction with the group in bringing out its potential for achieving shared goals. Hence work activity should be purposeful, and the leader should be the catalyst making it purposeful as well as meaningful.

In sum, recreation executives, supervisors, and activity leaders hold leadership posts. Their skill in performing leadership tasks is related to their skill in bringing out the best in others through cooperative effort to achieve the goals established by the recreation agency.

**Some Conclusions.** It should be clear to the recreation and park executive that his or her leadership responsibility is great and that his or her attitude and role in developing a permissive climate for leadership development can do much to improve the operation of the entire recreation and park system. However, it is important to keep in mind that much of what is designated as democratic ac-

[14]Dorwin Cartwright and Alvin Zander, eds., *Group Dynamics, Research and Theory* (New York: Harper & Row, 1953), p. 538.

tion or group leadership is merely a facade for a dominant figure to reach his or her own goals. This is true in many recreation and park systems as well as other organizations. Such leadership is not adequately covered in the theories as listed.

For example, there is little doubt that few groups operate without at least one status person being present, such as the boss, manager, or central figure in the structure of the group. Such a person is perceived as the leader. In fact, he or she is the leader even though he or she may do little in assisting the group toward desirable goals. As the central figure, all discussion is unconsciously pointed toward his or her approval; disapproval is tantamount to rejection of any idea presented. Further, he or she formally or informally colors every decision, even those of little significance. Moreover, his or her dominance is usually unknown to the group. Actually, members of a group are not conscious that they are merely "puppets" acting under a guise of democratic action. In fact, this central figure would be the most surprised if he or she were accused of manipulation or dominance.

That such administrative leadership situations exist in many organizations is a well-known fact. But perhaps groups or staff are oriented toward this pattern and not only accept it but even expect it. Actually, this central figure is a leader because he or she is accepted by his group as such. To call him or her a formal or appointed leader, as discussed earlier, does not clearly settle this issue, for in one instance the individual is recognized by the group as the status leader because of appointment or otherwise, while in this case the individual performs a leadership role in a manner that is not apparent to the group members. Resulting adjustment is identifiable in that members protect themselves; they keep quiet or else only identify their thoughts that are in keeping with known ideas of the central figure. They refuse to "stick their necks out." Creativity is stifled, and the group becomes stabilized with individual roles accepted. Some members never speak out; others only identify with the central figure in all matters of concern to the group or organization. In actual practice, the group accepts this institutionalized concept of leadership and in most instances it goes unchallenged. Thus to keep the status quo is important, or at least it is desirable not to go too far off course from present practices. Of course, this is not the desirable pattern of leadership in action, but since it exists, it should be viewed and changed as new understandings of the nature of leadership come forth. After all, about all that can be done at this juncture is to point out that leadership is a complex of social relationships and variables that involves the human element, the personality factors, the needs and structure of the group, and the situation. In sum, to view the known elements of the interaction of the leader and his or her group working as a unit toward goals in a specific environment is to open new horizons for future insights.

## LEADERSHIP TYPES

These questions are often asked: "How does a leader accomplish the objectives of the recreation system?" "How does he or she perform his or her function?" "How does he or she inspire individuals to attain goals?" Such questions imply a method or technique that is an activating force in getting the job done. But all techniques merely provide motivations that are conducive to reaching the goals of the organization. And, of course, differences of opinion exist as to how this can be done. Perhaps the easiest and simplest means of "typing" or grouping behavioral processes in reaching organizational objectives is that of using what may be described as a negative or positive approach, or of using no approach at all. Translated into workable terms, what is meant here is that one can motivate a group toward goals through the use of fear, threat to security, denial, or punishment or, on the other hand, one can motivate by making work meaningful, satisfying, and worthwhile to the worker. The former position is maintained by those who believe that the only way goals can be reached is through command and threat of punishment; the latter stresses the point that, given a chance and the incentive, an individual wants to do good work. Hence one type of leader uses a negative, or autocratic approach; the other, a positive, democratic, or participative method. If no direction is given whatsoever and the group is left to work out its own problems in complete freedom, a leaderless style exists, which is identified as a "laissez faire," anarchic, or free- rein approach. A closer look at these three types of leadership provides some illumination of human behavior traits in administrative action.

### Autocratic Leadership

The autocratic leader, sometimes known as the authoritarian-type leader, centralizes authority in himself or herself. The leader dominates, stresses discipline, and looks on human errors as signs of weakness. He or she does not have confidence in subordinates and hence considers their opinions as worthless.

Unfortunately, the autocratic leader rules through fear. Therefore followers are constantly on guard and anxious to please in sundry ways. This leader enjoys subservience because, after all, this is an outward indication of status and power over others. Those who do not succumb become the target of acts of retribution. Secretive in communication and poor in interpersonal relationships, the autocratic leader keeps all decision making to himself or herself. In turn, this further strengthens power, since subordinates, anxious and fearful, cannot chart a course of action alone. Such a leader seldom delegates responsibility or gives authority to others. Acting as the dominant or central figure, the leader doles out instructions as needed to his or her "puppets."

Subordinates run to him or her for decisions of the most inconsequential kind. The cringing, bootlicking actions of employees are reflected in poor morale, constant complaining, and rationalization.

This description of the autocratic leader is in sharp focus. However, there are degrees of authoritarian action; some are more negative and rigid than others. Generally, all such central figures have deep-rooted feelings of inadequacy and profound doubts about themselves, which are reflected in distorted action. Indeed, few feel any responsibility for their own failures or are consciously aware of their own shortcomings.

## Democratic Leadership

The democratic leader does not believe in unilateral action in decision making. Instead, he or she feels that his or her job is to help employees or groups arrive at decisions and determine purposes. In short, he or she conceives a leadership role to be one of stimulating staff by sharing and working as one of them in collective action toward goals. This type of leader does not have a power complex like that seen in the autocratic leader but, instead, encourages followers to give of their best. He or she encourages interaction of staff members, builds a feeling of security and self-confidence in them, and extends opportunities for their leadership development. His or her followers turn to him or her for help, have confidence in his or her ability and feel secure in their relations. Indeed, the democratic leader leads best by bringing out the staff member's greatest potential.

The democratic leader is "people oriented" and likes them as they are, with all their weaknesses and human frailties. He or she is flexible and tends to see as well as encourage the best in others. It is easy for him or her to relate with them. And not being too greatly concerned with discipline, he or she tends to give followers a chance to work under guidance toward goals in their own way. This leader is not dogmatic, nor does he or she feel that he or she is omniscient. Indeed, with open-mindedness toward ideas and belief in judgment of his or her subordinates a work climate is encouraged that makes activity meaningful to them. Truly, he or she represents democracy at its best.

## Laissez Faire Leadership[15]

By "laissez faire leadership" is meant the action of a central figure that gives complete freedom to a group in the solution of its problems. Actually, this

[15]The term laissez faire refers to maximum personal freedom or noninterference of government in the actions of people. In Adam Smith's *The Wealth of Nations*, the term is used to advocate less government interference. As used here, it refers to freedom of action, few if any restraints, or a leaderless situation in which the group establishes its own goals.

type is at the opposite pole of the leadership continuum from that of the autocratic leader. Where an autocratic leader takes full authority with no participation on the part of followers, the laissez faire leader gives maximum freedom with little or no central direction. In fact, this type of leader merely provides the resources and materials to do the job and participates only when asked.

*Laissez faire leadership* is often used interchangeably with the term *anarchic leadership*, although semantic practice differentiates between the meanings of these two terms. However, it should be clear at this point that the intent of this leadership style is to give subordinates or a group free rein in making decisions and reaching goals.

In summing up the kinds of leadership philosophies, it is important to point out that each has certain advantages and disadvantages. The democratic type of leadership, in which the leader works through people in a sharing relationship, offers perhaps the greatest opportunity for a recreation and park organization to reach its objectives. On the other hand, autocratic leadership centralizes decision making and makes possible greater utilization of a mediocre staff that has little initiative or need for decision making. However, this type of leadership is not the kind that leads to a high quality of work nor does it bring out the best in a staff, although it is perhaps the prevailing method used for getting the job done. In fact, it is regrettable that this style of leadership is stereotyped as a common and acceptable image for supervision.

Laissez faire leadership has many good features, although it is usually interpreted in a bad light. There are degrees in the amount of freedom given a group under this form, and in some situations the lack of formal leader may be what the situation needs. However, one must come to the conclusion that shared leadership is most productive, particularly in areas of service, such as community recreation and park development. In sum, a discussion of leadership leaves much to be desired, but one thing is certain: A leader must be looked on as such by his or her group and must satisfy its needs in terms of goal attainment.

## RECREATION AND PARK LEADERSHIP

There is no question about the need for dynamic democratic leadership in a recreation and park system. The tempo of our times makes it a necessity if such systems are to grow, develop, and compete by means of their services. Actually, the leadership within a recreation and park organization determines its effectiveness, for, after all, services to meet the recreational needs of people are its product and the quality of this product is determined by the vision, imagination, enthusiasm, and abilities of the employees that create it.

The leadership problem of motivating people to greater achievement is a big one. One solution is to develop an atmosphere that encourages democratic leadership. This involves a sharing at all levels of the administrative hierarchy, not just at the executive or supervisory level. Further, this means staff involvement in policy-making matters of concern to them, such as program making, long-range planning, working conditions, or personnel practices. Of course, it is not proposed here that staff should be responsible for the execution of policy. This task rests with the executive. However, it is suggested that formulation of policy by a legislative body can very well be shaped by the ideas and thinking of a dynamic staff.

## The Role of the Administrator

It is the job of the administrator and his or her leadership staff to perform those functions that lead to involvement of people in a variety of recreation and park activities. To do this the administrator must be constantly aware of the basic purpose of the recreation and park system and the changing social, economic, and political environment in which it operates. This need constantly to appraise the organization's function and to adjust to changes requires a conceptual skill that too few recreation and park administrators possess. To be able to see overall relationships, forecast future needs, develop creative plans, and implement them with the consent of his or her staff requires almost a charismatic quality of leadership. It is obvious to any administrator that change or even reappraisal of an organization's function usually meets with resistance or negative reactions unless a proper foundation is laid to convince people of the need of change. Usually, staff members develop a vested interest in a routine that is intrinsically satisfying to them, and to change this work pattern for reasons that are not clearly apparent only brings dissatisfaction.

It is a sad commentary that some administrators lose sight of their organization's goal and, instead of inspiring their staff to greater efforts, follow a pattern of holding the status quo until the organization becomes anachronistic. In this situation the recreation and park system develops operating procedures that are not focused on the purpose of the organization, and service to people becomes a necessary evil to maintain the several special interests of the staff.

The administrator must constantly be on guard against actions that chart a course away from the organization's purpose. He or she must think creatively, never letting the means control the ends but, instead, must develop those human relations and conceptual skills that project the organization into new and expanding areas of community service.

The recreation and park administrator has difficult leadership roles, depending on the size of his or her operating organization. In all departments, large or small, however, the executive and his or her administrative staff are expected to have technical and/or specialized skills as well as skill in working with people. However, in the small department the need for technical and specialized skill is much greater than in the large department. For example, in the smaller organization, the executive and staff are expected to perform many of the specific jobs themselves; there is no place for delegation or specialization of function. It is not uncommon for the recreation and park executive and his or her supervisory staff to perform such tasks as conducting activities, maintaining areas and buildings, or laying out designs for areas and facilities. On the other hand, as the executive ascends to the top of the of hierarchy of a larger organization, he or she is expected to be more of a generalist than a specialist. Duties are of an administrative nature, as was discussed in an earlier chapter. He or she is expected to spend most of his or her time in interpersonal relationships, planning, directing, organizing, and the like. Actually, little effort is spent in technical matters, except in matters of decision making.

This type of situation is rather paradoxical. For, indeed, as an individual becomes outstanding as a specialist, it is difficult to attain those qualities needed to be a generalist, which is so necessary in executive leadership. On the other hand, many appointing authorities have not recognized this viewpoint and have recruited their top recreation and park executives from positions of specialization. Regardless, however, specialization tends to limit executive leadership, for, too often, the specialist finds it difficult to cast off the cloak of his or her specialty to don the robes of the generalist. Perhaps the answer to this conflict is to recognize the effect of specialization on executive leadership and counteract it by providing opportunities for the development of leadership talent in the lower echelons of an organization's hierarchy. Furthermore, it should be recognized that not all individuals are seeking executive posts.

Much has been said about the need for human relations skill in administrative leadership. Ability to work with people in an atmosphere of respect and confidence is basic to successful recreation and park operation. One cannot have too much of this quality, since, goals are only reached through people; and human relations is only the effective interaction of people in a work situation that brings out their best. Yet, nothing does as much to achieve goals as getting people to work together harmoniously and with satisfaction. Hence the administrator must develop group leadership and develop his or her role as one who inspires rather than commands. Clearly, the effective administrator must have this skill.

Finally, skills that are many times overlooked but that are needed in administrative leadership are those of creativity and perceptivity. The higher and more responsible the position, the greater is the need for such skills. Strangely, these skills, or qualities, are most difficult to develop. Some of the most effective administrators have been average in intelligence but high in perceptive abilities. And the reverse is also true.

Actually, these skills relate to ideas and the almost intuitive ability to see relationships in their proper perspective. A person having this ability works best with unstructured situations. He or she is able to cut through the maze of detail or confusion and reach the heart of a problem, is not bound by conformity or tradition, but, instead, he or she perceives the major task and does not get sidetracked with detail. However, additional research is needed in analyzing this ability and the means by which it can be developed.

### Questions for Discussion

1. What is meant by "leadership"? How does leadership in a recreation and park system differ from other forms of "leadership"?
2. How does the role of leadership in a small recreation or park system differ from that in a large system? What skills are needed in each?
3. Explain briefly three theories of leadership. Which do you prefer? Why?
4. How would you proceed to develop leadership in your organization?
5. What are the pros and cons of the *autocratic* and *democratic* leadership? Can you make a case for *laissez faire* leadership? Explain.
6. How do individuals become leaders? Cite examples of leadership acquired in each of the ways you have enumerated.
7. Is it possible to have effective leadership in a group if goals are not well defined? Why or why not?
8. Is it possible for a group to reach objectives in a nonleadership situation? Explain.

## SELECTED REFERENCES

Browne, C. G., and Cohn, Thomas S. *The Study of Leadership*. The Interstate, Danville, Ill., 1958.

Dubin, Robert, *Human Relations in Administration*. Prentice-Hall, Englewood Cliffs, N.J., 1968, Chap. 16.

Edginton, Christopher R., and Williams, John G., *Productive Management of Leisure Service Organizations: A Behavioral Approach*. Wiley, New York, 1978, Chap. 6.

Fiedler, Fred E., *A Theory of Leadership Effectiveness*. McGraw-Hill, New York, 1967.

Gibson, James L.; Ivancevich, John M.; and Donnelly, James H., Jr., *Organizations: Structure, Processes, Behavior*. Business Publications, Dallas, Tex., 1973., Chaps. 8–10.

Massie, Joseph L., and Douglas, John, *Managing: A Contemporary Introduction*. Prentice-Hall, Englewood Cliffs, N.J., 1973, Chap. 15.

Newman, H. William; Sumner, Charles E.; and Warren, E. Kirby, *The Process of Management*. Prentice-Hall, Englewood Cliffs, N.J., 1972, Chap. 20.

Reitz, Joseph H., *Behavior in Organizations*. Irwin, Homewood, Ill., 1977.

Stogdill, Ralph M., *Handbook of Leadership: A Survey of Theory and Research*. Free Press, New York, 1974.

Stoner, James E., *Management*. Prentice-Hall, Englewood Cliffs, N.J., 1978, Chap. 17.

# CHAPTER 5

# Departmental Organization

The organization of public recreation and parks and their position within the framework of government constitute a significant chapter in the social progress of America. Indeed, no other country in the world has progressed in its public recreation development as has the United States.

While park systems have had long acceptance as a function of government, it has only been since the turn of the century that we have seen the emergence in government affairs of what is known as "play and recreation." Today, with the tremendous growth of the recreation and park movement and its significance in community life, greater emphasis is being given to this important phase of social living by all government bodies. However, its position in the organizational structure has not always been clear-cut or well defined, although the combining of recreation and parks into a single functioning body has become a reality.

## BASIC FACTORS

There is no one basic organizational structure for the administration and operation of public recreation and parks. All patterns of government organization will vary with local conditions and other contingent factors, some of which are the provisions of state enabling laws, existing legal powers, size of the community, adequacy of funds, community traditions, internal structure of the government body, and other features that are unique to a given situation. Furthermore, public opinion and the attitude of government authorities toward the inclusion of recreation and parks as a part of the family of government services are also important factors. Therefore any organizational plan must be compatible to situations as they exist. Keeping this in mind, we see that the task of government authorities is to evolve an administrative organization that meets their needs. Hence this chapter will deal primarily with an analysis of the administrative structure of recreation and parks. Since most services of this nature are provided at the community level, greater attention will be given to the municipal and district level. Expanding concepts at the county level will be probed more thoroughly in a subsequent chapter. Administrative structures at the state and national level are concerned primarily with services through the development and use of areas and facilities and are not covered, except as they were discussed under the role of government in Chapter 1.

## ORGANIZATIONAL PATTERNS

There are several common patterns or means by which the function of recreation and parks can be placed into the framework of a governmental structure: (1) A pattern that is still in common use involves having an independent recreation department in addition to an entirely separate unit devoted to parks. In other words, each is independent of the other and controls its own facilities. (2) Recreation and parks can be combined into a single functioning unit. (3) Recreation may be provided under the auspices of the public schools. (4) Authority for recreation and parks may be vested in a district system. (5) The entire recreation service including parks, libraries, and programs may be provided by a county authority. (6) Miscellaneous organizational patterns may be established to provide these services.

### Recreation and Parks as Separate Functions

Today one frequently finds the separation of parks from recreation in the general governmental structure. This is due in large measure to the early tradition and history of these two functions. At the turn of the century, park authorities were concerned basically with horticulture, formal garden designs, and general landscape features. On the other hand, recreation was developing as a play movement, and each seemed incompatible with the other. Hence independent playground and recreation departments were launched to fill the needs of an expanding industrial society. But in the last decade, expanding concepts and broadening points of view have shifted this focus from independent action to unified planning and control. In fact, so many public jurisdictions have combined these two functions that an independent poll would show this unified organizational approach to be most popular.

Many arguments have been given pro and con concerning the advisability of administering recreation as a separate function. Proponents have stated that separation gives greater recognition to a generally overlooked area of service by people and government; that programs do not lose their identity and, consequently, more adequate budgets prevail; that more capable personnel can be recruited; and that each is important enough to justify its own separate identity. On the other hand, those that oppose separate recreation organizations say that they make for a mere duplication of effort; that a park department is, in the final analysis, a recreation organization; and thus that it is ludicrous to have two separate units focused toward the same end.

Strictly speaking, advocates of separation believe that it strengthens the efforts of their independent unit of service, be it parks or recreation. Some recreation authorities claim that park departments are only concerned with engineering detail, property acquisition, and maintenance problems; that services to people are not given priorities as they should; and that they are not "people oriented" but are "facility oriented" instead. However, some park authorities say that this is not true, claiming that recreation personnel have little perspective or balance in dealing with the need for recreation areas and facilities and being so activity or play conscious they lack a sense of value as to the need for area and facility design, beautification, planning, and development. Actually, there is much to be said on both sides but, in the final analysis, whether separate organizations are desirable or undesirable depends on local tradition and other factors unique to a given community or region.

Far more controversial than whether parks should be merged with recreation is whether a combined department of parks and recreation should include in its operational policy the provision of cultural and educational services that involve the use of the public schools. Strictly speaking, parks and recreation should be combined, unless tradition or some other local factor makes it unwise. After all, a park is merely a recreation area and, as such, should be used for the enjoyment of people. Other services embraced within a park department are also aimed at recreation, or they have no reason for being. Furthermore, combinations are taking place in greater and greater numbers and are operating efficiently and harmoniously. To continue debate on this issue is only to belabor a point about something that is already in the process of being accomplished. But this in no sense means that there is no place for the separate and independent recreation department if this is the pattern that is needed. Too many of the outstanding recreation programs in the country are being administered independently for this to be cast aside as an archaic organizational pattern.

## Combined Recreation and Park Administration

It is now more of an academic question as to whether parks and/or recreation should be separate or combined, since such combinations, in fact, have taken place. However, it is important for a student to have some historical background concerning this issue, since it is one that was discussed during the 1920s, 1930s, and 1940s.

As early as 1940, George Hjelte, an eminent figure in the recreation and park movement, stated that consolidation of these two functions ". . . offers an opportunity for the development of a more comprehensive and at the same time more diversified program. . . . It is conducive to economical administration and enhances the services of the city in recreation."[1]

This combination of functions has become the most common pattern of administrative organization. Usually, the title of the person who heads this combined operation is "director of parks and recreation."

In addition to the fact that there is no inherent reason that parks and recreation should not be combined into a harmonious unit, this structure has many advantages. First, it makes possible a far more comprehensive program of activities and services that can be integrated under operational procedures that prevent duplication and overlapping. Indeed, the centralization of these two spheres of services not only will accomplish this end but will also foster efficiency and economy. Second, it makes possible the enlargement of recreation offerings into park areas that traditionally were thought of as inappropriate for activities—areas that were not to be touched by the vulgarizing influences of use by people. Third, it makes them far more attractive to the general public than would be the case if they were exercise yards, or merely functional areas for activity only by the addition of plants, shrubs, and trees and the general beautification of active recreation areas. Actually, integration of landscape features with utility of operation does, indeed, provide a setting that is conducive to relaxation and play. Fourth, it makes possible greater unity in planning and more specialization of functions. Larger operating units provide greater opportunities to specialize and develop areas of program that would not prove economical in the smaller unit. Fifth, it provides for better understanding of the full scope of recreation and, thus, eliminates fragmentation of services. Finally, it brings the operational, maintenance, and developmental functions together into a harmonious unit that makes for cooperation in planning as well as utility in use of people and materials.

Some claim that a combined department of recreation and parks has certain disadvantages. It is cited that there is the possibility that the head of the department will give greater emphasis to one phase of operation, either recreation or parks. This is possible. For example, this could occur if the department director is oriented toward parks and accords greater emphasis to this phase of operations at the expense of the activity program. Likewise, the same could happen if the head were recreation or activity oriented. But in the final analysis the same situation can occur in any given organization if its director is an inad-

[1]George Hjelte, *the Administration of Public Recreation* (New York: Macmillan, 1940), p. 64.

equate administrator or leader who brings to the job petty prejudices and lack of administrative skill.

The advantages of a combined park and recreation operation must be viewed in relation to the needs of the community to be served. Historically, the two operations were separated; currently, they are combined and function together effectively. For the public, there is every advantage to be gained by joining these two forces.

The organizational pattern in which recreation is provided by a division or bureau under a department of parks is disappearing from the American scene, and the few departments that claim such an organizational pattern are, in reality, park and recreation departments. Actually, no case can be made for this type of organizational pattern.

## School Administration of Recreation

A form of administrative organization for recreation that is used to a greater or lesser degree in many parts of the country involves providing this service under the auspices of the public schools. Many educators advocate this pattern of organization. Others do not.

Actually, the public schools are neither a recreation nor a park agency. Hence trends in the last decade have shown a diminishing role played by the schools as the primary agency for the direction of community recreation, although they do cooperate in most instances in the sharing of their facilities.

The major functions of the schools with respect to recreation have been education in the worthy use of leisure, provision of school areas and facilities for community recreation use, training of recreation leadership personnel, and training in recreation skills. The role of recreation in relation to the public schools will be probed more fully in Chapter 9.

School authorities have been given the major responsibility for the conduct of public recreation in a number of American communities. Notable instances of this organizational pattern can be found in Long Beach and Pasadena, California; Newark, New Jersey; Flint, Michigan; and Milwaukee, Wisconsin.

Still other school districts or local boards of education operate their own independent recreation program as a part of their education responsibility, and the programs provided are independent of the recreation opportunities offered by the public agencies in the city or county. In these instances the responsibility is primarily for school recreation. Programs include recreational activities after school, on weekends, and sometimes during the summer months.

In still other communities the schools coordinate their recreation efforts with that of the public recreation authority, through formal agreement that provides for joint use of school facilities, equipment, and leadership. In other instances the school authorities share with the city in paying the salary of the recreation director or make other contributions to the program. Under this pattern the schools contribute to the community recreation program, although they are not the major agency administering it.

Advocates of schools as the major recreation authority in a community have pointed to Milwaukee, Pasadena, Long Beach, and other cities as examples of highly successful patterns of organization. Actually, this pattern has grown out of local community situations where liberal understanding existed on the part of school administrators and board members as to their responsibility for this phase of community service.

The conditions under which the schools have been ready to accept their recreation responsibility have been dependent on a number of factors, some of which are as follows:

1. The availability of progressive educational leadership that gives full play to the role of recreation in developing "worthy use of leisure" as an educational objective.
2. The legal provisions that permit educational systems to become involved in community recreation programs and services.
3. The community tradition as to the role of education and its relation to recreation.
4. The financial status of school districts.
5. The adequacy and availability of school buildings and facilities that can be used for community recreation purposes.

Although excellent recreation programs are being conducted by some school authorities, the trend has been for boards of education to assume less direct authority and responsibility for community recreation.

In sum, we can observe coordinated organizational plans linking school and recreation. The recreation organization may be separate from the schools but use extensively school property and facilities; schools may conduct their own independent recreation program on school properties but cooperate with the municipal recreation authorities; schools may be the sole recreation authority; or schools may integrate with the municipal body to provide a coordinated recreation program involving both bodies.

The advantages of having schools as the major administrative body for recreation have been summed up as follows.

1. Community recreation is educational, therefore it is logical that programs be administered under school auspices.
2. Schools are staffed by teachers who are professionally trained to handle children, youth, and adults in their many recreational enterprises.

3. Recreation buildings and other facilities are already available for recreation purposes.
4. Education today, particularly in the adult education field, is largely recreational.
5. General public confidence in recreation would increase if it were under school auspices.

On the other hand, the arguments are many for not giving the function of community recreation to the public schools. The opponents to this form of organization stress the following points:

1. Public schools are concerned primarily with education, not recreation, and thus to give this important community function to a body that only looks on it as a secondary task is not in the best interest of the community.
2. Educational authorities with an exploding population and large school enrollments, have enough to do in financing their own curricular programs and facilities without taking on the function of recreation.
3. Recreation budget in periods of financial stress, is the first one to be cut.
4. Recreation encompasses much more than an activity program, and to place parks, playfields, swimming pools, beaches, golf courses, museums, and the like under school auspices is stretching the educational function.
5. Public school authorities have little interest in recreational activity for preschool, adult, and elderly people.
6. Schools are largely governed by educational policies, many of which are not related to recreation.
7. Many school boards feel that their recreation programs should be oriented only to the curricular school program.

Actually, argument pro and con could be given endlessly on this issue, but here again, the specific local situation determines whether this pattern of organization should be followed.

## The Separate Recreation and/or Park District

It is not unusual to see the recreation service in a community, county, or region administered under what is known as a "recreation district." Sometimes called a "park and recreation district" or "park, recreation, and parkway district," this administrative pattern has grown in popularity in many parts of the country. To avoid confusion, the term *recreation district* will be used here to describe any district relating to recreation and parks.

As a rule, the recreation district is entirely autonomous, that is, independent of other government bodies. This separation is provided under provisions of general state enabling acts passed by state legislatures. Strictly speaking, a recreation district is a civil subdivision of the state. As a legal government entity established under state law to provide the function of recreation for an area that wishes to use this organizational pattern, it has only those rights and privileges that the state delegates to it. In other words, a recreation district is a political subdivision of the state just as is a county, city, township, or village. It can be likened to a school district and, furthermore, is completely independent of a city, even though its boundaries may be coterminous with it. The only exception to this rule is when state law intentionally gives authority over a district to a city or county.

All states make provision for the establishment of special districts, one type of which is the recreation district. Scott and Bollens state that,

> . . . special districts usually have five characteristics: (1) a resident population and a defined area, (2) a governing body, (3) a separate legal identity, (4) the power to provide one or more public services, and (5) a degree of autonomy, with power to raise at least a part of its own revenue.
>
> The creation of a district, an action taken under general state enabling acts, follows three steps: petition by residents, examination of the petition by a local governing body, and approval at an election in the proposed district. Districts are governed by four types of boards: (1) elective, (2) appointive, (3) ex officio (city councils or county boards of supervisors), and (4) those comprising selected members of the governing bodies of constituent cities and counties. Most district laws permit the annexation of new territory, and a number provide for district dissolution and consolidation.[2]

While recreation districts can be found in most of the states, California leads all others in its use and variety of patterns. As was stated previously, there are numerous types of recreation districts. They may vary in geographical area from a few acres to many thousands of square miles, but all have specified boundaries. Likewise, some districts may employ no personnel or as few as one or two professional employees, while, on the other hand, some may have hundreds of employees.

Recreation authorities agree that the district pattern, when used, should be large enough for performance of the many varied services required. Some districts are so small and have such a limited tax base that they are inoperable. Research has still not brought forth the answer to the question as to how large the district should be, except as it relates to a specific function. Metropolitan

[2]Stanley Scott and John C. Bollens, *Special Districts in California Local Government* (Berkeley: Bureau of Public Administration, University of California, 1949), p. i.

recreation districts,[3] for example, usually encompass a designated metropolitan community. Seashore districts are limited to definite beach or shore areas, as are regional park districts, which are fairly circumscribed to include wilderness, mountain areas, lakes, streams, and the like. In general, however, the size of a recreation district that serves the recreation needs of an urban environment should be determined by the following criteria:

1. The district should include within its boundaries enough property to provide a tax base that is not burdensome to support a minimum of recreation and park services and programs under the direction of a professional leadership staff.
2. The district should be of sufficient size for people within it to feel an identification with the unit and exercise a voice in choosing the board of directors and developing programs.
3. The district should be large enough to permit the development of a variety of areas and facilities or use of lands and buildings that meet minimum program needs.
4. The district should serve at least one identifiable community or a group of interrelated "clusters" of people having community characteristics.

There are a number of reasons for the growth of recreation districts, particularly those that have been established to serve urban areas. They are:

1. *Need for service.* Some urban areas, particularly those that are not incorporated, feel a strong need for a recreation service but are not interested in annexation or incorporation.
2. *Lower costs.* Property owners feel that taxes will be made lower by establishing a district to provide a needed service.
3. *Referendum feature.* Local governing units' lack of interest in providing a recreation or park program has been offset by referring the issue to a vote of the people.

It should be stressed that the growth of recreation districts has resulted largely from the growth of decentralization, the expansion of people into peripheral areas of urban centers, known as "fringe areas." In other words, people have moved into unincorporated areas that border on municipalities or metropolitan cities. This process of decentralization has continued for a number of reasons—cheaper land values, availability of good transportation routes, more desirable residential areas, and lower tax rates. Furthermore, this process has resulted in the growth of unincorporated communities that, to all intents and purposes, are "unofficial cities." Living beyond the city limits of

[3]Sometimes the term *authority, ad hoc authority*, or *special district* is substituted for *district*.

the community in which they work and trade, residents are not bound to its tax structure, but they think in common terms and want the same government services that are available to those living within the municipal boundaries. Hence rather than incorporating or annexing, which would mean a heavier tax burden, they seek needed services through creation of special districts. Thus the area may be served by a fire-control district, water district, recreation district, and the like. Actually, the securing of a few basic community needs in this way reduces the tax load, since additional services are available within the adjoining city. Unquestionably, taxes are less, but the financial burden of the adjoining city is all too apparent. It is not uncommon to see the "fringe area" have a greater population that the core city.

Rather than have each separate urban "fringe" unit establish its own recreation district, some recreation authorities have fostered organizational patterns that include an entire metropolitan area within a recreation district. This unifies the planning problem, creates a larger administrative unit, and gives autonomy to recreation and park development. While this solves the "fringe area" problem for a single unit of service, the overall problem still exists. How can municipal boundaries be expanded to encompass those that live in what today is called the "fringe area"? This is one of the real issues that face municipal authorities today, and the solution will have profound effects on district operation.

There is little question that recreation districts as an organizational pattern will continue to grow. Expanding populations and growing metropolitan centers create urban clusters of people who seek solution to their need for recreation services. The district structure offers one solution. Authorities who support the district plan cite the following arguments to strengthen their case.

1. Recreation districts provide for fiscal independence and thus for better continuity of services.
2. District boards and their professional staffs are free from political influences.
3. District boards can devote all their time and energy to recreation and park problems.
4. Flexibility in establishing and changing programs is greater under an organizational pattern that is focused on one basic function.
5. District operation makes it easier to interpret needs directly to the people.
6. Unified long-range planning for programs and facilities can be given to a geographical area encompassing all the people, rather than being restricted to artificial political boundaries of a city.
7. Large districts that encompass a number of urban areas make for economy of operation and less duplication of service.

On the other hand, opponents to district organization stress the points that

1. Establishment of a recreation district only aggravates government problems by adding another overlapping taxing jurisdiction.
2. Independent district organization fragments government services and creates problems for unified planning.
3. One governing body providing all services is more economical than having independent boards plan separately for single functions.
4. Multiplicity of small recreation districts can neither plan effectively nor enlarge their services, owing to their limited tax base.

Thus the arguments go, pro and con. Actually, the district pattern is a problem that has not been solved at the municipal level. In urban communities that are faced with "fringe areas," where the proportion of people living in the unincorporated portions of the county outnumber the resident population of the incorporated city, the district pattern certainly presents a possible solution. Likewise, to establish a district to encompass areas that are homogeneously circumscribed, such as regional parks, seashore areas, and the like, also seems desirable. Furthermore, the problems of metropolitan government, unless solved, will only result in the establishment of additional metropolitan recreation districts.

## The County

In some instances the recreation and park services that are provided for local inhabitants are administered by a county recreation and/or park department. In other words, the county is the primary administrative body. Because of the growth in number and importance of county recreation and park departments, Chapter 10 will be devoted to the role of recreation administration at this level of government.

## Miscellaneous Organizational Patterns

Additional organizational patterns for recreation and parks can be found in various parts of the country. For example, cemeteries and memorials are added to the jurisdiction of some departments. Yacht harbors, stadia, auditoriums, and the like may be the major concern of others. In some cities recreation as well as parks may be structured as a division of another department, the department of public welfare, for instance. Or in the specific case of parks they may be administered by a division of the department of public works. These are exceptions to the rule.

## New Developments in Organizational Structure

The question of the right name by which to identify the services provided by a governmental unit involved with leisure, recreation, parks, cultural services, sports, and the like has faced the profession for decades. Because so many departments have not been all embracing as to the scope of their services, the word "leisure" has gained popularity and is now being used extensively. Proponents have stated that leisure answers the question of identity. On the other hand, those that are less enthusiastic about its use believe the name is confusing to the general public as to its scope when used as the title for a department.

Actually, advocates believe it adds to the stature of services provided as well as to the stature of the profession. But this is questionable if a title is not clearly understood by the users of a service. The layperson in a community does not know what to expect in the way of services from a Department of Leisure Services, and thus it would be far better to establish a name with meaning and identification. A department of parks, recreation, and cultural services would be far more meaningful in title than one called a department of leisure services. Furthermore, by emphasizing the cultural service aspect of departmental operations, we open the door to services that have not always been identified with parks and recreation. For example, a number of cities, including Phoenix, Arizona, and Tampa, Florida, have added public libraries as a part of their department operations. Others have included concert halls, auditoriums, dance and art studios, and theaters for development of the arts. This is a progressive move that should be encouraged. It is not too farfetched to see more departments also include within their sphere the operation and management of publicly owned museums, stadiums, skating rinks, and the like.

There has been some discussion on the part of professionals in the field that it might be wise to consider a department of local government identified as one relating to human services or resources. In other words, park and recreation leisure services would be under a unit of government called the "Department of Human Services." Although some isolated cases that have this form of organization may exist, it is not one the authors would recommend since it encompasses too many facets of local government under a "general umbrella."

Proponents of this proposal have suggested that human services would include parks, recreation, cultural services, libraries, concert halls, art studios, and the like, as well as other local social and welfare services. To have these types of services under a Department of Human Resources would tend to identify the welfare field with that of parks, recreation, and leisure services. This would be a mistake. Clear identification needs to be made between services classified under a welfare category and those in a leisure setting. The dichotomy of services would be too broad. Many local welfare services have little in common with leisure services, and combining them would only lead to confusion. For this reason, the authors firmly believe that parks, recreation, leisure, and cultural services identify more clearly the setting for leisure and should not be confused with a broader base.

## PARK AND RECREATION BOARDS

It is not uncommon to see the recreation and park function administered under a board or commission that is independent of the municipal government. In other words, the board or commission is the policy-making authority for recreation and parks and its work program is independent of the administrative authority of the political jurisdiction in which it operates. Of course, when recreation is administered under the "district" pattern, administrative power rests with a board. But administrative boards can also be found in those cities wherein state-enabling laws and municipal charters give policy-making responsibilities to the recreation and park body.

The traditional literature on recreation and park administration makes a strong case for boards or commissions that have full administrative authority over parks and recreation, but before review of this issue, consideration must be given to the current status of such public bodies as they now exist in American communities.

Actually, all recreation boards and commissions may be classified as follows: (1) separate and independent bodies with full responsibility for policy and administration of recreation services, (2) semi-independent boards with policy-making and administrative authority over the recreation function, and (3) advisory bodies with no final authority or responsibility for policy or administration.

### Separate and Independent Recreation Agencies

Some recreation and park organizations are so constituted that they have little or no relationship with the municipal government of the city wherein they are located and which they serve. Many recreation districts are of this type. It will be recalled from the previous discussion that many local recreation and/or park districts are legal entities or civil subdivisions of the state and as such have their own taxing power and policy-making function. Furthermore, all recreation districts have duly elected or appointed policy-making boards, usually of three to seven members.

On the other hand, some recreation agencies or departments are separate from the local government because they are so constituted through provisions of city charter or state law. In other words, the structure and responsibilities of the recreation agency are vested in a recreation board that is autonomous—independent of the municipal government. Provided with a mill levy or other finance base, these recreation and/or park commissions[4] carry on the recreation function much as if they were members of the family of government agencies of a city or county. Of course, these boards have final policy-making authority.

In discussion of "final authority, policy-making and administrative authority," it should be made clear here that final responsibility and authority for decision rest with the duly elected or appointed board or commission of the agency. Final authority to hire and fire, acquire property, approve budgets, and plan for facilities and programs rests with this body. Decision making is not passed to another body; it rests here, just as decisions on school matters rest with the school board, or matters of general municipal affairs rest with a city council. This does not mean that the board performs an executive function—it means that the board is responsible as the final legislative body for the affairs of the agency. A city manager or city council would not have any direct control over the board or its staff. The responsibilities of boards or commissions and their relation to the executive function will be probed more fully in Chapter 6.

## Semi-independent Boards

Many recreation and park departments of local or state government operate under policymaking boards, but in this case they are not independent of the public jurisdiction of which they are a part. For example, at the municipal level the department would be a part of the city government but its operation would be under the control of the board.[4] In other words, the board may be appointed by the mayor and the city council but be delegated policy-making and administrative authority to manage its own affairs in providing community recreation service. The board would submit reports to the city council and present its budget to this body for approval.

Actually, it can be seen that this body would have full authority and responsibility over the recreation programs, services, and facilities of the department, as well as its personnel. Most recreation and park authorities favor this type of semi-independent organization, although many favor independent status.

## Advisory Boards and Commissions

The advisory board also exists to serve recreation and park authorities in many American communities. Just as its name implies, this board is advisory to the governing body of the public jurisdiction that appoints it, and although it

[4]The terms *board* and *commission* will be used interchangeably in this discussion.

works closely with the recreation agency, it acts in an advisory role. Consequently, these advisory recreation bodies do play an important part in the affairs of the organization if the administrator through his personal leadership provides a role for the group that gives them a feeling of contributing to the work of the agency.

## The Issue of Integration

The advocates of the integration school of administration claim that to have anything but advisory bodies interferes with their over all administrative direction and coordination and violates the principle of having a single executive responsible for the affairs of a city. Furthermore, they claim that independent or semi-independent bodies having their own special tax levy violate the concept of the executive budget and thus make control of spending impossible. Nonetheless, much can be said in favor of boards, in addition to the traditional argument that they keep the agency free from politics and assist in protecting the budget in times of emergency or periods of retrenchment.

Boards do give a degree of "representativeness" and involve citizens in the affairs of government. But even more important, is the tradition behind them. These boards have been the bodies responsible for the growth of existing recreation and park services; they have attracted outstanding civic leaders to serve on them and have provided a direction for an important aspect of community and regional development. In no way have they worked against the interests of the community or the governing body of the public jurisdiction. In effect, these boards are not completely independent of the local government, except in those cases just explained. Therefore it is imperative that these bodies cooperate in every way with the unit of government that created them. Otherwise, they would soon be abolished.

Furthermore, while the administrative management school advocates only the use of advisory or *ad hoc*[5] citizen boards, it is generally believed that such bodies are effective. Actually, it is not in the American tradition for groups to meet on problems without being given a degree of power to do something about them. Hence if boards are to be used, they should be given the authority to do the job.

It should be stressed again that many units of local government do give the policy-making function to a designated recreation and park body but still keep administrative direction over its operation. Such departments are, of course, parts of the local unit of government, and the recreation and park executive heads his or her own department, just as the heads of other departments do, except he or she works through a body that has final authority on

[5]An *ad hoc* board is a citizens' body that is established to delve into special problems and then disband after a report is rendered.

matters of recreation policy. It should be clear that those who recommend policy-making boards are not trying to separate the recreation function from local government; instead, they are merely advocating that such bodies be given a responsible role, rather than one that is merely advisory.

## INTERNAL ORGANIZATION

Since recreation and park services are provided through differing forms of managing authority and patterns of organization, it is presumed that the internal structure of the department will be organized to fit the needs of the local situation. There is no one best way to organize a department. Indeed, the uniqueness of each community and its particular managing authority calls for varying kinds of internal structures. But basically, the organizational principles outlined in Chapter 2 can be applied regardless of the kind of administrative form that exists.

### Policy and Executive Level

Every recreation and park department has two basic phases of operation—the policy-making phase and the executive phase. The policy-making phase gives overall direction to the agency, and the executive phase carries out this direction through its departmental operation. In other words, the latter phase is concerned with the execution of policy. The policy-making function is relegated to the recreation and park board or commission. Because of the importance of boards and commissions and their relationship to executive responsibilities, a full discussion of these bodies will be given in Chapter 6.

The execution of policy is carried out by the executive and the department staff. In carrying out the policies, rules, regulations, and objectives promulgated by the managing authority, the executive must structure the department to reach goals in the most efficient manner possible.

### Internal Structure of Staff and Services

The internal organization of a department is designed basically to provide the means of reaching departmental objectives through division of responsibility and unification of effort. Thus responsibility must be divided and integrated into functioning units.

In every recreation and park department the major areas of responsibility can be grouped into the following five major divisions.

1. *Program (Recreation).* The program division is related to the general recreation service. The activities and programs that are provided under direct leadership are involved here. The operation of com-

munity centers, playgrounds, playfields, and the like, as well as special acitivity services including arts and crafts, athletics, drama, music, social activities, nature, and special services would be included. In fact, all general and special recreation and park services would be classified under this heading.

2. *Special Facilities*. The special facilities division may be a single unit in which all services relating to the supervision and operation of special recreation and park areas, buildings, and other structures are grouped together. Or it may be necessary to divide the various kinds of facilities into separate divisions. For example, the operation of camps, beaches, yacht harbors, civic auditoriums, municipal piers, museums, golf courses, concessions, conservatories, and the like may have an identity that requires separation into divisional units.
3. *Construction and Maintenance (Parks)*. This division is concerned with the planning, development, construction, and maintenance of recreation and park lands, buildings, and other structures used in the program of services by the department.
4. *Business and Finance*. This division involves the financial record keeping of the organization, the receiving and disbursement of moneys for the operations and improvements, and the business function relating to personnel, record keeping, and office management.
5. *Public Relations*. The work of this division involves the public interpretation, publicity, and information aspect of the system.

In the structure of the department the responsibilities for planning and development, research, personnel, finance, and public relations do not usually appear as a division in the charting of the system, except in the large department. Personnel matters are usually allocated to a central personnel division of the unit of government, except that leadership training is usually absorbed as a responsibility of one of the supervisors or even of the superintendent of recreation.

Except in the large department, planning and development are absorbed by the executive as are the public relations, research, and finance functions. As readily seen, the larger the system, the greater the need for specialization of function. In the small community it is not unusual to see the executive perform personally many of the required leadership functions. Hence in allocation of responsibilities the scope of services required to meet the recreational needs of the community determines the number of staff members required. All of the basic elements outlined in Chapter 2, on administrative integration, would be considered here in structuring the department.

Keeping in mind the preceding list of major responsibilities, one can readily see that charting begins with making a diagram that shows the areas of responsibility. The following indicates how the policy-making and executive function would appear.

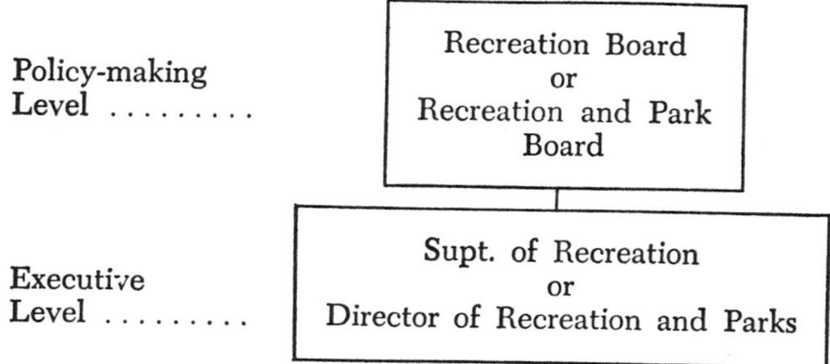


The next step in structuring the organization would be to determine the major divisions of work or areas of responsibility that would be headed by a person carrying the title of supervisor or superintendent or some other appropriate title. Structuring now becomes somewhat subjective, but one of the simplest divisional breakdowns could be shown as follows.

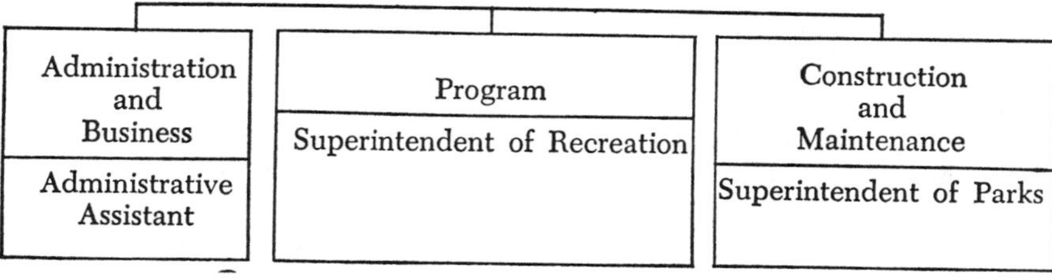


Another way of charting the divisional organization would be

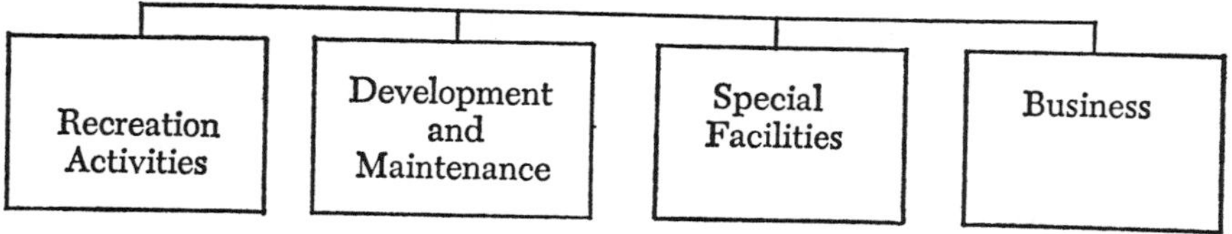


As a recreation and park system grows in size and complexity, it is usual practice to add divisions of specialization. For example, park police, special services, public information, planning and development, engineering, horticulture, and the like are found in the larger metropolitan recreation and park department.

In the separate recreation department it is not unusual to see a divisional breakdown like that found in the Tacoma, Washington, system as follows.

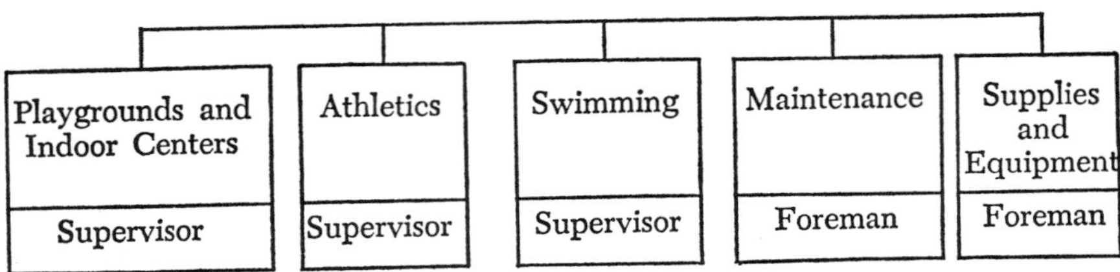


It is also common to see the following divisional breakdown of the separate recreation system.

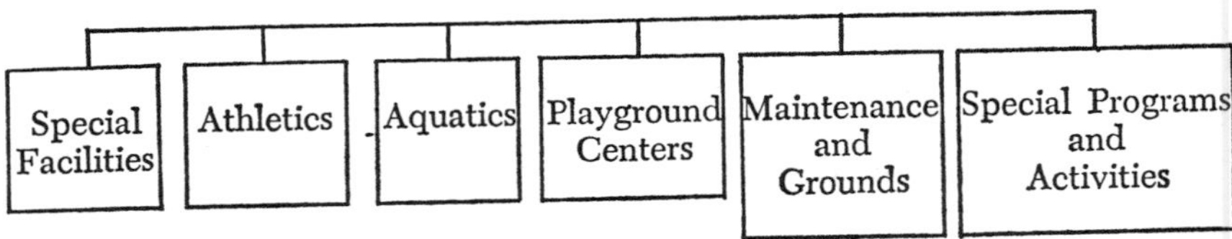


It is important to note that each division may be further divided into groupings by areas of specialization. For example, the division of special programs may include units of arts and crafts, dance, drama, music, and the like, or the division of athletics may include units such as golf, archery, trap and skeet grounds and rifle ranges, in addition to general athletic activities. Some departments also include aquatics under the athletic division.

Under special facilities can be found such units as camps, auditoriums, beaches, marinas, golf courses, museums, riding stables, garden centers, conservatories, street trees, and zoological parks.

Actually, the allocation of responsibilities to the various units and the establishment of an organization chart are not difficult if the basic principles of administration outlined in Chapter 2 are followed. Common divisions found in recreation and park departments are:

1. Special facilities.
2. Aquatics.
3. Playgrounds and centers.
4. Construction and maintenance.
5. Special programs and activities.
6. Administration.
7. Engineering.
8. Horticulture.
9. Police.
10. Forestry.
11. Botanical gardens and arboretums.

As the divisions and their subunits are placed together, an organization chart emerges that indicates the scope of departmental operation (Fig. 5-1).

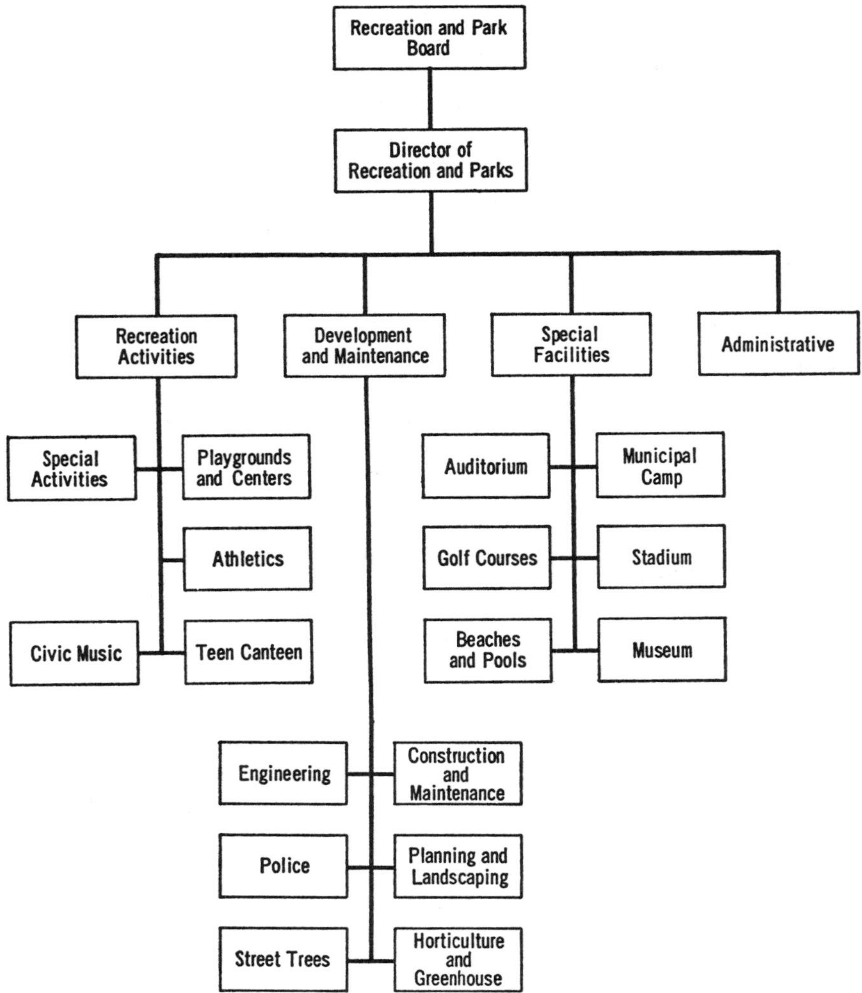


**Figure 5-1**

Because some of the divisions are found principally in a separate park department or a division of parks of a recreation and park department, a closer look will be given to the responsibility of this unit.

## Park Division or Department

A park department or division performs a recreation function, although its identity is largely associated with special kinds of recreation areas and facilities. In division of the park function into its component parts a common core of services prevails. All are concerned with the planning, construction, and maintenance of recreation areas and facilities, the business function, and the care of plants and shrubs as found in public parks. Nearly all are also concerned with special types of areas and facilities such as golf courses, swimming pools, beaches, parkways, greenhouses, and zoological parks. Others include, in addition to those mentioned, responsibility for law enforcement in the park system, shade trees, botanical gardens, museums, and landscaping.

The functions of these units are implied in their titles, but further explanation should be given to the divisions of engineering, horticulture, forestry and street trees, botanical gardens, police, and miscellaneous categories.

**Engineering.** This division of responsibility relates to the location, acquisition, and construction of recreation and park properties. It is concerned with topographical surveys and work with architects in carrying out the construction detail.

**Horticulture.** As used here, *horticulture* is the art of growing plants and shrubs for public parks and recreation areas. Propagation, planting, and care of plants require training and skill, and although this function is sometimes found in the maintenance division, it is an area of specialization that is basic to sound park operation.

**Forestry and Street Trees.** In many communities the care of shade trees is a division of the park department. In others, it is a separate department. Nonetheless, this unit of operation is a responsible one that requires specialized education and training in forestry and tree care.

**Botanical Gardens.** It is not unusual to see in many park and recreation departments a division devoted to development of highly specialized types of gardens. Rock gardens, Japanese gardens, rose gardens, and other specialized types are fairly common, but the genuine botanical gardens are more unusual. Such gardens are found in a number of our larger cities, and some are world famous. The objective of such gardens is to give people a chance to enjoy native and foreign trees, shrubs, and herbaceous plants, as well as to use them for educational and research purposes.

**Park Police.** This division is concerned with the enforcement of rules and regulations governing the use of park properties and facilities. A park guard or police system is usually found only in a system that has extensive acreage.

**Miscellaneous.** Some departments have established divisions for units of service such as aquariums, zoological gardens, refectories, cemeteries, piers, and the like. Again it must be pointed out, many of the large park systems have units that may be unique to their specific department.

The diagrammatic outlines and listings of units of recreation and park service presented in this chapter should be considered as hypothetical only, since nearly every city will modify its divisional organization to meet its own needs. In Appendix B, to show the actual organization of recreation and park systems, there are charts from various cities or districts. These are followed by a system of charts that were developed by the Recreation Division of the California Park and Recreation Department.

## SUMMARY

Many patterns of departmental organization exist for providing public recreation and park services, and the internal structure of each will vary with the scope of program offerings, but in the final analysis all departmental organization must be structured on the basis of sound principles of organizing.

The main issues involved in departmental organization center around the function of boards or commissions and whether they should be advisory or policy making, the role of schools in the recreation settings, the place of the "district" as an organizational pattern, and the current problem as to whether recreation and parks should be combined into a single operation.

It should be stressed that the joining of parks and recreation into a single unit is a sound move, and solid logic supports this combined type of organization. After all, a park is only an area that is developed for the recreational use of people, and to say that one type of recreation should be separate from another is merely to create conflicts in jurisdiction and add to operational problems.

In sum, an organization that deals with recreational pursuits of people should be unified and structured to give the broadest program of services possible. The new focus on cultural developments and leisure pursuits having broad educational implications should be fostered by the recreation and park system. Only as we expand into new areas of programs and provide the areas and facilities for people who are highly educated, exceedingly mobile, and living in an affluent society can we keep abreast of changing times. Thus a departmental structure is not static but must change with the varying recreational needs of our society. Indeed, a look to the future may show recreation and park departments changing direction toward the development of extremely large cultural centers and programs, and organizational patterns must remain fluid to keep pace with these changes.

### Questions For Discussion

1. What is a "recreation district," and how does it differ from a municipal recreation and park department?
2. Why were recreation departments ever separated from the park function, if both have similar objectives?
3. Explain the kinds of organizational patterns for recreation and parks. Which do you prefer and why?

4. Explain the principles of organizing you would follow in structuring the internal organization of a recreation and/or park system. What divisions would you use for a separate recreation department? A combined recreation and park department?
5. Do you think the role of the schools in recreation will expand? Why or why not. Specifically, what is this role?
6. Explain the spheres of responsibility of the different kinds of boards and commissions. Which do you prefer and why?
7. How would you integrate parks, recreation, and schools into a cooperative working unit?
8. Make a sound case for the integrated park and recreation department. The school-managed recreation agency. The separate recreation system. Why should they all not be included under a county park and recreation department?
9. How does the county recreation function differ from that of a "district"?

## SELECTED REFERENCES

Edginton, Christopher R., and Williams, John G., *Productive Management of Leisure Service Organizations: A Behavioral Approach*. Wiley, New York, 1978, Chap. 5.

Gibson, James L.; Ivancevich, John M.; and Donnelly, James H., Jr., *Organizations: Structure, Processes, Behavior*. Business Publications, Dallas, Tex., 1973, Chaps. 3, 4.

Hjelte, George, and Shivers, Jay S., *Public Administration of Recreation Services*. Lea and Febiger, Philadelphia, 1972.

Krause, Richard G., and Curtis, Joseph E., *Creative Administration in Recreation and Parks*. Mosby, St. Louis, Mo., 1977, Chap. 3.

Lutzin, Sidney J., Ed., and Storey, Edward H., Assoc. Ed., *Managing Municipal Leisure Services*. International City Manager's Association, Washington, D.C., 1973, pp. 60–68.

Massie, Joseph L., and Douglas, John, *Managing: A Contemporary Introduction*. Prentice-Hall, Englewood Cliffs, N.J., 1973, Chap. 6.

Newman, H. William; Sumner, Charles E.; and Warren, E. Kirby, *The Process of Management*. Prentice-Hall, Englewood Cliffs, N.J., 1972, Chaps. 5, 6.

Pfiffner, John M., and Fels, Marshall, *The Supervision of Personnel*. Prentice Hall, Englewood Cliffs, N.J., 1964, Chap. 10.

Stoner, James E., *Management*. Prentice-Hall, Englewood Cliffs, N.J., 1978, Chap. 9.

# CHAPTER 6

# Board and Executive Relationships

Much of the success of a recreation and park department depends on the direction provided by its policy-making board and the administrative skill of its executive. Indeed, the governing bodies that control the recreation service, named commissions or boards, have a traditional and important part to play in determining the type and quality of public recreation and park programs and services. Moreover, the role of the executive and his or her relationship to the board provide a backdrop for administrative practice that should be closely scrutinized. Therefore this chapter will examine the operation of these bodies and the role of the executive in relation to them.

## BOARDS AND COMMISSIONS

It was indicated in Chapter 5 that the terms *boards* and *commissions* are used interchangeably. Also, it was stated that such bodies have degrees of authority, some having complete and independent policymaking authority and others having only an advisory function. Since in the final analysis those bodies usually have a degree of policymaking and administrative authority that give significant direction to the recreation service, it is this type that will be examined here.[1]

## Functions of Boards[2]

Boards can only receive their power through legal enactment, whether by general state enabling acts, provisions of city charters, special legislation, or provisions of local ordinances. Hence the duties and responsibilities of these bodies will vary with the specific legislation that created them. But regardless of this difference, there is general agreement that the more important and significant functions of all can be summarized as follows:

1. To define the objectives of public recreation services and make plans and general policies harmonizing with them.

Every recreation and park body must ascertain its purpose and direction. Not knowing what is to be achieved makes program planning an impossible task. Statements of objectives can be used as a frame of reference for policy making.

2. To maintain the highest quality and standard of recreation and park service.

The recreation and park governing body not only charts the direction but also establishes a standard of quality it hopes to achieve. Such guidelines assist the professional staff in achieving its goals.

3. To interpret the importance and need of recreation and park services to the general public.

The governing body of the recreation and park system needs to keep the general public aware of the progress being made in achieving its goals and of the problems that confront it.

4. To keep public officials informed of the status and progress of recreation and park services.

[1]Some advisory recreation and/or park boards or commissions provide excellent direction with regard to an organization's objectives. This is due to the dedication of those that serve on these bodies and the excellent leadership that is given them by the executive officer. Furthermore, these bodies, because of their excellence, exert influence that carries much weight with the governing authority, and their role is more than a passive one. Consequently, while such bodies are called "advisory," in a real sense, they exert genuine community leadership and provide significant direction in vitalizing the recreation service. No effort is made to eliminate this type of board from our discussion. Instead, only those bodies that assume a completely passive role and function only as rubber stamps for executive decision making are not given any place in this discussion.

[2]The term *recreation and park board* will be used interchangeably with *park board*, and *recreation board* or *commission*.

The governing authority of the public jurisdiction wherein the recreation and park services are being provided should be aware of the scope of the programs being provided and the current and long-range problems that confront the recreation and park body.

5. To select the recreation and park executive and define the duties and responsibilities for that position.

The selection of the executive who will manage the recreation and park system is one of the most important tasks of the board. Whether the recreation body is a district board, a recreation commission, or a recreation and park board, the selection of the chief executive officer by this body is perhaps its most important single duty.

6. To establish a sound fiscal plan to achieve recreational goals.

Every recreation and park body must be concerned with its income source and must plan its expenditures. The board must give final approval to the budget and exert final control over all expenditures.

7. To provide for an adequate system of recreation areas and facilities.

The scope of recreation and park services requires the use of a variety of areas and facilities designed to meet specific needs and perform specific functions, and the recreation and park body must be concerned with these facility problems.

8. To give stability and continuity to general operations.

An alert and dynamic board provides stability to departmental operations, through calm judgments on needs for program expansion and leadership.

9. To evaluate the services of recreation and park system in relation to its objectives.

Once programs are inaugurated to carry out policy, the task of the recreation and park body is to appraise services in relation to policy and recommend change as needed. In other words, evaluation relates to the policies that are formulated and changed as experience dictates.

**Example of Board Responsibilities and Functions.** As already stated, a recreation and/or park board only receives power through legal enactment—by charter, special law, or local ordinance. An example of the duties and responsibilities of an advisory park and recreation commission are those cited in Section 907 of the city ordinance that established the Anaheim Park and Recreation Commission. The ordinance specifically states that this body will

> (A) Act in an advisory capacity to the city council in all matters pertaining to parks and recreation.

(B) Consider the annual budget for parks and recreation purposes during the process of its preparation and make recommendations with respect thereto to the city council and city manager.

(C) Advise in the planning of parks and recreation programs for the inhabitants of the city; promote and stimulate public interest therein; and to that end solicit to the fullest extent possible the cooperation of school authorities and other public and private agencies interested therein.

(D) Recommend policies for the acquisition, development and improvement of parks and playgrounds and for the planting, care and removal of trees and shrubs in all parks and playgrounds, subject to the rights and powers of the city council.

(E) Perform such other duties as may be prescribed by ordinance not inconsistent with the provision of the Charter.[3]

The legal status and primary responsibilities of an administrative and policy-making board are illustrated by the provisions of an ordinance of the Park District of Oak Park, Illinois. Chapters IV and VI of the ordinance cites the legal status and functions as follows.

## LEGAL STATUS OF THE PARK BOARD

Section 1. The state constitution, statutes, attorney general's rulings and court decisions on matters relating to public parks, park boards and park districts constitute the foundation of the legal status of park boards.

Section 2. Park boards are constitutional state agents created by the legislature for the purpose of effectuating, within the respective park districts, the state laws pertaining to public parks and recreation.

Section 3. Park boards are empowered to make contracts, employ persons, sue or be sued, make rules and regulations governing their own procedures and those of the parks under their jurisdiction, rent or lease property, and, in most instances, purchase and hold title to property in the name of the park district as well as to sell and give title to property.

Section 4. Park boards, like cities, counties and other units of local government, have no inherent or original governing powers. Being subject to state control, park boards can neither add to nor subtract from their responsibilities, powers and limitations, as defined by state law.

Section 5. Park boards are not subject to the authority of any governing unit other than the state, except in such special areas as the legislature may determine.

[3]City of Anaheim, California, Anaheim Park and Recreation Commission Role and Function, rev. 1975.

Section 6. Park boards are elected by the citizens of the local park districts to represent and act for the state in performing the legal function of providing their district with the kind of park and recreation program and facilities required or permitted by state law.

Section 7. Park boards are corporate bodies, or "artificial persons," and can act officially only in duly authorized and legally held meetings.

## PRIMARY FUNCTIONS OF THE PARK BOARD

Section 1. The Park Board's major function is policy making in nature. The Board shall formulate and adopt broad policies regarding the employment of staff personnel, programs, physical plant and equipment, finances and public relations.

Section 2. The Board shall employ its chief administrative officers as its professional advisors and properly delegate to them the authority and responsibility to execute its policies, enforce its rules and regulations and administer the parks. The Board shall exercise its supervision primarily through its chief administrative officers and shall not deal directly with individual subordinate staff members on specific problems.

Section 3. The Board shall approve sound, realistic annual budgets developed and recommended by its administration and provide the financial resources necessary for executing its policies.[4]

**Policy-Making and Executive Functions.** As stated previously, the recreation and park board is the policy-making body and that the executive staff carries out the policies. Therefore boards make policy decisions and the executive uses technical acumen to execute policies. Hence two elements are involved here that should be clarified.

First, policies reflect the aims to be achieved by the recreation and park system. They provide the guidelines of what is wanted and thus give positive direction to the executive and staff in the discharge of their duties. Policies do not give specific directions; these are labeled as rules and regulations. Instead, they reflect value judgments on issues related to the purposes of the recreation and park system. On the other hand, rules and regulations are more specific and tell precisely how, where, and when things are to be done. They spell out in greater detail the specific course of action to be taken within the general framework of policy. Examples of policy questions are whether to use school buildings as recreation centers, whether to build a museum or nature center,

[4]Park District of Oak Park, Illinois, Revised Code of the Park District of Oak Park, Ill., 1977.

whether to adopt a graduated salary schedule, whether to extend the summer recreation season, whether to extend the recreation service, and whether to add street tree planting to departmental operations.

While extension of the program into use of buildings and grounds would be a policy matter, the decisions on how the school will be used, by what age groups, and the time of use would be governed by rules and regulations. Clearly, then, it can be seen that policy provides guidelines while rules and regulations deal with policy application.

Second, the execution of policy reflects the means used by the recreation and park executive and staff to reach those ends as outlined by the board. Thus policy execution is left to those professionals whose factual judgments and knowledge make it possible for them to perform the technical work needed in reaching the goals of the organization.

Boards have neither the time nor the knowledge to carry out policies. In fact this is not their function. Therefore it can be seen that policy-making bodies provide the guidelines, the purposes, and the goals while the implementation of these purposes rests with the executive and his or her staff.

Some confusion between the policy-making and executive functions will exist if the relationships between the recreation and park board and its executive officer are not clearly drawn. It is not the job of the recreation and park board to meddle in administrative detail, nor the job of the administrator to perform the tasks reserved for the board. Therefore clear understanding and agreement should exist as to the respective spheres of interest of each.

Some writers have suggested that policy making and execution can be differentiated by the kinds of practical decisions involved. For instance, policy matters involve value judgments while administrative decisions involve factual judgments.[5] Value judgments reflect values and purposes that have meaning in charting the direction of the recreation agency. But factual judgments reflect the practical means and technical knowledge needed to carry out the value judgments. Hence proponents of this point of view claim that value judgments involve policy decisions while factual judgments involve administrative decisions.

Much could be written on this point. But in truth, no clearly defined statement separating these areas can be forthcoming. Administrators must be involved with policy. They do much in the way of guiding policy through their technical knowledge of problems. Likewise, boards must be involved in factual judgments in order to make proper value judgments. Thus a clear division of authority is not possible, but this should not confuse the issue as to board–executive relationship.

[5]This point is examined thoroughly in *Managing the Modern City*, (International City Managers' Association, Washington, D.C., 1971), pp. 85–86.

## Need for Policy

Every recreation and park system has need for written policy statements to guide its actions. Such statements give positive direction, clarify relationships, assist in evaluation, reduce criticism, assure fair treatment, provide uniformity of action, and give a clearer picture of recreation objectives.

Policy should cover decisions of what has been done and what should be done. Each system must develop its own set of policy statements, but areas to be included would consist of all phases of operation that would concern the board. Specifically written statements would cover such matters as objectives and purposes, board responsibility and relationships, personnel, program, business matters, public relations, areas and facilities, budgeting and finance, and the like.

All decisions of the board relate in one way or another to policy. Therefore either all board action should conform to established policy or changes should be made as needed.

While some recreation and park boards are consistent in following an unwritten policy such as one of securing the lowest price in purchasing goods, it is obvious that confusion and lack of continuity will not arise if such procedures are placed in writing. There is no reason that statements of policy cannot be published and distributed throughout the recreation and park system. A maxim to follow would be "do not keep departmental policy a secret." However, written policy statements should not become so detailed that the executive and staff have no latitude in carrying out their administrative function.

## Rules and Regulations

Rules and regulations outline the administrative procedures and relationships to be followed by recreation and park officials and staff. In other words, they spell out the directions of what things are to be done, when, how, and where. Every recreation and park system should adopt a sound set of rules and regulations to guide action in the conduct of its business. Rules and regulations that are generally found in recreation and park agencies cover topics such as affirmative action and hiring, emergency procedures, environmental controls, fiscal controls, gifts and donations, personnel conduct, purchasing, use of facilities and equipment, and the like.

> The personnel rules and regulations of the Department of Parks of the City of New York spell out its provisions in a booklet that covers the following.
>
> Article I. RULES OF CONDUCT FOR EMPLOYEES
> Section 1 General Provisions
> Section 2 Behavior and Conduct of Employee
> Section 3 Injuries and Accidents Sustained by the Employees or the Public

Section 4 Violation of Park Rules and Regulations and Vandalism on Park Property
Section 5 Uniforms

Article II. RULES GOVERNING VACATION, ABSENCES, LEAVES AND OVERTIME
Section 6 Notification of Inability to Report for Duty
Section 7 Leave of Absence
Section 8 Vacation
Section 9 Hours of Employment and Overtime
Section 10 Tardiness

Article III. GENERAL INFORMATION FOR EMPLOYEE
Section 11 Transfers
Section 12 Probation
Section 13 Service Ratings
Section 14 Salary and Salary Increments
Section 15 Pay Check and Deductions
Section 16 Health Insurance Plan
Section 17 Veteran's Preference
Section 18 Grievance Procedure

### Bylaws of the Board

Many boards establish a system of what are called "bylaws" for the conduct and guidance of their business affairs. Bylaws are merely the enactment of rules and regulations to provide guidelines for board action. They spell out in greater detail the legal provisions that established the recreation system. For example, the time and place of board meetings, committee organizations, and the like are given, and it is not unusual to see matters of general policy of the board incorporated in a written statement of bylaws. In fact, some bylaws incorporate board organization, statements of policy, and general rules and regulations. A good example of bylaws is provided by those of the Topeka, Kansas, Recreation Commission, shown in Appendix C.

## BOARD ORGANIZATION

A recreation and park board or commission is either appointed or elected. An appointment to a board at the municipal level is usually the responsibility of the mayor and city council; at the county level, it is the responsibility of the Board of County Commissioners. Recreation and park boards are elective if the legal provisions creating them provide for this manner of selection. Boards for district operation are usually elective.

Generally speaking, boards composed of five or seven members are most numerous, although it is not unusual to see the number vary from three to eleven members. Boards of five or seven members are the most workable units.

In selecting citizens for a recreation and park board, the appointing body should keep in mind the qualifications of individuals that make for successful service. These include (1) a fundamental belief in public parks and recreation as a vital social force in the community, (2) a desire for personal service with no thought of remuneration or advantage to self or friends, (3) the ability to deal with issues and policies without becoming involved in administrative details, and (4) the courage to uphold standards of service and, if necessary, to fight against misunderstandings and misrepresentations. An understanding of the functions and responsibilities of a board by its members is as important as the finding of qualified citizens who will be willing to serve.

In most public jurisdictions, board members serve for overlapping terms of three or five years without compensation except for expenses. Furthermore, it is sometimes specified that school and other public interests should be represented on this body.

## Internal Organization

The internal organization of the board is sometimes spelled out in law, but this will vary with the public jurisdictions. Nonetheless, the three officers most commonly designated are chairman, vice chairman, and secretary, or president, vice president, and secretary. Some policy-making bodies also designate an office of treasurer.

It should be stressed that the executive of the recreation and park system should not be burdened with the task of keeping minutes or performing other secretarial chores. Likewise, a board member, even the one designated as secretary of the board, should make use of stenographers to assist in these chores. It is impossible to participate in the business at hand and perform the detailed task of record keeping. Furthermore, it should be noted that all minutes of meetings and records of the board are legal documents and should be kept in the business office of the recreation and park system where they are available to the public.

## Committees

Committees when they are used by park and recreation agencies are usually classified as standing committees or special committees. A standing committee is usually authorized in the organization's bylaws and is a permanent part of the organizational structure. Special committees are usually temporary in their existence and are appointed for a special task. Actually, recreation boards are small bodies concerned with legislative or policy-making matters and as such do not always utilize committees. However, when used, the committees most frequently appointed are finance or business management, personnel, program, public relations, building and grounds, and equipment and supplies. Standing committees are composed primarily of board members with the executive and at times other staff members assigned to work with the committee.

Special committees are usually appointed by park and recreation agencies to undertake a specific task such as helping to sell a bond referendum, advising on a particular park location and development, advising and evaluating a special program service such as a sports league or a program for persons with handicaps. In addition to board and staff members, these committees very often include citizens in the community.

**Advantages and Disadvantages of Committees.** The chief argument for standing committees is that they give definite responsibility for a phase of board operation to certain members, who in turn advise the body as a whole. This provides more opportunity for individual members to become informed and involved. Another argument for standing committees is that it provides a better sounding board for staff recommendations. It affords more in-depth study of a proposal and the discussions can generally be more frank as the work of committees is not so much in the public view as are the official board meetings.

Arguments against the use of standing committees are that the assignment of specific activities or problems to committees tends to dissipate full consideration of these matters by other members of the board. The assignment to a committee takes extra time on the part of the board member, which many are not willing to give. Also, it requires that the staff use additional time to prepare and present information two or more times to the committee and the full board. Furthermore, evidence substantiates the claim that such committees encroach on the professional areas of recreation and park administration that are the responsibility of the executive.

These same arguments can be made for the use of special committees. In addition, the involvement of citizens on the committee, when used properly, can develop strong support for the recreation and park agency within the community. Special committees can also utilize community strengths by calling on citizens whose expertise relates to the proposal under discussion. Many citizens are quite willing to volunteer their time and knowledge as a public service, thus saving considerable expense that otherwise might be needed to obtain a consultant.

One good rule for boards to follow when utilizing committees is not to appoint the number of board members to a committee that would constitute a majority of the board. If this were the case the committee could in fact take action that would be equivalent to action by the board. For example, four members of a seven member board are in effect making a decision for the board if they unanimously agree to a course of action by the committee.

## Board Meetings

Board meetings should be held at a regular time and place. Since such meetings take care of public business, they should be open to the public except in instances when discussion is focused on problems of a confidential or personal nature. Many states have an open meeting law that limits closed meetings to matters of personnel, land negotiations, and litigation. Even though executive or closed meetings are discussion, not decision-making sessions, they should be used sparingly and only in those instances where an open meeting would prove either costly to the board or embarrassing to individuals. However, all actions of the board should be taken in an open meeting.

**When and Where to Hold Meetings.** Recreation and park board meetings are usually held at least once a month, at a time convenient to all members. The first or second Tuesday of each month is a popular time. Also, most boards hold their sessions during the evening hours, although other times could just as well be used. While the number of meetings will vary with the press of business, too few or too many meetings may be an indication that some study and adjustment is necessary to see that the board functions as specified are being properly performed.

The place of meeting is usually the board or conference room in the administration office, or in the small recreation and park system it may be in the office of the executive. Needless to say, the meeting place should be large enough to accommodate the needs of the board and the public that may attend.

**Arrangements for Meetings.** The physical arrangements for the meeting can be very important in setting the meeting's tone. A sloppy arrangement can lead to a poor meeting, but a sharp physical arrangement can help to ensure a good meeting. Tables should be arranged so board members can easily talk to each other and to persons attending the meeting. The seating should be arranged in order of seniority and name plates used to identify the board members. Paper and pencils should be at each place, with special folders to hold the meeting material. The president or chairperson should have a gavel. Glasses of water and microphones are items that might be helpful. Audio-visual materials including maps, drawings, and slides are very important to provide additional information and should be used as much as possible.

Although the recreation and park executive is not a member of the board, he or she is its executive officer and, as such, attends all board meetings. Indeed, it would be difficult for a board to act on matters without explanations or counsel from this person. The executive speaks for the department, even though members of the staff may on occasion be called on to explain specific matters of concern to the board. The executive does not vote on issues but does, as a general rule, prepare the agenda for board meetings.

**Rules of Order.** The business of the board should be handled efficiently and expeditiously. Thus a definite order of business should be instituted and a definite parliamentary procedure followed. The order of business typical for many recreation and park systems is:

1. Call to order.
2. Roll call (quorum must be present).[6]
3. Approval of minutes of last meeting.
4. Report of communications and petitions.
5. Approval of bills.
6. Report of officers and executive.
7. Remarks from visitors.
8. Unfinished business.
9. New business.
10. Miscellaneous and announcements.
11. Adjournment.

It is important that the board agenda be carefully prepared and distributed together with supporting data to all members well in advance of the meeting. The agenda is the guide for board action at its regular meeting. Thus board members should have a chance prior to the scheduled meeting to scrutinize carefully all proposals to come before the board. In many cases the executive will also prepare additional notes, providing information on agenda items, to send to the board.

**Minutes of Meetings**. It is almost unnecessary to say that the official record of board actions is of the utmost importance. Such records of actions are known as the "minutes of the board," and they should show in an accurate, complete, and clear fashion the actions of the board. A sample set of minutes is shown as follows.

Smithville, Oregon
October 6, 1980

| | |
|---|---|
| Roll Call | The Smithville Recreation and Park Board met in regular session at 7:00 P.M., in the board room at City Hall with George Hall, President, presiding. Those present were Tom Brown, Ella Christiansen, Sam Bell, Harold Jones, and George Hall. |
| Approval of Minutes | The minutes of the previous meeting were approved as read. President George Hall and Secretary Tom Brown attested to and signed the minutes. |
| New Business | It was moved by Harold Jones and seconded by Ella Christiansen that general fund warrants 16350 with 16343 void; general fund payroll warrants 15 through 69 with 36 void; Building and |

[6]A quorum usually consists of a majority of the members.

Construction warrants 19985 through 19989; Trust Fund warrant #93, be approved for payment. Motion carried.

It was moved by Sam Bell and seconded by Tom Brown that the change orders as presented by George Smith, pool architect, for anchor holds, repairs to pool ceiling, and change in wiring in the pool office be approved, the cost not to exceed $450.00. Motion carried.

It was moved by George Hall and seconded by Harold Jones that the board accept the bid of the Eastman Company for pool paint according to specifications for $870.44 and the bid from the Hillyard Company for the electric cash register for the pool according to specifications for $230.00. Motion carried.

Report of the Superintendent

George Hill, Recreation and Park Superintendent, gave a report on the Eastside Park extension and recreation needs for the fall and winter season. He also reported that the Rotary Club was interested in helping with the development of Foothill Park and it is estimated that the cost would be approximately $2300. He further reported that President Hall and he met with the school authorities and accomplished the following:

a. Agreement on the design and development of the River Road school–park site.

b. The architect will meet with the board on the Eastside school site.

c. There was agreement to the recreation and park department using areas of school property in the northeast section to extend program needs.

The superintendent reviewed the program for the coming recreation and park conference.

It was moved by Ella Christiansen and seconded by George Hall to allow $250.00 for conference expenses for the superintendent to attend the Pacific Northwest Park and Recreation Conference in Seattle, Washington, October 17 and 18. Motion carried.

Communications

A letter was received from Senator Ed Smith advising the superintendent that he has directed a request to the Administrator of the Housing and Finance Agency urging him to give our request for open-space land thorough and prompt attention.

A letter was received from the Power and Light Company that the request for additional lights in Hill Park has been approved and that work would begin on installation by November 1.

Adjournment

Since there was no further business to come before the board, it was moved by Tom Brown and seconded by Sam Bell that the board adjourn. Motion carried and meeting adjourned at 8:50 P.M.

George Hall, President
Tom Brown, Secretary

**Approval of Minutes.** As a general rule the minutes of the meetings should be sent to board members in advance of the meeting. This allows each member to read the minutes and note corrections at their leisure. At the next regular meeting the chairman has the option of having the minutes read or simply asking if there are any corrections or additions to the minutes "as mailed." The second option saves time, which can be used for other business matters. In either case, after corrections are made, the minutes are approved.

**Parliamentary Procedure.** It is essential that members of a board know the rudiments of parliamentary procedure. Although much more goes into the planning of a good business meeting than the following of rules or order, knowledge of how to conduct a meeting is basic if the business before the board is to be accomplished.

The chairman may say, "We now have a report of the finance committee." The report is given, and the chair may ask, "Is there a motion that the report be accepted?" Another member may then say, "I move for the acceptance of the finance report." Another member will remark, "I second the motion." The chair then states, "It has been moved and seconded that the finance report be accepted. Is there any discussion?" (The report is now open for discussion, not before.) Discussion then follows, after which the chair may bring the question to a vote by stating, "Are you ready for the question?" (This means "Are you ready to vote on the business before the house?") A member may say, "Question," and the chair then states, "All those in favor of the question say 'Aye,' opposed, 'Naye.' " If the majority vote in favor, the chairman says, "The motion is carried."

It is not unusual, before the final voting, for a member to ask for the wording of the question, in which case the chairman states the issue that is being voted on. Finally, any motion to adjourn must be seconded, but it cannot be debated.

The agenda of the meeting is followed as a guide by the chairman, and the business at hand is disposed of in each case by the procedure of bringing it to a vote, as shown in the preceding example.

## BOARD AND COMMUNITY RELATIONSHIPS[7]

The services of a park and recreation department are unique among those of local government. Participation in its program and the use of its facilities are

[7]The discussion on "Board and Community Relationships" is based on material used in the first writing of this book that was prepared by John Collier, Director of Parks and Recreation, Anaheim, California, for use by his commission. The original material has been revised and additional items added.

matters of free choice, as far as the individual is concerned, and are absolutely dependent on his or her attitude toward them. Unless there is respect for the quality of the leaders responsible for the recreation program in the neighborhood, parents may not permit their children to attend. The participants must be convinced that what is offered is of good quality and that they will gain satisfaction and enjoyment, or they may not take advantage of the opportunities offered.

### Criteria for Commission Success

The ultimate success of an advisory park and recreation commission depends on the following four factors.

1. The commission must maintain a clear-cut division of responsibilities between itself, its parent body, and the professional staff. The commission has the responsibility to advise the city council, the various school boards of trustees, and the administration on various policy matters that have been referred to them in regard to the provision of park and recreation services to the total community. It has no administrative authority or responsibility, making only general recommendations.
2. The commission is responsible for advising on the basic policies that guide the department. The administration of the policy—the actual provision of the park and recreation service—is a technical problem that requires a technically trained staff. This division-of-work principle serves the important and practical purpose of enabling the commission and staff to recognize their respective functions and to achieve the most effective working relationships to fulfill the department's purposes.
3. The commission must recognize that its recommendations to the city council and administration will not always be followed but should not take this rejection as a personal affront. Many times, proposals are rejected because of other factors not yet public knowledge. However, with few exceptions, if the commission secures the pertinent facts on matters referred to it, deliberates carefully, and acts objectively, its recommendations will become the basic policies for the operation of the department.
4. The commission gives advice that the city council and administration must evidence a willingness to accept, even though they are under no legal obligation to act on it. Without such support, the commission becomes a mere observer, serving no useful purpose other than to ratify decisions already made.

### Delineation of Duties and Responsibilities

The following delineates some of the ways the commission can carry out its duties and responsibilities as outlined in an ordinance for the city of Anaheim, California.

1. *Give due attention and study* to recreation and park services as they affect the welfare of the people.
2. *Interpret* the recreation and park services of the Department to the community.
3. *Interpret* community recreation and park needs to each of the public jurisdictions participating in the joint cooperative program.
4. *Take initiative* in planning for future recreation and park areas and facilities, as well as in determining means of bringing present areas and facilities up to an acceptable standard.
5. Serve as a *sounding board* against which the department administrator and staff may test their plans and ideas.
6. *Recommend policy*—keeping in mind that the adoption of the policy is the prerogative of the city council and that the park and recreation administrator must have a free hand to carry on the work of the department within the framework of these policies.
7. *Enable* civic and service organizations to accomplish results through cooperation that they could not possibly accomplish alone.
8. *Encourage* individuals and citizen groups to give funds, property, and manpower for the development and operation of the park and recreation facilities.
9. *Generally enlist* community interest in parks and recreation.[8]

## CODE OF ETHICS

The following code of ethics prepared and adopted by the Citizen-Board Members Branch of the National Recreation and Park Association is a good guide for all boards and commissions.

> As a Park and Recreation Board-Commission member, representing all of the residents, I recognize that
>
> **1.** I have been entrusted to provide park and recreation services to my community.
>
> **2.** These services should be available to all residents regardless of sex, race, religion, natural origin, physical or mental limitation.

[8]City of Anaheim, California, Ordinance No. 1300, November 25, 1958.

**3.** While honest differences of opinion may develop, I will work harmoniously with other Board or Commission members to assure residents the services they require.

**4.** I will invite all residents to express their opinions so I may be properly informed prior to making my decisions. I will make them based solely upon the facts available to me. I will support the final decision of the Board or Commission.

**5.** I must devote the time, study and thought necessary to carry out my duties.

**6.** The Board-Commission members establish the policy and the staff is responsible for administering the policies of the Board-Commission.

**7.** I have no authority outside of the proper meetings of the Board-Commission.

**8.** All Board-Commission meetings should be open to the public except as provided by law.[9]

## A Guide for Community Relationships

No board or commission can be isolated in the community and remain effective. The board or commission must maintain a working relationship with the city council, the school board, and other community agencies. The influence that the board or commission has in the community is dependent on the manner in which these relationships are maintained. The following are some specific suggestions as to how these relationships can be most effectively maintained.

### Relationship of the Commission with the City Council.

1. Know your ordinance and the role and function of the commission.
2. Meet with the council from time to time in formal session when policies are needed in order to ensure a smooth-running operation.
3. Remain warm, friendly, and understanding toward the council, but make your decision on all the facts.
4. Back up recommendations made to the council with written facts, oral communications, and action, as necessary.
5. Support the over all program of the council, not just the park and recreation services.

### Relationships with Boards and Commissions of Other Agencies.

1. Recognize that other agencies have an important role to play in the provision of park and recreation services. Joint planning and use of

[9]Citizens-Board Members Branch, Code of Ethics, National Recreation and Park Association.

facilities with school districts, for example, can save the taxpayers thousands of dollars each year.

2. Maintain a cooperative and mutual trust relationship with the boards of other agencies in the community.
3. Meet with the board of other agencies in formal and informal sessions to discuss matters of joint interest.

**Relationship of the Commission with other Commissioners.**

1. Attempt to understand their point of view.
2. Do not speak for other commissioners unless formal action has been taken by the commission as a whole.
3. Maintain a level head during consideration; do not "go overboard" with your personal recommendations or allow the commission as a whole to become too radical in its suggestions.
4. Admit a mistake, when made, as a unit without trying to fix the blame.

The chart shown in Fig. 6-1 illustrates the relationships of the park and recreation commission with other official municipal bodies.

## THE EXECUTIVE AND THE BOARD

It has already been stated that a close and cooperative relationship should exist between the executive and the recreation and park board or commission. Indeed, the lack of an effective and cooperative relationship will seriously cripple the work of the organization. Board members are laypersons; the executive and departmental staff are professionals in the affairs of recreation and park operation and administration. Therefore it is necessary for each group to understand its relation to the other. The following are some expectations between the board and executive.

**What the Executive Can Expect from the Board.**

1. The board gives the executive a clear understanding of the responsibilities of the position.
2. The board gives the executive the authority to carry out the responsibilities of the position.
3. The board acts as a group in working with the director. Directives are not issued by individual members.
4. The board evaluates the director's performance, but if negative criticism is necessary, the evaluation is given in a closed session. The board will not ridicule the director or staff in public meetings.
5. The board gives the director's recommendations a fair hearing.

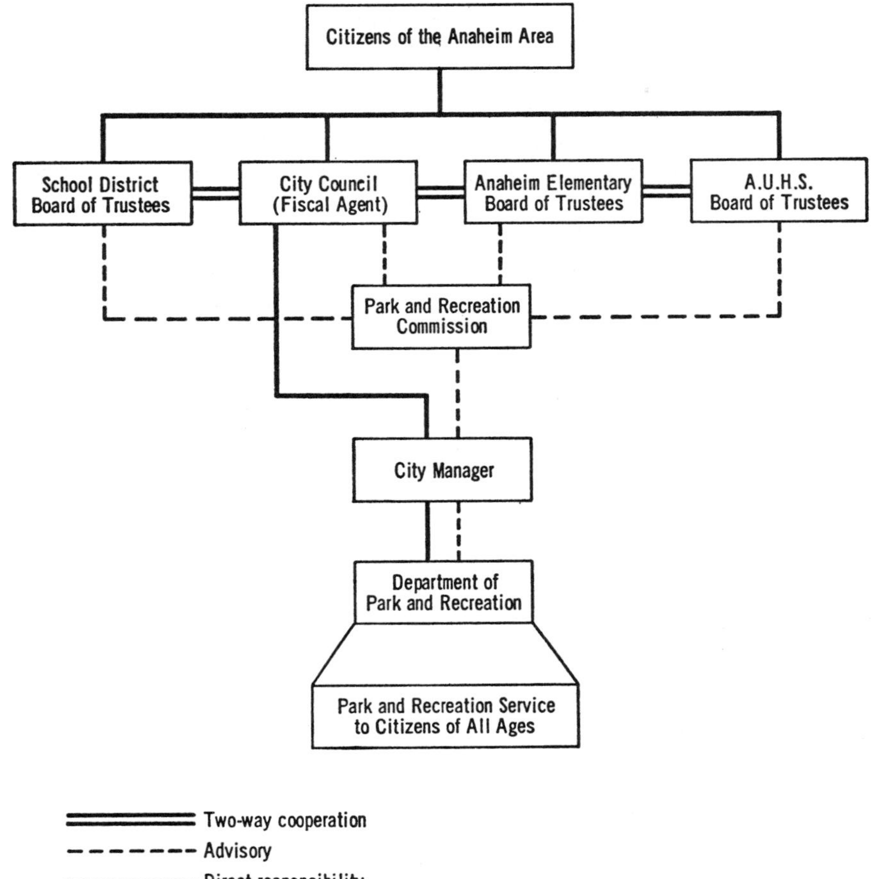


**Figure 6-1.** Organization chart of Anaheim area park and recreation services, showing relationships of commission.

6. The board assures the director of a fair hearing and supports the director when necessary.
7. The board does not bypass the director in dealing with other employees.

**What the Board Can Expect from the Executive.**

1. The executive will set an example for the staff and maintain an honest administration.
2. The executive will carry out the policies and enforce the regulations as set by the board.

3. The executive will direct the operation of the agency.
4. The executive will provide for the hiring, training, and supervision of the staff.
5. The executive will prepare an annual budget, take fiscal control of the agency, and present fiscal reports to the board.
6. The executive will recommend policies and various courses of action.
7. The executive will keep *all* members of the board informed of the agency's operation.

In all matters of recreation and park operation the board should look to its executive officer for leadership and guidance. But a reciprocal relationship exists in that the executive looks to the board for the direction the recreation and park system will take. Hence a team effort is essential. The board makes the policy, and the executive is charged with executing it. Obviously, unless there is good feeling and cooperation between the executive and the board, not only will the system suffer, but the executive will be looking for new employment. Consequently, the executive who is dogmatic in his or her viewpoints may have real difficulty with his or her interpersonal relationships. There must be a "give and take" attitude, a spirit of cooperation, and complete trust and faith of each in the other's work. The executive is not omnipotent; neither is the board.

## Questions for Discussion

1. Examine carefully the legal provisions that establish recreation and/or park boards in your state. What powers do these boards have? Are they advisory or administrative?
2. Distinguish between an advisory and an administrative board. Which do you prefer? Why? Can you make a case for an advisory board as well as for an administrative one?
3. How are board members selected? How can good candidates be encouraged to seek a post on the board?
4. Explain fully the difference between a policy-making decision and an administrative decision.
5. How do rules and regulations relate to policy? Cite examples of each.
6. Explain fully the relationship that should exist between the recreation and park executive and the board.
7. Assume that you are hired as a recreation and park executive and that the board has been performing the executive function. How would you proceed to take over your rightful function as recreation executive without offending the board?
8. Should recreation and park board meetings be open to the public? Should they ever be closed? Why or why not?
9. During board meetings, what is the role of the recreation and park executive?

**10.** If a staff member is disgruntled and does not like the way matters are being handled, should he be allowed to air his grievances before the board? Why or why not?
**11.** Why have a board? Would it not be better to eliminate them entirely? Explain.

## SELECTED REFERENCES

Butler, George D., *Introduction to Community Recreation*. 5th ed., McGraw-Hill, New York, 1976.

Edginton, Christopher R., and Williams, John G., *Productive Management of Leisure Service Organizations: A Behavioral Approach*. Wiley, New York, 1978.

Hjelte, George, and Shivers, Jay S., *Public Administration of Recreation Services*. 2d ed., Lea and Febiger, Philadelphia, 1978.

The International City Managers' Association, *Municipal Recreation Administration*. 4th ed., The International City Managers' Association, Chicago, 1960.

Kraus, Richard G., and Curtis, Joseph E., *Creative Administration in Recreation and Parks*. Mosby, St. Louis, Mo., 1973.

# CHAPTER 7

# Personnel Practices and Policies

Personnel practices and policies determine in large measure the effectiveness of any recreation and park system. Indeed, the success of any recreation and park program is dependent on the quality of its staff. But high quality is only attainable if an organization attracts and retains outstanding leadership and provides a climate for its development.

In a real sense, personnel administration is concerned with people. Over half the planned expenditures of recreation and park systems are marked for personal services or employee salaries. Hence those conditions and techniques used to attract, place, classify, train, promote, and challenge employees in an environment that is conducive to bringing forth their maximum effort are the essence of personnel administration. Actually, personnel matters cut through all phases of administration, for the human resources of an organization make possible the achievement of recreation goals.

This chapter explores those personnel practices and policies that are essential for effective recreation and park operation. Discussed here will be the three phases of personnel management: (1) the recruitment, selection, and placement phase, (2) the employment policies phase, and (3) the job separation phase. This will be followed by a discussion of position classification and wage standardization, which are essential to effective personnel operation.

## RECRUITMENT, SELECTION, AND PLACEMENT

Of all the tasks confronting a recreation and park administrator, none provide more benefits to the organization that that of selecting a competent and effective leadership staff. Such personnel determines the tone of operation and the quality of the program. Indeed, the ability of staff enhances the success and efficiency of the recreation and park service. It is a truism that "leadership is the key to a successful operation."

Any program of recruitment and selection is concerned with the process of finding and selecting the best qualified individuals to fill the openings in the recreation and park organization. This process involves the preparation of job descriptions, the advertising of job openings, the reviewing of applications, the testing of applicants, the selection of personnel, and the testing of new employees on the job by means of a probationary period.

### Job Descriptions

An effective recruitment or employment program must have a written job announcement that provides information about the duties, responsibilities, salary, and qualification requirements of the position to be filled. The National Recreation and Park Association recommends that the following basic information be used in the job description format.

IDENTIFICATION OF EMPLOYING AGENCY:
(Mailing address) (telephone number)

CLASS OF POSITION TITLE:
Should be descriptive of the job

FUNCTION:
General statement of major duties

SUPERVISION:
To whom employee is responsible and employee's responsibility for assigned personnel.

EXAMPLE OF DUTIES:
Listing distinguishing features and examples of work (illustrative only, not all-inclusive). Should be broad enough to include anticipated supplemental assignments.

QUALIFICATIONS:
Required knowledges, skills and abilities.

SPECIAL REQUIREMENTS:
In addition to above where need such as technical or professional registration or certification by a National Agency in the specified field.

EDUCATION AND EXPERIENCE:
Required education, training and experience.

SALARY RANGE:
Add pertinent information (car furnished, etc.)

CLOSING DATE FOR RECEIPT OF APPLICATIONS:

CONTACT:
Full name, title, agency and address for submitting applications.[1]

Figure 7-1 is an example of a job description.

## Advertising the Position

If a position is to be filled by the best qualified person available, it is essential that every effort be made to advertise or otherwise publicize the opening, and every communication channel should be used to reach the greatest number of prospective candidates. Placement organizations, colleges, universities, and professional societies do an excellent job of communicating with qualified applicants through use of personnel bulletins, newsletters, professional journals, convention postings, and direct correspondence. The National Recreation and Park Association; the American Association of Health, Physical Education and Recreation and other service and professional organizations have up-to-date rosters of job openings and records of qualified personnel seeking employment. Civil service commissions have extensive mailing lists for the sending of job notices, as do state recreation associations and societies. Actually, these service and professional recreation and park bodies do the major task of advertising the openings in the recreation and park profession. Some publicizing, however, is also done by word of mouth and through personal correspondence by the staff of the department in which the opening is located.

## Reviewing of Applications

Most recreation and park organizations state in their job announcements the manner in which applications are to be received and the personal data that are required. In most instances application forms provided by the department are to be filled out by the applicant and returned before a closing date. On the other hand, some small departments only require a written letter of application that gives pertinent personal data, references, and professional qualifications.

[1]National Recreation and Park Association, *National Personnel Guidelines for Park, Recreation and Leisure Service Positions* (Arlington, Va.: National Recreation and Park Association, 1978), p. 4.

**Employment Opportunity City of Smithville, Oregon,**
**Director of Parks and Recreation**

| | |
|---|---|
| **Duties:** | Directs and coordinates a city-wide recreation program including the operation of recreation centers, swimming pools, parks, golf course, and activities held in the schools. Supervises department heads and other staff, prepares budget, plans recommendations, and prepares administrative reports. Responsible for recruiting department heads, testing employees, and preparing publicity, in-service training manuals, and other publications as required. |
| **The Examination:** | Will be competitive and consist of three parts: (1) evaluation of training and experience, (2) assembled written test, and (3) oral interview. A character investigation will be conducted on each qualifying candidate. The three divisions of the examination will be weighted 3 and 3 and 4, and an aggregate rating score will be given each applicant to determine his or her respective order on the eligibility roster. Arrangements will be made for applicants to take the written examination outside the city of Smithville if necessary. Failure in the written examination will automatically disqualify an applicant. |
| **Minimum Qualifications:** | Thorough knowledge of the objectives and ideals of public park and recreation programs. Good background in the planning, operation, and maintenance of park and recreation areas and facilities. Understanding and experience in park and recreation administration, and supervision, personnel procedures, and program planning.<br>Must have four years of progressively responsible experience in a public recreation program with at least two years' experience on a supervisory level. A B.S. degree in parks, recreation, or related field required. |
| **Salary Range:** | \$19,500–26,000 annually. |
| **Method of Application** | Closing date for applications is October 5, 1979. Applications should be mailed to John D. Brown, President, Parks and Recreation Board, 910 South Street, Smithville, Oregon, 12345. |
| **Equal Opportunity:** | The City of Smithville is an equal opportunity employer. Employment will be based on merit as it relates to position requirements without regard to race, color, religion, sex, or national origin. |

**Figure 7-1.** Job description.

Once these applications are received, an informal screening takes place and those applicants not meeting the qualifications specified in the job announcement are eliminated and so notified. Those who meet the requirements are admitted to the testing or evaluation program.

## Testing of Applicants

A system of testing, examination, and evaluation is used to select the best qualified person for the position. Probably the most used method of testing is selection based on merit, which is objectively evaluated on the basis of (1) education and experience, (2) personal interview, (3) written or performance tests, and (4) character investigation or reference check. At times a physical examination may also be required. A weight may be given to each of the areas under which the candidate is evaluated, except for the physical examination and character investigation, where it is a pass-or-fail situation. The candidate receiving the highest score on these items is placed first on the eligibility list.

A new system for evaluating candidates for a position is the assessment method. It has had great use in the business community and many park and recreation agencies are beginning to use the technique. Under the assessment method the top four to six candidates, following an initial screening, are invited to participate in an assessment center. The candidates are put through a series of tests, exercises, management games and interviews all designed to test the "behavioral dimensions" of the candidates. A team of trained assessors observes the candidates and prepares a report on the strengths and weaknesses of each candidate.

The person responsible for making the selection utilizes the assessment reports in addition to methods he or she used in reviewing the candidates, as a major aid in the final selection process. This method is normally used only as an aid in filling top positions.

It should be emphasized that the purpose of a testing program is not only to select the most efficient person for the position to be filled but also to select the person having the greatest potential for development while on the job. Too often, this second objective is overlooked. With more and more emphasis on affirmative action and promotion from within, it is essential that the individual's potential be carefully considered, as well as his or her technical and professional competence.

**Education and Experience.** Most professional testing programs consider education and experience in their rating score for job candidates. Although it is difficult to evaluate these experiences objectively and to give them an assigned weight, much progress is being made by personnel agencies in standardizing the scoring of this item.

**Written Examination.** Many park and recreation agencies, particularly those under civil service, consider a written examination as an important part of any testing program. The written examination attempts in a short period of time to test the candidate's professional competence for the job to be filled. In other words, it is a sampling of a candidate's knowledge. Tests known in civil service parlance as an "assembled" test are given to all candidates in a central location at a specified time. This type of test is used primarily to fill professional recreation positions.

On the other hand, the "unassembled," or "nonassembled" examination is frequently used to fill high-grade executive and supervisory positions where intangible qualities and competences are sought. For these examinations the candidates are not assembled but, instead, are given the test individually at their place of residence. Naturally, this type of test is different in content from the assembled examination. In most instances it consists of a statement of education, experience, written publications, community service, honors awarded, and professional offices held. Rarely does it consist of a formal written test. A written statement of one's philosophy or attitude toward current issues is sometimes requested, however. This test is then scored, and the most promising candidates are called for a personal interview.

**Personal Interview.** The purpose of the personal, or oral, interview is to see if a candidate has the personal qualities needed for the job to be filled. Strictly speaking, it is a personal evaluation of a candidate's appearance and personality.

The personal interview is used extensively in the testing of candidates for professional recreation and park positions. And rightly so, for it is a very important step in the selection of persons needed to fill those positions that require a high degree of knowledge of the community, intelligence, initiative, tact, resourcefulness, and ability to get along with others. In no other way than through the face-to-face appraisal of the candidate can these intangible qualities of leadership be assessed to the same degree.

A distinction should be made here between the oral interview and the oral examination. The purpose of the oral examination is to test orally the professional competence and fitness of a person for the job to be filled. It either replaces entirely or supplements the written examination. Questions are asked on technical administration or professional matters, and the candidate is evaluated on the basis of his or her answers. Likewise, he or she is rated on personal qualities.

The oral examination is difficult to administer and difficult to rate. Probably, the best practice is to assess a candidate's technical knowledge or ability through a formal assembled or unassembled examination and to size up the candidates intangible personal qualities through the personal interview. In effect, the oral examination could very well be eliminated as a testing device. It is low in validity and reliability. For top-level administrative positions, the unassembled examination and oral interview might well be used.

**Physical Examination.** Some recreation and park systems require a medical physical examination before a new employee can be placed on the payroll. The purpose is to reveal any physical defects that can impair the work efficiency of an employee and, further, to provide a check against specious claims of work-connected disability in the future.

**Character Investigation.** Another step in the testing program is that of conducting a character investigation of the job applicant. Such investigations are used by nearly all public bodies responsible for personnel selection, but the questionnaire method that is so prevalent today in securing such information about an individual's character or moral standards has definite weaknesses. Such inquiries are usually sent to selected references supplied by the applicant. Invariably, such references give only the outstanding traits of the individual or his or her job skill. To get reliable personal data about the candidate, it is necessary either to conduct a confidential investigation or to seek such information from selected individuals through a confidential questionnaire or preferably by a phone call. It is generally agreed that more accurate information can be obtained from a personal phone call than by written questionnaires.

## Selection of Personnel

The actual appointment of the recreation and park worker to the position that is open is the next-to-the-last step in the recruitment process. The selection-and-appointment procedure follows a ranking of the candidates on the basis of test scores. An eligible roster is developed, and the highest three are usually certified for the position.[2] The recreation and park executive then has the choice of selecting or hiring one of the three.

The actual certification procedures begin with the recreation and park executive notifying the personnel body that a position is open in the department. The personnel department follows the examination procedure outlined and certifies those eligible for appointment. The executive then fills the vacancy. Of course, names remain on the eligible roster and additional appointments can be made from it as vacancies occur. Such rosters are usually dissolved after one or two years.

[2]In some instances, the top person on the eligible list must be appointed. In other cases five names may be certified but, in general, the "rule of three" is favored by personnel bodies and civil service commissions.

### Probationary Period

The last stage in the selection-and-placement process is that of providing for a probationary period before an appointment becomes final. This trial period usually varies in length from one recreation and park system to another. However, three to six months are commonly provided. While the probationary period is a "working test" to determine whether the employee can satisfactorily perform the functions of the position, in actual practice it is tantamount to final appointment. Little effort is made by most recreation and park authorities to use this period as it was intended to be used—for the final testing of the worker on the job.

## EMPLOYMENT POLICIES

The employment policies phase deals with those on-the-job personnel practices and policies that are in effect for all professional staff. In this category fall matters such as promotional policies, hours of work, leaves, professional development, evaluation, and health and welfare provisions.

### Promotional Policies

Most employees desire opportunities for advancement or promotion. Here is one of the vexing problems for many recreation and park systems. Although it is desirable to promote from within the organization, such a policy is desirable only if those in the lower ranks meet the qualifications of the higher position. In no way does this mean that the door for advancement should be closed to those within the department. But it does mean that mediocrity in one's present position does not automatically provide a gateway to a higher position. Here is where the testing program has its effect. If those in the lower ranks were initially selected on the basis of their potential development in addition to their job skill, promotion from within would create few problems.

There is also much to be said about the morale factor involved in having a haphazard policy that discourages advancement of those within the department. Such a policy creates dissatisfaction, poor worker morale, loss of initiative, and high job turnover. Hence it seems desirable to have a flexible promotion policy that adjusts to the needs of the positions to be filled. If a position is such that excellent work in it is a good criterion of successful performance in a higher position, a policy that rewards such effort is desirable. On the other hand, there exist in many recreation and park systems top level positions that call for unusual ability, and to restrict applications only to those in the department could seriously impair the efficiency of the organization if none of the staff have the abilities needed for the job to be filled. Actually, every effort should be made to reward meritorious services. Therefore policies should be developed that make it clear that appointments will be made through promotional examinations if at all possible.

## Hours of Work and Leaves

Every recreation and park system should establish a clear-cut policy as to hours of work, annual and sick leave, and leaves of absence. Too often, no fixed number of working hours is specified for professional recreation and park employees. Instead, work schedules are arranged to meet the demands of the job. This frequently requires many additional hours of work beyond a normal workload. Actually, there is no excuse for an employee to spend his or her mornings, afternoons, and evenings on the job. Recreation leaders have responsibilities to their families and their own recreation leisure pursuits, just as do other people. No profession asks that its members have a duty to work 12 to 18 hours a day because the job demands it, but many recreation and park staff do just this. On the other hand, work schedules may call for many odd hours of work, in the afternoon and evening, for example, or even work on weekends and holidays. However, a uniform policy on the number of hours of work required of employees is highly desirable, and if an individual is called on to spend additional time in preparing program features he or she should be given compensatory time off.

The granting of an annual vacation to employees in the recreation and park system is a widely accepted practice. At present, most departments provide a minimum of two weeks with pay after being on the job for one year, and many provide additional time after a specified number of years of continuous service. However, in some instances recreation and park executives are given as much as a month of annual leave with pay, but this must usually be taken at a time other than the busy summer season. A liberal vacation policy is highly desirable. Not only does it create a climate that aids morale and job satisfaction, but, even more, it helps prevent staff members from stagnating and going stale. It provides vigor and renewed energy to those creative and imaginative staff members who give maximum effort to their jobs.

Most public recreation and park systems provide for special leaves. Generally, the rules governing such provisions are specifically outlined in the personnel policy of the department. This policy usually includes sick leave, personal leave days, provision for leaves without pay, and for authorized absences due to holidays, illness, or death in one's family and maternity.

Every recreation and park system should establish a sound sick-leave policy. Failure to provide one creates a hardship on employees who are ill but still report to work. Furthermore, it endangers the health of those who come in contact with them on the job. Wherever possible, it is desirable to make sick leave cumulative for a number of years. Thus an employee can accumulate many weeks or months of such leave to be used as needed. Experience has shown that this is a sound practice; however, close supervision must be exercised in preventing abuse of the privilege.

### Health and Welfare Provisions

A great variation exists among park and recreation systems as to the types of health and welfare services available to members of their staff. The trend today is for agencies to provide more and more of these services. It is generally mandated by the federal and state governments that worker's compensation insurance be carried on all employees and that a retirement program be provided the employees. In most instances employees are provided with health, accident, and hospital benefits and in some cases the insurance is provided for the employee's family. Generally, however, the family insurance coverage is optional to be paid for by the employee. Life insurance coverage is also provided by many agencies. Mention should also be made of cafeterias, coffee breaks, rest periods, staff lounges, credit unions, cooperative purchasing plans, legal aid, and recreation services that are available to recreation staff members through employee organizations or as a part of public services.

### Professional Development

Most professional staff members seek opportunities for continued education that will increase their effectiveness and open new channels for promotion. The recreation and park system that overlooks continued education programs is neglecting one of the best means available for improving the efficiency of its staff. Indeed, no philanthropic motive is implied here. It just makes practical sense to educate for improvement.

Provision of opportunities for professional growth while on the job may take the form of directed reading, training institutes, registration in formal courses of study, training sessions in special activities, individual instruction on the job, staff meetings, or attendance at district, state, or national conferences. Many universities in cooperation with the National Recreation and Park Association and/or other organizations are conducting excellent executive development programs on management workshops. Regardless of the method used, every recreation and park system should establish a program that fully encourages the development of every employee. Today some state park and recreation societies require evidence of participation in a continuing education program by a member if he or she wishes to continue a professional membership in the society or the state registration plan.

Very few recreation and park systems provide opportunities for, or encourage, their professional staff members to take leaves of absence to further their education through study or travel. This practice is usually associated with the education field, but if recreation leadership and management have the educational overtones claimed, a case can very well be made for the need of sabbatical leaves.[3] With educational institutions providing new horizons for advanced study and research and with communities experimenting with exciting adventures in program planning and facility development, the desirability of utilizing such leaves for the professional improvement of staff and thus enriching the recreation life of the community is obviously evident. In any number of instances, private industry is financing training periods at graduate and research centers for its promising staff members. Moreover, participation in either short- or long-term training institutes, conferences, graduate programs, research sessions, or even travel through leaves of absences or sabbaticals will pay such dividends to the recreation and park service that every encouragement should be given to its utilization.

## Evaluation of Personnel

The systematic evaluation of an employee's performance is one means used by recreation and park executives to determine the relative effectiveness and efficiency of the employee. In effect, such evaluation provides the guide for additional training, promotion, demotion, salary raising, or transfer, and further, it acts as a stimulant for the employee to seek the rewards provided for meritorious service. Performance evaluations commonly called "service ratings" or "performance ratings" are basically judgments of the work of an employee. By their very nature, such ratings have come under repeated attack. Most employees and supervisors dislike service ratings.[4] They believe them to be rather crude instruments for measuring intangible qualities and work habits of professional staff. Furthermore, many believe such ratings are too subjective, imprecise, and ambiguous, and that they reflect the bias and prejudices of the rater. Broadly speaking, service ratings have not been too satisfactory, and there is much to be said for these criticisms. The danger of error is inherent in judgement ratings of a person's work. In fact, some recreation and park supervisors and administrators have given up in despair trying to evaluate performance objectively on the basis of checking points between "poor" and "out-

[3]A sabbatical leave is usually granted to a faculty member of an institution of higher learning after a specified number of years of service. It requires the member to use the period for his or her own professional development. The institution usually pays part or all of the salary while he or she is on leave.

[4]John M. Pfiffner, *The Supervision of Personnel* (Englewood Cliffs, N.J.: Prentice-Hall, 1958), p. 309.

standing'' for the qualities possessed by the staff member being rated. Items commonly listed for rating include such factors as quality of work, personality, dependability, disposition, leadership capacity, resourcefulness, personal appearance, tact, relationships with others, intiative, performance skills, enthusiasm, attitude, organizing ability, open-mindedness, and ability to take criticism. The actual form of the rating instrument is usually as follows.

| *Quality* | *Excellent* | *Very Good* | *Average* | *Below Average* | *Poor* | *No Basis for Judgment* |
|---|---|---|---|---|---|---|
| | *1* | *2* | *3* | *4* | *5* | *6* |
| 1. Employer plans and organizes work | | | | | | |
| 2. Works well with individuals and groups | | | | | | |
| 3. Is resourceful in planning programs | | | | | | |

Other rating forms provide descriptive phrases denoting the various degrees of the quality, performance, skill, or trait being rated. This gives the rater the opportunity of choosing the phrase that most nearly represents the item under review. For example, the several degrees of the quality or trait being rated would be listed in place of very good, average, poor, and the like as shown in the preceding table. Included instead might be the following gradations of a quality such as resourcefulness.

| | *1* | *2* | *3* | *4* | *5* |
|---|---|---|---|---|---|
| Resourcefulness | Extremely resourceful and imaginative | Moderately resourceful | Resourceful when directed | Requires guidance and imagination | Lacks resourcefulness and imagination |

Too often the qualities being rated are not the items needed most by the incumbent of the job being rated. In other words, the abilities needed by a recreation and park executive would be different from those required of a recreation leader or a park gardener. Therefore it is advisable for recreation and park systems to establish rating scales that carefully analyze those traits and skills that are most important for quality performance in each of the classes of work. Executive positions, for example, would require for rating such factors as personality, human relations, organizing ability, resourcefulness, ability to plan, work with others, leadership, and the like, while those in the lower echelons of leadership or maintenance positions might be rated on such items as amount and quality of work performed, cooperativeness, desire to learn, attitude, disposition, initiative, personal appearance, punctuality, ability to accept criticism, and the like. Examples of rating scales developed by various recreation and park agencies for evaluation of their staff are in Appendix D.

It is true that ratings have their imperfections, but some means must be developed to recognize excellence of service as well as poor performance that needs correction. While the present image of evaluation through service ratings is negative, some personnel authorities have suggested that this problem be approached through new evaluation procedures that will help an employee with his or her weaknesses, encourage cooperative working relationships between the supervisor and the employees, and reward merit and excellence.[5] This new philosophy abandons the competitive and punitive type of rating in favor of an employee-development pattern that emphasizes the full development of the individual through cooperative superior–subordinate relationships.[6]

Along these same lines some organizations use a management by objectives or modified MBO process as a part of the evaluation process. After all, the long-range aim of evaluation is to bring about improvement in the recreation service. What better way is there to accomplish this than through a positive team approach that motivates the employee to realize his or her full capacities?

## JOB SEPARATION

All employees of a recreation and park system will at some time or another take permanent leave of their positions through resignation, dismissal, retirement, or death. Each of these methods of job separation should be covered in an organization personnel policy, for, indeed, staff members feel deeply about their positions and the separation policies that govern them.

### Resignations

Every department will receive resignations from employees who leave for personal reasons or take positions of higher responsibility. But a clear-cut policy should exist as to the type of notice required from those who wish to terminate their service. Most recreation and park systems require a written resignation with a minimum of two weeks notice. However, this will vary with the kind of

[5]John M. Pfiffner and Robert V. Presthus, *Public Administration*, (5th ed., New York: Ronald Press, 1967), pp. 291–94.

[6]Ibid.

position being vacated and most professional positions will require 30 days or more notice.

### Dismissals

Civil service rules and regulations, the Civil Rights Act, and local personnel policies generally cover all matters relating to dismissal of employees. No employee should be separated from his or her position without just cause; security in tenure is a basic premise of the merit system. On the other hand, dismissal is sometimes necessary for those who fail to measure up to the needs of the position. In most instances an employee has the right to appeal the dismissal before an impartial reviewing body. This acts as a safeguard against capricious and arbitrary dismissal for reasons that are not sound. Therefore it is wise for administrators to have concrete evidence to substantiate their reasons for dismissals. After all, employees have a proprietary feeling toward their jobs and will almost without exception want the dismissal action reviewed. Some situations have arisen where an administrator has been overruled by a reviewing body because he or she could not substantiate the claims against an employee of inefficient work or violation of department policy. Hence it is very important that adequate performance records be kept for each employee, as well as proved statements or written reports of disloyalty, dishonesty, warnings, offensive behavior, violations, or insubordination. Likewise, records and statements of commendable service and excellent performance should be kept.

### Retirement

Most recreation and park authorities provide a retirement plan, usually of a joint contributory type in which both the individual and the employer contribute a specified amount, which is usually expressed in terms of percentage of salary. For the most part, retirement plans are provided and payments made from actuarial reserve funds. In other words, the employee and employer contributions build up in a fund over the years to pay the retirement benefits when they are due. Very few recreation and park bodies provide pension plans to which the employees make no contribution. Strictly speaking, a pension plan requires no contribution on the part of the employee, as opposed to a retirement plan that is supported by joint contributions from employee and employer.

A liberal retirement plan attracts capable young men and women into a recreation and park system and keeps them there. Furthermore, it relieves an employee of anxiety as retirement approaches, and thus the employee can give more to the job. On the other hand, a department without a retirement plan will find poor work morale and general inefficiency. Retirement benefits should not be looked on as charity or gratuities. Rather, it is a planned insurance scheme that gives an employee an opportunity to start saving early in life for later years of leisure.

Although most retirement plans will vary in their benefit provisions, it is desirable to specify the qualifying requirements for eligibility. Generally speaking, these requirements are based on a specified number of years of service or else on reaching a certain age, usually 65 or 70, as now specified by federal law. In some instances, an employee may retire at an earlier date—age 62, for example—or may extend his or her employment above the specified age through special request to, and approval by, the recreation and park body.

Some public authorities provide for retirement after a number of years of continuous service. However, it is not in the best interest of a recreation and park department to use length of service as the basis for retirement. Those who enter a department early in life are rewarded by early separation, and it is not unusual to see a person retire at age 45 or earlier. On the other hand, those who enter a department in middle life may be required to remain many years beyond their period of usefulness.

Although some hardships are occasionally inflicted on employees by holding to a rigid age limit for compulsory retirement, it is good practice to do so if the retirement policy includes both optional and compulsory features. For instance, it is not unjust to permit optional retirement at age 60 and to require compulsory retirement at 70. Instead, it is fair and in the best interest of the department and its staff.

### Death Benefits

There is wide variation in personnel practice of recreation and park systems relative to death benefits. This aspect of job separation is such that little can be said about it except insofar as it relates to benefit provisions. Yet it is of concern to recreation authorities, since all types of employee benefits, including those relating to death, improve staff morale and make entry into the department more attractive to prospective employees. State retirement insurance programs that include death benefits are provided in many instances and if not, group insurance plans to take care of this contingency have proved popular. In many cases the recreation agency pays for this insurance but even if the employee pays through withholdings from salary, this fringe benefit adds greatly to the attractiveness of employment.

## POSITION CLASSIFICATION AND SALARY STANDARDIZATION

A discussion of personnel policies and practices would be incomplete without considering position classification and wage standardization. Actually, a classification-and-pay plan is the foundation of a satisfactory personnel program.

Without one, a recreation and park department gropes in the dark in trying to provide equitable treatment and pay schedules for its employees.

## Position Classification

A position classification plan attempts to group similar positions into classes on the basis of common elements or characteristics. Stated in another way, it is merely a means of systematically arranging the various positions that are alike into units or classes on the basis of the similarity of job responsibilities, likeness of duties or functions, and levels of difficulty. By establishment of such groupings, the similarities and dissimilarities among them can be clearly seen and the foundation laid for the recreation and park body to carry out its personnel function. Without such a classification, it would be almost impossible to equate pay schedules or handle such matters as recruitment and selection on a sound basis.

Reference is sometimes made to a "series of classes," a "group of classes," or the term *service*. Actually, a "series of classes" is a grouping of two or more classes that are similar but differ principally in degree of responsibility or rank. For example, the classes of senior recreation center director, recreation center director, and junior recreation center director would constitute such a series. On the other hand, a "group of classes" consists of two or more classes that are related as to responsibility, educational qualifications, and difficulty, but that differ with respect to area of specialization. For illustration, the classes of recreation supervisor (general) and recreation supervisor (special activity) would represent such a group, or the classes of supervisor of landscape, supervisor of construction, and supervisor of maintenance might also constitute a group of closely related classes in this category.

Finally, the term *service* as used by personnel authorities refers to those groupings wherein are lumped all classes having the same or similar general job characteristics and entrance qualifications. Thus a public jurisdiction might have a service such as custodial, subprofessional, clerical, technical, administratve, or the like. The number of services will vary with different public authorities. Some may have four or five; others may have ten or more. The city of Houston used for many years the following seven basic services: (1) legislative, (2) executive, (3) controller, (4) administrative and supervisory, (5) technical and professional, (6) clerical and fiscal, and (7) labor. Each of these services, in turn, was broken down into its respective classes.

## Salary Standardization

Position classification grew out of the need to equate the pay of public employees doing the same type of work. It was based on the premise that all employees doing the same or similar tasks should be given equal pay—or, expressed in another way, "equal pay for equal work." Actually, it was not unusual to find a clerk receiving a salary of $7,500 while in the next office a clerk doing a similar job would be receiving $8,500. Thus the need for a basic classification plan that grouped similar positions into classes was very much needed if any kind of salary standardization was to be achieved.

It is obvious that positions of higher responsibility in the structure of the recreation and park department will command more pay than the lower ones. But positions must be compared and a consistency developed in pay schedules for the various levels of job responsibilities. Consequently, the compensation plan must work closely with the classification plan. Only in this way can standardization and uniformity of pay rates be achieved for these levels. Generally, most public recreation and park systems have classified their positions, developed uniform job titles, and established uniform pay rates. Much of the pioneering effort was due to the efforts of the National Recreation Association working through public personnel systems and more recently The National Recreation and Park Association in the publication on personnel guidelines.[7] Very few instances exist, if any, of public jurisdictions not having plans for position classification and compensation.

## Determining Pay Schedules

Each class of position is given a salary schedule. The amount of pay for each is usually determined by the personnel agency of the public jurisdiction wherein the recreation and park department is located. In general, the steps for establishing a pay schedule after positions are classified are as follows. (1) A salary study is conducted by the personnel agency, budget office, appointive committee or outside research agency. (2) Recommendations are made to the chief executive and civil service commission for a proposed salary schedule for each class of employees. (3) The proposals with adjustments are passed on to the elected governing body for approval or disapproval. (4) The new pay schedule goes into effect, if approved.

In study and determination of the amount of money to be paid to each class of positions in a recreation and park system, attention is given to the prevailing wage rates for the same or similar positions in the community or in other public recreation and park departments. Also, comparisons are sometimes made with the salaries paid to public school employees. In the final analysis, salaries are competitive and pay schedules reflect the wage scales of the region.

[7]National Recreation and Park Association, *National Personnel Guidelines for Park, Recreation and Leisure Service Positions* (Arlington, Va.: National Recreation and Park Association, 1978).

The salary ranges of many recreation and park positions are too low. Not enough attention has been given to interpreting the professional competences needed to fill the positions in question. Actually, if recreation and park authorities cannot interpret the importance of the responsibilities and the high qualifications needed to fill a leadership position to an impartial salary study or reviewing body, perhaps the importance of the job should be questioned. Experience has shown, however, that high quality work performance by professional staff tends to upgrade salaries.

Mention should be made of the prevailing practice of having a step plan for salary increases. Such a plan could have a number of levels or steps—five, for example.[8] This means that an employer would begin his or her job tenure at the first step and progress at regular intervals to the fifth, or last, step, where maximum salary is paid.

The time interval between steps will vary with the many recreation and park systems. In some instances the second step is reached after the period of probation; this could be from three to six months. Then the time between steps may be one year or longer. Furthermore, some jurisdictions base the step increase on meritorious service; others make the increase automatic. Still others will induct new employees at any step level depending on the qualifications of the individual. This tends to defeat the purpose of the step plan, since bringing an employee into the department at the last step means he or she is starting at a maximum salary with little chance for salary increase unless the basic wage plan is amended or revised.

The purposes of the step plan of salary increases are to (1) provide an incentive to employees to remain with the organization, (2) reward employees for meritorious service, (3) be competitive with private concerns that pay on a step basis, and (4) let prospective employees know pay levels and how they can be reached.

## COLLECTIVE BARGAINING

Today over half of the states have legislation supporting collective bargaining on the part of public employees and the trend is for more states to pass collective bargaining legislation. Most park and recreation managers are unfamiliar with collective bargaining, therefore they are afraid of it and tend to act negatively. The authors do not wish to propose or oppose collective bargaining but do feel that it is important to present some basic information.

The passage of a state collective bargaining law does not necessarily require that a public employees' union be formed. It does mean that employees

[8]A step plan could have any number of levels, and the number will vary with recreation and park systems. However, the most prevalent practice is to have five steps.

have the right to form a union if they desire. However, in many cases where the agency has established good personnel practices, the employees may choose not to form a bargaining unit. Collective bargaining is essentially a process of shared decision making by two parties. However, this process is restricted to a limited area—wages, working hours, and working conditions. Management still retains various rights such as the right to hire and fire, the right to discipline, the right to determine the number and location of facilities and the type and number of operations and services to be provided.

It is important that every park and recreation manager become acquainted with the laws and procedures of collective bargaining. But at the same time the manager must put forth special effort to see that the existing personnel system is equitable and adequate including a grievance or complaint system that allows employees some say in their working conditions and benefits.

## SUMMARY

A basic objective of a recreation and park organization's personnel policy is to recruit, develop, and keep an outstanding staff. To do this, constant study and evaluation of all personnel practices and policies are needed. Furthermore, the findings of research in human motivation, leadership, supervision, psychology, and the like should be constantly funneled into the organization, if this means greater efficiency and a higher degree of worker satisfaction.

Many recreation and park systems are constantly on the alert for new personnel practices. To open new channels of study and research and to provide opportunities to observe changing leisure-time habits of people may provide new opportunities for employee development.

In addition, the developing requirements of certification and registration for professional employees need further exploration and study.[9] Although such licensing is in effect in a number of states on a voluntary basis, a great expansion in this direction is foreseen. At the present time, certification rests with professional recreation and park societies or associations. Mandatory or legal certification is still not in effect. But progress is being made, and the decade ahead may see this requirement mature.

Rating scales and evaluative techniques need additional study and research. The theoretical value of such ratings breaks down in practice. Likewise, all testing techniques and processes need careful scrutiny. So do health and welfare provisions. But, most of all, flexibility and open-mindedness to the needs of an expanding recreation

[9] Certification, sometimes known as "registration," is a license to practice a profession. Medicine, architecture, law, and many other fields of study require that practitioners be certified or registered.

service through liberal personnel practices and policies open the door to vast developments in the future.

## Questions for Discussion

1. Explain why the recruitment process is so important to a recreation and park system. As an executive, how would you proceed to fill vacancies in your department?
2. What is the difference between an assembled and an unassembled examination? For which positions would you use each? Why?
3. Should a recreation leader in a community center have as much education and professional preparation as a supervisor? Why or why not?
4. What are the pros and cons of rating scales? If they are not as effective as they should be, how can they be improved?
6 Explain fully the basic elements of a good personnel program for a recreation and park system.
7. How may a recreation and park administrator encourage his or her staff to engage in programs of study that lead to professional development?
8. On what basis would you limit promotion to only those from within your department? Outside your department?
9. How could the probationary period for a new employee be made more effective? Should it be lengthened or shortened? Why or why not?
10. What information should be included in the job description?
11. Would you pay recreation workers for overtime work? Why or why not?
12. What are the advantages of having a good retirement program?
13. How does an oral interview differ from an oral examination? Explain the function of each.

## SELECTED REFERENCES

Brademas, James D., *Parks and Recreation*. 24 February (1979).

Butler, George D., *Introduction to Community Recreation*. 5th ed., McGraw-Hill, New York, 1968, Chaps. 7, 9.

Crouch, Winston W., *Local Government Personnel Administration*. International City Manager Association, Washington, D.C., 1976.

Edginton, Christopher R., and Williams, John G., *Productive Management of Leisure Service Organizations, A Behavioral Approach*. Wiley, New York, 1978, Chap. 11.

Hjelte, George, and Shivers, Jay S., *Public Administration of Recreational Services*. 2nd ed., part IV, Lea and Febiger, Philadelphia, 1978.

Illinois Department of Local Government Affairs, *Collective Bargaining/Labor Relations. A Guide Book for Local Government*. Illinois Department of Local Government Affairs, Springfield, no date.

Kraus, Richard G., and Curtis, Joseph E., *Creative Administration in Recreation and Parks*. Mosby, St. Louis, Mo., 1973, Chap. 5.

Lutzin, Sidney G., and Storey, Edward H., eds., *Managing Municipal Leisure Services*. International City Manager Association, Washington, D.C., 1973, Chap. 14.

National Recreation and Park Association, *National Personnel Guidelines for Park, Recreation and Leisure Service Positions*. National Recreation and Park Association, Arlington, Va., 1978.

Stahl, Glenn O., *Public Personnel Administration*. Harper & Row, New York, 1971.

# CHAPTER 8

# The Recreation and Park Staff

Every recreation and park system provides services to people. Indeed, services are its product. But these services cannot be provided without a leadership staff that is specialized and knows its function in the framework of the organization. Hence it is imperative to select qualified staff that can most effectively implement the purposes of the recreational undertaking. This chapter will set forth the desirable personal, professional, and educational qualifications of the personnel engaged in leadership endeavors in recreation and park systems.

## PERSONAL QUALIFICATIONS

While professional and educational qualifications will vary with the many positions of specialization in a recreation and park organization, certain personal qualities are essential for all positions of leadership. These were discussed at length in Chapter 4 on "Leadership," under the trait theory. While it was brought out that traits alone do not make for leadership, they are important in the total picture of influencing and motivating people. Honesty, integrity, sincerity, sense of humor, sense of service, and faith in people are essential qualities for those who have leadership roles, and there are many more. Such personal qualities are the foundation of successful job performance.

## TYPES OF POSITIONS

The need to provide people with the many varied recreation services found in large and small recreation and park systems results in specialization of personnel. This is natural, for a varied program of activities or services can be carried on in no other way. The number and kinds of recreation and park positions found in local public agencies will vary from one jurisdiction to another, depending on the size of the public authority and the areas and facilities it operates. Other variables also condition the quantity of staff as well as the scope of operation. In many recreation and park departments of small cities the administrator is the only full-time employee and is required to be a combined executive, teacher, activity leader, and maintenance worker. However, as the functions and responsibilities of a department increase, two clear-cut areas of work occur—that of administration and of program operation. In other words, the administrator must perform a definite executive and supervisory function that takes much time and specialists or leaders must be obtained at the operating level to care for program needs and park maintenance. Such work specialization begins in the medium-size community.

As can be seen, the larger the community and the department, the kinds of positions and the skills needed to fulfill them will become greater. Also, the larger the recreation and park agency, the more specialized the functions. For example, in the larger non metropolitan communities the recreation and park agency begins to separate administration from supervision. The addition of general and functional supervisory positions is seen here.[1] Of course, additional operating personnel are also needed. As a recreation and park system gets still larger, as in the large metropolitan communities, additional positions of a highly skilled and technical nature are added, as well as those of district supervisors, general and functional, who are each responsible for a geographical area of the metropolitan community.

Various methods are used to classify the recreation and park positions. In some cases they are classified according to size of the agency and in others according to function or location. A generally accepted classification follows.

> *Executive* administrative staff are considered to be responsible for overall direction and operation of the agency within policy guidelines and agency resources.

[1]So that the student does not confuse a general supervisor with a functional one, a distinction needs to be made between the two. A general supervisor is one who is responsible for any or all phases of departmental operation, while a functional supervisor is responsible for a specialized field or area such as athletics, drama, music, camping, and the like. (In the park area, it could be street trees, arboretum, nursery, planning, etc.)

*Supervisor-facility directors* refer to those staff who provide front line operational direction of administrative policies, guidelines, resources, and supervision of programs and facilities within an agency.

*Operating staff* are considered to be those on the production line of the agency including program leaders and park operations personnel.

*Trainees and volunteers* are those staff who are serving the agency in a learning situation or a nonpaid basis. While these positions are not normally listed in a table of organization, they can be very valuable members of the organization's staff.

## POSITION NOMENCLATURE

In the discussion of many types of positions in the area of recreation and parks a standardization of job titles is increasingly apparent. This is due largely to the initial efforts of the National Recreation Association, the American Institute of Park Executives, and those other national, district, and state organizations that have issued reports on job classifications and standardization. More recently, the National Recreation and Park Association has completed extensive work in this area and published a manual of job descriptions and titles.[2]

Job titles are very important in any organization, recreation and parks being no exception. In a number of instances the term *manager* is being used to designate the top position in an organization. However, the term *director* is the most widely used title to designate the chief executive of a department of recreation and/or parks. On the other hand, the term *superintendent* is used most frequently to designate the head of a major division of a department, such as a division of recreation or a division of parks, although the term is still used widely by some bodies to characterize the head of a separate department. Actually, the term *superintendent* became widely used at the time that recreation and/or park systems operated under separate boards or commissions. But as this function is increasingly being added to the family of local public jurisdictions as a department of government, the term *director* is used to designate the top executive position of a department. Although a department head is commonly called a director, a division head is frequently called a superintendent and the head of a bureau is known as a chief. All positions can be classified under administrative, supervisory, operating, trainee, and volunteer staff.

[2]National Recreation and Park Association, *National Personnel Guidelines for Park, Recreation and Leisure Service Positions* (Arlington, Va.: National Recreation and Park Association, 1978).

## Administration

1. *Director of Recreation and Parks.* Chief administrative officer in an agency or department that has both recreation and park functions, which include the development of comprehensive recreation areas and facilities, playgrounds, and specialized areas and structures. The title "General Manager" is used by a number of recreation and park districts that conduct recreation and park services as a separate local function under an elected or appointed board of directors, in order to avoid confusion in reference to both a board member and the chief administrative officer as director.
2. *Administrative Assistant.* Assistant to the executive head of the department or division; responsible for certain management functions as assigned by the executive head including affirmative action, budgeting, grant writing, purchasing, personnel management, research and related staff services. Normally this is not a line staff position and there is no direct supervision of other personnel.
3. *Director (Superintendent) of Recreation.* Chief administrative officer responsible for recreation program functions, which include the planning, organizing, developing, and managing of a comprehensive recreation service. The title "director of recreation" is used most frequently when the executive heads a separate recreation department. The title "superintendent of recreation" is usually used when the executive heads a division within a department of a jurisdiction.
4. *Director (Superintendent) of Parks.* Administrator responsible for planning, developing, constructing, and maintaining recreation and park areas, facilities, and structures, as well as specialized areas and facilities. The title "director of parks" is used when the administrator is responsible for the park functions in a separate department of a jurisdiction. The title "superintendent of parks" refers to one who heads a division within a department.
5. *Assistant Director.* "Assistant Director"—sometimes referred to as "deputy director" in larger departments—might be found under any of the administrative positions. The assistant is responsible for delegated responsibilities in administration, planning, organization, and supervision as delegated by the director of parks and recreation, director of recreation or director of parks. The assistant also acts for the administrator in his or her absence.
6. *Business Manager (Director of Financial Management).* "Business manager" is a position that emerged from the growing complexity of fiscal management, in organizations that especially operate as a

separate governmental agency. Responsibilities include budgeting, purchasing, accounting, investment of funds, control of insurance programs, capital item inventory, and related staff services.

## Supervision—Facility Directors

1. *District Supervisor.* Under direction, responsible for the professional supervision of the work of others in all phases of the program for a district or a large geographical section of a community. Usually there are two supervisors in each district, the recreation supervisor for the recreation programs and the park supervisor or foreman responsible for the facilities and park areas.
2. *Functional Supervisor.* Under direction, responsible for the professional supervision of others in the planning and organizing of an assigned function or specialized field of the community wide program. Recreation supervisors would be responsible for specialized programs such as music, drama, arts and crafts, aquatics, sports and athletics, camping, and so on. Park supervisors would be responsible for specialized programs such as landscape design, street trees, arboretum and nursery, building maintenance, equipment repair, and so on.
3. *Facility Directors.* Under direction, responsible for the administration of a special facility and for the professional supervision of others in the planning, organization, and operation of that facility. Facility directors usually designated as director or manager would be responsible for golf courses, sports centers including tennis, racquetball and ice rink facilities, nature centers, zoos, museums, arts centers, arboretums, camp facilities, and so on.

## Operating Staff

1. *Senior Recreation Leader.* Under supervision, responsible for professional recreation work in directing a wide variety of activities at a large recreation center or major playground, plus the supervision of recreation leaders and other subordinates at a specific area or facility.
2. *Recreation Leader.* Under supervision, responsible for professional work in planning, directing, and conducting a wide variety of activities at an assigned small playground or center, or assistance at a larger playground or recreation center.
3. *Assistant Recreation Leader.* Under close supervision, responsible for the leadership of simple recreation activities with primary em-

phasis on organizing and leading groups of children and, on occasion, adults, in a variety of recreation activities. Serves as an assistant to a receation leader.

4. *Recreation Attendant.* Under the direction of a recreation leader, responsible for routine work at a recreation center, playground, camp, or swimming pool.
5. *Special Activity Leader.* Under supervision, responsible for the conduct of a recreation program around a special interest such as arts and crafts, dance, drama, music, athletics, or camping.
6. *Senior Park or Operations Staff (equipment operator, forester, gardener, grounds keeper, greens keeper, horticulturist, landscaper, mechanic, and so on).* Under supervision, responsible for a variety of skilled grounds and facility maintenance and construction work. Supervision, on assignment, of the work of park or operations crews.
7. *Park or Operations Staff.* (A second and third level of park staff generally with the same title designation as the senior staff.) Under supervision, responsible for performance of routine grounds and facility maintenance work.
8. *Custodian.* Under supervision, responsible for the care, cleaning and maintenance of facilities and buildings.

## Trainee and Volunteer Personnel

1. *Park and Recreation Intern or Field Student.* Under supervision, responsible for wide variety of administrative, supervisory, and operating level tasks in a rotated program of work, directed reading, conferences, and examinations.
2. *Park and Recreation Staff from Special Training Programs.* (Staff who are employed from a government or private training program such as Comprehensive Employment Training Act, "CETA"; and Operations Industrialization Center, "OIC.") Under supervision and responsible for a wide variety of assignments that promote job skills. Many of these programs are subsidized by the sponsoring agency and the local park and recreation department receives funds for the salaries of these staff persons.
3. *Volunteers.* Under supervision, responsible for assisting park and recreation staff in a variety of programs and operations. Local volunteers provide a tremendous assist in the operation of local programs, paticularly in the areas of sports and athletics, and programs for the handicapped.

More detailed job descriptions outlining the basic elements of administrative, supervisory, and leadership positions are contained in Appendix E.

### Questions for Discussion

1. Name the standard job titles for positions at the executive, supervisory, and operating levels of a recreation department. How do they differ from those of a park department?
2. What is the difference between a district supervisor and a functional supervisor?
3. Give examples of duties for the position of director of recreation. How do they differ from the duties of the position of superintendent of parks?
4. Give examples of duties for the position of business manager. Why are organizations utilizing the position of business manager?
5. How does a superintendent's position differ from that of a supervisor? Does a superintendent have supervisory responsibility? Why or why not?
6. What knowledge, skills, and abilities are needed to be the director of parks and recreation in a large city?
7. Give the reasons that experience is necessary for an executive position in a large city. Is it needed for a small city?
8. Which do you think is more important, experience or education? How is personality related to abilities?
9. Explain fully the competencies that are needed by a (a) park foreman, (b) supervisor of horticulture, (c) recreation leader, and (d) district recreation supervisor.

## SELECTED REFERENCES

Butler, George D., *Introduction to Community Recreation.* 5th ed., McGraw-Hill, New York, 1976, Chap. 7.

Hjelte, George, and Shivers, Jay S., *Public Administration of Recreational Services.* 2nd ed., part IV, Lea and Febiger, Philadelphia, 1978.

Lutzin, Sidney G., and Storey, Edward H., *Managing Municipal Leisure Services.* International City Manager Association, Washington, D.C., 1973, Chap. 14.

Meyer, Harold D., and Brightbill, Charles K., *Recreation Administration.* Prentice-Hall, Englewood Cliffs, N.J., 1956, Chap. 8.

# CHAPTER 9

# The Schools and Public Recreation

The role that schools have played in providing play and other recreation opportunities for people through the use of the educational plant and facilities has been of major importance in the recreation movement. The closeness of schools to the citizenry, the type of facilities, and the recognition of the importance of public recreation have been factors inducing acceptance of the principle that schools should provide the guidelines for leisure-time living.

The schools have a larger role than merely providing plant and facilities for the community recreation resource. Recreation is an essential part of education. As early as 1917 the National Education Association published its well-known *Cardinal Principles of Secondary Education*, in which "preparation for the worthy use of leisure time" was listed as one of its seven objectives.[1]

The prime goal of education is the full development of the individual—intellectually, physically, socially, and spiritually. Its purpose is the improvement of people; it attempts to develop people to their fullest, preparing them for citizenship and effective living. Hence education must extend beyond the classroom into the realm of leisure living; it must be vitally concerned with raising the cultural level of people through the medium of wholesome, creative, and socially sound leisure-time activities.

[1]National Education Association of the United States, *Cardinal Principles of Secondary Education* (Washington, D.C.: Bureau of Education, Department of Interior, 1917).

The place of the school in recreation is clear. After a period of expanding concepts of the relationship of education to recreation the public schools have identified their role in recreation as including (1) education in the worthy use of leisure, (2) education in the basic arts, skills, and appreciations that provide a foundation for self-fulfilment, (3) provision for joint facility planning between school and community, (4) mobilization of resources for an expanded recreation program, and (5) establishment of a community education program.

## EXPANDING CONCEPTS

The use of schools for public recreation purposes had its beginnings at the close of the nineteenth century. In the 1890s the schools were concerned more with the "three Rs" than with recreation or play, and the leaders in education during this period who urged play as a necessity for children had more effect on social reform than on general school practice.[2] Regardless of the general negative attitude on the part of most educators toward the use of schools for community recreation services, some progress was made. By 1900 several states had passed general legislation permitting school buildings to be used as civic or social centers, but the provisions were often indefinite and the use of school buildings for recreation at this time was meager and sporadic.[3]

As the play movement gained momentum the attitude of educators toward the use of the educational plant for purposes other than classroom work began to change. Several schools in New York City were opened in 1898 under leadership for evening recreation programs, and the number was increased until, in 1907, 26 schools were being used for recreation programs in that city.[4]

The movement for the wider use of schools for recreation purposes received its greatest impetus in 1907, when Rochester, New York, appropriated money to establish a school center demonstration. This demonstration was an attempt to establish a civic center in the schoolhouse for the purpose of encouraging better citizenship and training for democracy.[5] This experiment stimulated other cities to make wider use of their school facilities and plants for community recreation, and some states passed legislation that, in effect, declared that the school was a civic center. Wisconsin was one of the early leaders in this

[2]Elizabeth Halsey, *Development of Public Recreation in Metropolitan Chicago* (Chicago: Chicago Recreation Commission, 1940), p. 38.

[3]National Recreation Association, *Recreation and Park Year Book* (New York : The Association, 1950), p. 5.

[4]George D. Butler, *Introduction to Community Recreation* (New York: McGraw-Hill, 1959).

[5]George Hjelte, *The Administration of Public Recreation* (New York: Macmillan, 1940), p. 78.

movement when, in 1911, it passed legislation authorizing education authorities to levy a tax of two-tenths mill for community recreation. As a result of this legislation, Milwaukee established its school recreation center program, which has since become nationally known.[6]

The movement to make schools available for leisure-time use and the changing attitude of school authorities concerning their responsibility for play and recreation were due to a number of factors. One of the first was the evident economy of using the school buildings and facilities full time rather than duplicating this use by the construction of a separate recreation center.[7] Second, the schoolhouse was brought into use for community service during World War I, and this practice established a pattern for general use of the school plant. Third, the National Education Association, in 1918, issued a report on the "cardinal principles" of secondary education, in which it listed health and worthy use of leisure time as two of the "seven cardinal principles" of education. A statement taken from these principles emphasizes the need for schools to guide the leisure pursuits of youth.

> The school has failed to organize and direct the social activities of young people as it should. One of the surest ways in which to prepare pupils worthily to utilize leisure in adult life is by guiding and directing their use of leisure in youth. The school should, therefore, see that adequate recreation is provided both within the school and by other proper agencies in the community. The school, however, has a unique opportunity in this field because it includes in its membership representatives from all classes of society and consequently is able through social relationships to establish bonds of friendship and common understanding that cannot be furnished by other agencies. Moreover, the school can so organize recreational activities that they will contribute simultaneously to other ends of education, as in the case of the school pageant or festival.[8]

These principles influenced school authorities to include activities of recreational significance in the curriculum. Fourth, the evolution of broader thinking in education, in which emphasis was placed on the development of the child, further stressed the need for the school to develop extracurricular activities. Fifth, the trend toward urban living and changing family patterns placed more responsibility on the school for supervision of the child, as well as increased responsibility for recreation supervision. Sixth, school authorities stressed the development of the child's total personality as one of their major goals, and this emphasis has given acceptance to education for leisure and to

[6]Butler, op. cit., p. 65.

[7]Halsey, op. cit., p. 39.

[8]National Education Association of the United States, *Cardinal Principles of Secondary Education*, Bulletin No. 35 (Washington, D.C.: U.S. Government Printing Office, 1918), p. 15.

the belief that individual maturity can only be achieved by giving attention to a person's leisure and recreation needs.

These influences have been a factor in bringing about a better understanding on the part of school authorities of their responsibility for making wider use of school facilities in the administration of community recreation programs.

## EDUCATION IN THE WORTHY USE OF LEISURE

Education has the task of developing individuals to their fullest. To implement this task, education must extend beyond the formal classroom into the realm of recreation and leisure. Like the schools, recreation has as its purpose the development of wholesome habits, skills, attitudes, and appreciations. Actually, there is no conflict between education and recreation as to purpose. While methods differ, each complements and supplements the other.

The schools may justifiably feel that their primary function is not recreation. However, their education function must prepare young men and women to live effectively in an expanding world of leisure. Thus education in the worthy use of leisure is a prime educational objective. There is no other way to acquire skill in the product of formal education except through intelligent planning of the values for which our schools strive. Indeed, the ends of leisure and the pursuit of these ends are of the utmost importance to the educative process.

The educational approach by schools toward worthy use of leisure should relate instruction to experience. In other words, the student has the leisure-time laboratory of recreation in which to learn the meaning of democracy by practicing it, to develop leadership by progressively accepting responsibility, to learn social grace and wholesome relationships by participating in group life, to grow in insights and skills through teamwork, and to fashion attitudes, values, and behavior through peer relationships and club activity. Therefore, with many aspects of our cultural environment, the student learns through active experiencing. In fact, students are more aware of their environment and the satisfaction in doing than of their responsibility to learning. Thus directed and motivated learning through participation in human activity and group life is remarkably effective.

## EDUCATION IN BASIC ARTS, SKILLS, AND APPRECIATIONS

The learning of basic skills, attitudes, and appreciations is the foundation on which worthy use of leisure rests. Understanding is basic to appreciation; appreciation is the product of education. Therefore the schools must assume the major task of giving instruction, understanding, and experience in those skills that will broaden interests and appreciation for individuals in their present environment, future work life, and leisure. It is a truism that people engage in those activities and avocational life pursuits in which they feel secure and have a degree of skill. What better opportunity exists to form these broad foundations for life's leisure pursuits than in the schools through a program of work in basic arts, sciences, reading, music, drama, clubs, social activities, sports, games, and physical education.

The family and other social institutions have a responsibility to leisure education, but the schools have the greater task of intelligently planning for the learning of leisure skills. The school does not automatically develop these interests, skills, or appreciations through mere exposure or chance. Only through a program of artful leadership working toward their acquisition can the objectives be reached. In fact, their acquisition must be deliberately planned for in the school environment. History, science, literature, music, art, physical education, and the like can all be taught in a way that will contribute to lifelong interests. Truly, the years that children and youths spend in school provide the foundation for skills and life values that enrich personality and provide the seasoning for future adventure in leisure living.

## JOINT FACILITY USE AND PLANNING

School areas and facilities represent a major capital investment, and modern needs point to the urgency of using and planning these facilities for community recreation purposes. The public schools belong to the people; so do parks, libraries, and other public recreation properties. It is to the best interest of taxpayers and the rest of the public to coordinate, integrate, and consolidate public facilities when basic functions are not incompatible.

Joint planning and use of facilities by municipal and school authorities have been accomplished in many instances. In fact, public school buildings and their many facilities are almost universally used for community recreation. However, cooperative and joint planning of school plant and recreation areas is a modern innovation. Actually, until very recently, few school districts gave much consideration to planning of buildings or grounds for other than formal class use and extracurricular school activity. While the present trend is toward cooperative planning, this development has evolved as the result of several historical steps. These are

1. The period of original requests by municipal recreation departments for use of school facilities at which time limited use was allowed by school boards to test the reliability of municipally employed leadership.
2. The use of school faculty personnel as paid leaders or supervisors of programs conducted by municipal recreation departments on school properties.

3. The drawing up of written agreements between school boards and city recreation authorities.
4. The payment of compensation to school janitors for additional work involved in community use of schools.
5. The interpretation to the school board of its responsibilities regarding community use of schools.
6. The inclusion of school board representatives on city recreation boards.
7. The joint planning of school building programs with other community recreation agencies.[9]

Today educators have accepted the need for extending the role of schools into the leisure hours of their students and also for relating their schools to community recreation purposes. School buildings and grounds are the only public facilities distributed widely over a city. Indeed, the schools are the largest youth-serving agencies in a community. Therefore, the need for extending the use of school property for community recreation programing, for planning the school plant for community education and after-school use, for developing outdoor areas and school parks, and for formalizing cooperative policy statements for joint use and planning as well as contractual agreements between city and school have the effect of welding community cooperation into a workable plan of action.

## Use of Schools

Reports show that thousands of school recreation centers are now in operation in the country. In some instances use of schools for recreation means only use of outdoor play areas. Such arrangements are entered into by informal or formal agreements between recreation and school authorities. School restroom facilities may or may not be included. This use of school property does extend the outdoor physical activity program of the community. However, the use of indoor school facilities gives rise to complications, problems, and negotiation. The development of a community recreation program making use of the indoor school plant involves questions of cost and administration. Most school buildings are not planned so that a portion of the plant can be separated and used for community recreation programs. Hence the cost of heating the entire structure, the problem of supervision, and the extra cost for heat, light, maintenance, and repairs create a major problem, not to mention the animosities engendered by recreation leaders disturbing the tranquility of the day school faculty through evening use of school equipment and classrooms.

[9]California State Department of Education, *The Schools and Community Organization for Recreation* (Sacramento: Bureau of Health Education, Physical Education, and Recreation, 1953), p. 2.

In many instances, actual costs of maintenance, including custodial service, heating, and lighting are established by a fixed fee and are assessed against the recreation agency. In other cases the schools may only charge for the extra time involving the custodian. In rarer instances, the schools absorb this entire cost.

It would be a mistake to give the impression that schools only look with favor on allowing their outdoor play areas to be used for community recreation. While it is true that this limited property use provides the fewest problems, many schools do make provisions for use of gymnasiums, swimming pools, auditoriums, classrooms, and the like. Here again, the pattern for indoor use is often limited to the athletic plant for administrative and supervisory reasons. Next comes use of auditoriums, and finally, libraries, general classrooms, and special study rooms such as shops, music and art rooms, cafeterias, and home economics facilities. In short, school authorities look realistically on the use of school plants in light of their basic responsibility to formal education. If extended use does not interfere with basic school instruction and extracurricular programs, cooperation is excellent; if interference exists, the priority is for class instruction; other purposes are secondary. Establishment of a plan that minimizes interruption of classes and school programs is basic to good school–community relations.

## Planning of School Plant

As stated previously, the modern school plant lends itself to the function of recreation as well as to that of education. While school facilities are distributed widely throughout a community, these facilities are, generally, not designed for general community use. Hence cooperative effort between school and community in designing school plants for combined use is highly desirable. New school buildings should be planned so that the portion of the building that is to be used by the community can be separated from the rest of the plant. Included in this section of the building would be the gymnasium, auditorium, shops, swimming pool, music and art rooms, community room, and cafeteria, as well as storage space and restroom facilities. Other kinds of rooms that could be used for general community purposes might also be included. These rooms and facilities could be grouped, insofar as this is practical, in one wing or area of the building, with ready access to outdoor play areas.

The best results in planning will be secured if the following basic principles are accepted as the criteria for action.

1. All school building planning should be an integral part of the overall plan used to guide the physical development of schools, community recreation, and parks.
2. Schools should be planned as community structures with ample recognition given to the function of recreation in which adults and

children are equally served in creative arts, music, drama, and other activities.

3. Planning of school plants should be a cooperative venture among educators, recreation officials, city planners, and parents.
4. School plants should be planned to provide the most economical operation for school and community by giving consideration to the following points: Does the part of the school that is to be used for community purposes
   (a) Provide for economical maintenance?
   (b) Have ready access to street and parking areas?
   (c) Provide means for cutting off access to other parts of the school plant?
   (d) Provide for true multiple use with a minimum of interference?
   (e) Have provision for separate heating?
   (f) Make for ease of circulation and access?
   (g) Provide for adequate restroom facilities, drinking fountains, and circulation of people?
   (h) Have classrooms and facilities located for ease of supervision and efficient control?

## Planning of Outdoor Areas

Schools and community recreation properties tend to have roughly the same service areas. Moreover, the considerations affecting the locations of school sites apply generally to recreation and park sites as well. Hence the school–park, or school–recreation, concept seems a logical one to follow in modern community planning.

This concept envisions the development of neighborhood and community recreation areas or parks on lands adjoining the school plant. The importance of including school grounds in recreation planning becomes more apparent with each decade of increasing land values and disappearing open spaces. Miles explains this concept as follows.

> The park-school is a modern school on a park-like site which is designed to function as a center for programs of education and recreation throughout the year. The park-school for the neighborhood serves the children of the elementary grade age level during the day throughout the year and is used during the evening for youth, adult, and family education-recreation programs and related neighborhood activities.
>
> The park-school for the community likewise serves the youth of the secondary grade level during the evening hours for youth, adult, and family education-recreation programs and related community activities.[10]

[10]M. A. Gabrielsen and C. M. Miles, *Sports and Recreation Facilities: for School and Community* (Englewood Cliffs, N.J.: Prentice-Hall, 1958), p. 47.

The school park, with its attractive landscape, the functional design, and its healthful environment, adds to the character of the neighborhood, the value of the property. and the civic pride of the community.

School-recreation planning of outdoor areas and facilities should be a part of the long-range planning process of all localities. Indeed, the basic elements of community planning lie in cooperative effort. For example, the elementary school, by its very location, is strategically situated to serve as a neighborhood recreation center and play area if indoor and outdoor facilities are properly designed. In general, a neighborhood is identified as the service area of an elementary school, just as the community has similar identification with the area served by the junior and senior high schools.

The justification for school-recreation planning can be both practical and reasonable. The elementary school is centrally located within the neighborhood it serves. Ideally, children should be able to walk to school without having to travel excessive or unnecessary distances and without having to cross busy streets. These same considerations apply to the location of recreation areas and facilities; school, park, and recreation properties all tend to have the same service radius.[11] The elementary school and the neighborhood playground and park serve the neighborhood preferably from a central location; the junior and senior high schools and the larger park facilities serve a group of neighborhoods, preferably from a central location. The community or junior college serves the region, just as a regional park or facility serves this same area. In actual practice, school sites are planned and buildings constructed in relation to the area to be served. What better way can be found for economy in spending the "community dollar" than through integration of the park, the recreation center, and the school. A planning study in California aptly describes this policy as follows.

> The cooperative planning process offers opportunities to present-day recreation and park agencies and school districts to achieve functional groupings of properties and facilities that were seldom possible in earlier days. The trend toward inclusion of the neighborhood school, playground, and park on a single site receives particular stimulus from this type of collaborative planning in which the common interests of school districts and recreation agencies become more readily discernible.[12]

School planners should also not overlook the place of outdoor education in the overall design of school sites. To use the school–park as an outdoor

[11]The effect of the school busing decisions by the U.S. Supreme Court have tended to take students away from their neighborhood school, however, the recreation areas and facilities continue to serve the neighborhood children after school hours and on weekends.

[12]California Committee on Planning for Recreation, Park Areas and Facilities, *Guide for Planning Recreation Parks in California* (Sacramento: California Recreation Commission, 1956), p. 21.

laboratory for conservation and natural–science education, to develop a school forest, or to develop other aspects of outdoor life as mediums for enrichment of education is one means of broadening the horizons of learning. Furthermore, the development of additional facilities and areas for outdoor school laboratories known as school camps opens up entirely new vistas of exploration. Regional parks, forests, seashores, mountains, deserts, and valleys all have an educational implication for cooperative planning with other jurisdictions. Indeed, the realms of learning are as vast as the minds of humans.

## Cooperative Policy Statements and Contractual Agreements[13]

The working relationships between school and recreation authorities in the acquisition, use, development, or maintenance of their respective properties range from simple, informal arrangements to formal contracts. These agreements attempt to spell out and clarify the areas of responsibility of each public jurisdiction for providing the desired recreation service and thus eliminate those points of difference that might develop friction. Fortunately, education and recreation authorities have a strong feeling of concern for the task at hand. Both have similar objectives; both serve the same group of taxpayers; both see the need for properties designed for functional instruction and leisure education. In fact, both even see the need for expanded working agreements in the areas of joint program planning, financing, and leadership.

Many communities, school districts, and counties have initiated formal working relationships or contractual agreements for joint acquisition, development, and use of their respective properties. In reviewing the relationships that exist, we see several patterns emerge. These are:

1. Informal, unwritten agreements between school and recreation authorities on use of school properties, usually play areas.
2. Written agreements and resolutions endorsing the principle of cooperation between school and community.
3. Formal agreements or contracts for the joint use of a specific facility or area.
4. Formal agreement as to policy governing joint acquisition, use, development, and planning of school and recreation properties.
5. Review of proposed joint action by coordinating committee and recommendation to governing bodies.
6. Joint employment of planning staff, landscape architect, or other personnel in developing a cooperative plan.

[13]For examples of agreements between school and recreation and park authorities refer to Appendix F.

## COMMUNITY RECREATION PROGRAMS

What is the function of the school in providing a recreation or leisure program for the people in a community? It has been stated that the schools are not primarily recreation agencies. This is true. State statutes clearly indicate that the responsibility of the public schools is to provide an education for all the people. But individual mental, physical, social, and moral development encompasses more than formal classroom instruction. Hence the development of the individual must involve attention to leisure as a part of daily living, and therefore the schools have an obligation to provide for recreation.

Should the public schools, then, act as the public recreation authority in a community? This is a debatable question, but usually, local school districts are not charged solely with this responsibility, although in some instances school boards have been given this task. In other cases the schools conduct their own independent program on school properties, and the municipal recreation agencies conduct other activities on city or park areas, but by informal agreement, each cooperates with the other. In still other instances the schools will join hands in a cooperative and coordinated venture with other public bodies. Such a coordinated program may be under a recreation commission, constituted in part by representatives from school and municipal government, and by agreement, the existing program makes use of school and city facilities. Variations of this coordinated pattern exist when school districts contribute financially or otherwise to the independent program.

Whether the school will become the principal recreation authority in a community depends on many factors. State statutes make it clear that the administration of community recreation programs may be undertaken by school or nonschool public bodies either singly or cooperatively. Each community is unique in its interests, needs, and resources. Therefore the administrative structure to provide recreation services must fit the unique characteristics of each community. The willingness of school districts to provide this recreation service is conditioned by:

1. The close relation between recreation and education as interpreted by educational leaders.
2. The adequacy of school funds to provide a recreation service.
3. The nature of school plant and facilities and their adequacy in providing needed indoor and outdoor space.
4. The intensity of public demand that the schools undertake this service.
5. The availability of professional trained recreation personnel who can meet school standards.
6. The desire of, and encouragement by, public authorities to let the schools lead the way in developing a community recreation service.

Many community recreation programs now operate wholly or partially under school auspices. Cities in Pennsylvania, New York, and California receive services from the schools, as do many cities of other states. Regardless of the role of the schools in providing or assisting in the function of community recreation, all school bodies may be classified as giving one of the following degrees of service:

1. The schools are the primary recreation agency.
2. The schools provide the community recreation service in partnership with the municipality or other public governing jurisdiction.
3. The schools are not the primary recreation authority but do assist in the service by making school plant, areas, and facilities available to the public jurisdiction that administers the recreation program.

Some school districts feel that recreation is not a part of the school function and take no part in this phase of community service. This point of view is rapidly disappearing, however, as the schools are more and more discharging their responsibility by making recreation a vital part of education.

### School Administration of Public Recreation

Advocates of schools as the major recreation authority in a community believe that school facilities can be best used and programs best developed if the board of education has administrative control. While the local community situation will determine the structure of administrative organization for this service, some notable examples of school-administered public recreation can be found in New York City, New York; Milwaukee and Madison, Wisconsin; Newark, New Jersey; and Pasadena and Long Beach, California, as well as in other cities.

It is not uncommon for many small cities to look to their schools for provision of a recreation service. These small communities have in many instances the only facilities available for such service, in addition to the potential for leadership. Furthermore, the close ties between school and community make for excellent cooperative effort in joining forces for financing such ventures on either a full-time or part-time basis. Because of this need in the small community, it is not unusual to see community education programs flourish here to provide recreation and other needed services. For a more complete discussion of the advantages and disadvantages of the management of recreation services under school auspices see Chapter 5.

## THE ROLE OF COMMUNITY EDUCATION

Community education is a new concept that encourages an expanded community role for the public school.[14] The traditional public school is concerned

[14]This entire section on "the role of community education" was prepared by and used with permission of Mr. William Boldt, Assistant Professor of Recreation and Park Management, University of Oregon, September 19, 1979.

with serving students in school from 8:00 A.M. to 3:00 P.M. Community education advocates opening the school to people of all ages within the community before and after the regular school day. This new type of school is called a community school and is the vehicle by which the community education concept is carried to the people.

### Early Development

The community education idea was developed in Flint, Michigan, in 1936 when Frank Manly noticed that schools were not being utilized after the regular school day was over. The financial support of philanthropist Charles S. Mott enabled Mr. Manly to become the first community school coordinator. His job was to facilitate the opening of five schools to community residents of all ages before and after regular school hours. The community education idea mushroomed from this humble beginning in 1936 to include over 5000 schools throughout the world in 1979.

### Goals of Community Education

The "community education" concept is best understood by examining the six major goals of community education.

1. *To Extend the Use of Public School Facilities.* Schools can be used by everyone from infants to senior citizens 12 to 18 hours a day, seven days a week. The school should become a community center for the school neighborhood.
2. *To Integrate Community Resources into the Required School Curriculum.* Every community has an abundance of valuable human and physical resources. Schools can identify and utilize the community to improve learning in the classroom. Examples of human resources utilized to strengthen the school's curriculum could include volunteers to help teachers personalize education in the classroom; guest speakers to present a variety of viewpoints to students; and volunteers to help in a variety of roles from coordinating a Christmas play to organizing and conducting an outdoor school. Examples of physical resources utilized to strengthen the school's curriculum could include fund-raising activities to build a community playground, utilizing a local park to study ecology, and obtaining a church-owned bus to use on field trips.
3. *To Coordinate Optional Enrichment Activities for Youth.* Activities for youth could include free breakfasts, sports, chess, puppetry, cooking classes, drama, gymnastics, tutorial assistance, community beautification, reading made fun, and many other activities.

4. *To Coordinate Optional Enrichment Activities for Adults.* Activities for adults could include lunches for senior citizens, as well as for students; adult basic education job assistance; cultural arts; single adult meetings; parenting, investment counseling; and many other activities.
5. *To Facilitate Involvement in Community Problem Solving.* The community school's major purpose is to involve the community residents in determining community needs and identifying qualifiable resources to meet those needs. Each community school appoints a community school advisory council that helps to determine the direction the school should follow. Surveys, personal visits, phone calls, and meetings are used to assess community needs and interests.
6. *To Facilitate School Cooperation and Coordination with Others.* There are numerous government and nonprofit agencies attempting to serve every community. Agencies have a difficult time in determining the needs for the many communities that they are required to serve. Community schools can help coordinate agency efforts in their school community so that the agency can best serve the school neighborhood. An example includes 4-H, Girl Scouts, Campfire Girls, City Parks & Recreation, Boy Scouts, Boy's Club and YMCA—all offering programs to the youth in the neighborhood. The community school can coordinate the time and type of agency programs offered to eliminate competition and, through collaboration, can best serve the community's youth. This process helps each agency to present efficiently its services to the school neighborhood.

The community education concept represents a new approach to the integration of local elementary schools with the neighborhoods that they serve. It allows neighborhood residents to use their school as a focal point for activities and meetings that they feel develop their community and make it a better place to live. It allows the school staff to tap into the breadth of community resources in the neighborhood (people and places) that enrich the school's curriculum. The schools can be opened for extended periods of time to serve the educational, social, cultural, and recreational needs of all citizens, regardless of age. The concept is based on the belief that local resources can be utilized to solve most community problems.

## Community School Personnel

The staff of the community school includes both paid professionals and community volunteers. The key person on the paid staff is the community school coordinator. The community school coordinator promotes and coordinates the use of the school and other community facilities for all age groups. He or she enlists community involvement in the community school and its programs through the advisory council surveys, and personal contact with residents. Other paid staff may include assistant coordinators, building supervisors, secretaries, and paid instructors for classes.

The volunteer staff is numerous. The key volunteers in regard to the direction the community school takes make up the community school council. This council is often composed of residents from the community, students and staff from the school, and representatives from cooperating community organizations. The community council draws together all interested individuals and groups to represent the community to local agencies, organizations, and governmental bodies. Through the council, all needs and issues of concern to the community may be discussed and dealt with in a community problem-solving manner. The development efforts of the community school are fostered through this council. Other volunteers include class instructors, tutors, fund-raising specialists, transportation helpers, lawyers, and many others.

## Programs and Process

The two prime ingredients of the community education concept include programs and process. The programs within each school are unique to that school community. Once community needs and interests are gauged, programs are limited only by the creativity of the planners and the neighborhood resources available. Most of the schools include programs shown in the following chart.

| *Community* | *Social* | *Cultural* | *Recreation* | *Skill* | *Academic* |
|---|---|---|---|---|---|
| Beautification | Games | Concerts | Sports | Vocational | Language arts |
| Traffic | Dances | Films | Aquatic | Social living | Social studies |
| Day care | Suppers | Exhibits | Parks | Typing | Science |
| Health | Picnics | Lectures | Playgrounds | Woodworking | Mathematics |
| Lighting | Movies | Theater | Outdoor | Automotive | Economics |
| Security | Parties | Literary | Hobbies | Sewing | Adult basic |

The process aspect of community education relates to the involvement of the community in determining common direction, identifying available resources and formulating appropriate plans of action to reach community goals. The process is the system for involvement of people in the identification and solution of their problems. The best geographic unit for the process to be most effective is the elementary school area. This elementary school area is small enough to allow for maximum community participation and is in walking distance to approximately 90 percent of the population to be served (see footnote 11, p. 186).

## Financial Resources and Support

The financial statements of the community school are low when compared to other social service agencies. Existing school facilities are used, reducing the capital cost of building new facilities. Staffing costs are reduced due to the extensive use of volunteers in the schools. The community school staff generally consists of a coordinator and one or two support staff members.

The funding sources for community schools vary greatly from school to school. Generally, community schools rely on the school district and one or more cosponsoring groups for financial support. These cosponsoring agencies include parks and recreation, community colleges and city or county government. Other sources of revenue include fees and charges from programs, state and federal aid, business and industry foundation grants, and local fund raising.

The community education concept is backed by federal and state legislation. The Federal Community School Act was passed in 1974 and authorized $17,000,000.00 for funding of community schooling nationwide. In addition to funding, this legislation defined minimum elements of a community education program and provided strong support for the community education concept nationwide. The community schools and Comprehensive Community Education Act of 1978 further strengthened federal support for community schools by appropriating $240,000,000.00 over a five-year period for local community education programs.

State support for community education has gained appreciably with nine states funding the program in 1979. These states include Michigan, North Carolina, Texas, Oklahoma, Utah, Minnesota, Florida, Maryland, and Alaska. The majority of the remaining states have passed legislation supportive of the community education concept.

This concept is further supported by over 80 centers for community education located within educational agencies throughout the United States. Basic services provided by these centers include:

1. Information dissemination regarding the community education concept.
2. Technical assistance to communities and agencies interested in implementing or expanding Community Education offerings.
3. Training all levels of personnel interested in community education; and
4. Support and evaluation of community education programs at all stages of development.

The potential of the community education concept and community schools is tremendous. Presently, there are billions of dollars worth of school facilities that are underutilized.

This concept would allow not only the full utilization of these facilities but the involvement of the citizens in grass roots decision making regarding the direction of their neighborhoods.

## Rationale for Community Education Concept

In summary, it is important to look from the community perspective of reasons for adopting the community education concept. The Salem, Oregon, community school program in their handout, "Why does Salem have Community Schools?" listed the following 12 reasons for supporting community school programs.

1. Community schools increase the use of one of the largest and most underutilized public investments, the public schools, by keeping them open on afternoons, evenings, weekends, and summers for community activities.
2. Community schools provide an opportunity for all district patrons to be involved in local community advisory councils that help identify neighborhood needs and strategies for meeting them.
3. Community schools enhance the required school curriculum by actively seeking out community resources that can help in educating children.
4. Community schools recognize that learning is a lifelong process and provides educational activities for all ages.
5. Community schools assist community agencies by identifying neighborhood needs and passing those along to the agency designed to meet them.
6. Community schools assist community agencies by making school facilities available to them for extended periods of time.
7. Community schools bring many community residents into contact with their neighborhood school for programs that are of assistance to them.
8. Community schools assist patrons in developing community programs to meet special community needs.
9. Community schools promote volunteerism at the neighborhood level.
10. Community schools provide local neighborhoods with a facility and staff that they can use as a community center.
11. Community schools strengthen family units by providing family-oriented activities.
12. Community schools build a sense of community in local neighborhoods by getting residents together in a variety of activities.

The key to community education lies in identifying and mobilizing community resources to meet neighborhood needs. This process approach of "helping people help themselves" is vital to education and American democracy.

### Questions for Discussion

1. What is the goal of education, and how is this goal related to recreation?
2. Do the schools have an educational interest in recreation? Explain.
3. Trace the history of the use of school plant and facilities for public recreation. What is the present thinking on such use?
4. Explain the four roles of education in relation to recreation. Do all school systems identify themselves with these roles? Why or why not?
5. Should public recreation authorities have free use of school areas and facilities? Why or why not?
6. If school buildings are planned for community recreation as well as for education, should the cost of construction be shared? If so, how? Explain your answers.
7. In general, how should the school plant be planned for recreation use? Outdoor areas?
8. What is meant by a school park? Explain the park-school concept.
9. Why should a city enter into contractual agreements with school authorities?
10. What provisions are usually found in city-school contractual agreements?
11. Should the schools administer the community recreation program? Why or why not?
12. What is meant by community education?
13. What are the six major goals of community education?
14. Explain the differences between a community education program and a community recreation program. Do they conflict? If so, where and why?

## SELECTED REFERENCES

Hjelte, George, and Shivers, Jay S., *Public Administration of Recreation Services*. Lea and Febiger, Philadelphia, 1972, Chap. 6.

Kraus, Richard, *Recreation and Leisure in Modern Society*. Prentice-Hall, Englewood Cliffs, N.J., 1971, pp. 62–63; 70–73; 93, 187, 190, 196, 199–200.

Kraus, Richard J., and Curtis, Joseph E., *Creative Administration in Recreation and Parks*. Mosby, St. Louis, Mo., 1977. p. 41–41, 60.

Sessoms, H. Douglas; Meyer, Harold D.; and Brightbill, Charles K., *Leisure Services*. Prentice-Hall, Englewood Cliffs, N.J., 1975, pp. 165–169.

# CHAPTER 10

# County and Regional Park and Recreation Systems

The recreation movement in America has been concerned primarily with providing recreation opportunities for people in urban areas, and only since the beginning of the twentieth century have counties taken an active part in public recreation and parks. To understand adequately the development and growth of county and regional recreation, one should understand the county as a unit of government and its role in providing park and recreation service, as well as the region as a governmental entity to provide this service.

## THE COUNTY AS A UNIT OF GOVERNMENT

The county as a unit of government originated in England and was transplanted to the American colonies. First instituted in Virginia, the county flowered throughout the South and became its principal form of local government. The colonists in the region found this form of government to their liking because they were agriculturalists. They lived far apart on plantations and did not develop local self-government around the town as did the merchants and other townspeople of the New England states.

When the colonies became states, this form of government was continued and copied by the new states to the west and the southwest, except for Louisiana, where this unit is the "parish." Today counties may be found in 46 states and range in number from 3 in Delaware to 254 in Texas. In organization, the county varies little from its colonial counterpart.

The importance of the county as a unit of government varies with the section of the country. In the middle states, the South, and the Far West, it is an important unit; in the New England states, it has little significance.

### Functions of Counties

Counties are merely convenient subdivisions of the state, created to administer state law and perform other functions such as are assigned to them by the state government. The county is, in fact, an adjunct of the state; it operates with direct reference to general state policy. It has little legislative power and operates under the statutes passed by the state legislature.

The functions counties perform have become somewhat traditional and are similar in character. In most instances counties build and maintain roads and bridges, care for the indigent, register land titles, preserve the peace and administer justice, collect taxes, and provide for state and county elections. In most states, they also act as administrative units of the state for education. In recent years, however, the county functions have become somewhat liberalized to include services of an urban nature, such as parks, recreation, libraries, drainage, sewage control, and the like.

The major difference between a municipality and a county as a form of local government is one of degree—the former being a cluster of people organized for their own interest and convenience under permissive state laws, the latter being established by the state to implement its own functions. All cities lie within county territory. Hence people in cities have two layers of government, one the municipal corporation, or "city," and the other the county. Each has its own sphere of service.

## URBAN VERSUS RURAL COUNTIES

Although all counties are similar with respect to organization and function, each county has its own unique characteristics. They vary in size from a few square miles (New York County, New York) to areas embracing thousands of square miles (Pima County, Arizona). Likewise in population, they vary from a few hundred to millions of inhabitants.

It is obvious that problems existing in heavily populated counties are in sharp contrast to the problems arising in rural counties. Indeed, the traditional pattern of county service has changed little through the years in the rural county. Since 90 percent of all counties in America can be classified as rural in character, the scope of county park and recreation responsibilities is somewhat limited. Actually, relatively few of the small rural counties provide a recreation or park service and, when provided, it generally consists of maintaining a few park sites. Only where counties serve large populations has emphasis been given to park and recreation development. In fact, a parallel can be seen between the population of a county and the scope of the park and recreation service.

## HISTORY OF COUNTY PARKS

The history of county parks and recreation in the United States had its beginning at the close of the nineteenth century, and while there was some development before this time, the only type of county-owned properties that might be classified as fulfilling the function of a park were those areas that provided a setting for the county courthouse or the county fairgrounds.[1] Early pioneering in this field led to the establishment of a metropolitan park commission in Boston in 1892 by an act of the state legislature, which was designed to develop a park system for the people in the 36 cities and towns around that city.[2] This progressive and farsighted move included many of the elements of present-day departmental planning, and one of the proposals in the plan pointed to the need of acquiring property of the following types.

> (1) spaces on the ocean front, (2) as much as possible on the shores and islands of the bay, (3) the courses of the larger tidal estuaries . . . , (4) two or three larger areas of wild forest on the outer rim of the inhabited area, (5) numerous small squares, playgrounds and parks in the midst of dense populations.[3]

Although Boston was the first city to establish a metropolitan park commission, the first county to establish a system of parks was Essex County, New Jersey. Several other counties had acquired parklands through gifts prior to that date, but as far as is known, no other county took steps to develop its parks or establish a park commission or board.

During the first decade of the twentieth century two additional systems were established: Hudson County, New Jersey, in 1902, which followed the pattern of Essex County, and the Metropolitan Park District of Providence Plantation, which was established as a result of the progress achieved in the Boston region.[4] In 1900 a 160-acre park was given to Orange County, California, and six years later, additional parks were acquired through gifts in Ven-

[1]National Recreation Association, *County Parks* (New York: National Recreation Association, 1930), p. 1.

[2]National Recreation Association, *Recreation and Park Year Book, 1950* (New York: National Recreation Association, 1950), p. 4.

[3]Clarence E. Rainwater, *The Play Movement* (Chicago: University of Chicago Press, 1922), p. 34.

[4]National Recreation Association, *Recreation and Park Yearbook, 1950,* loc. cit.

tura County in this same state, but no official authority was established to administer these areas.

Additional progress was made in the second decade. Milwaukee County, Wisconsin, purchased its first county park in 1910, and in 1915, Cook County and DuPage County, both in Illinois, began a program under the terms of a State Forest Preserve Act of 1913, to acquire forest preserve areas, which are similar in function to county parks.[5] County parks in Michigan were also established this same year in Muskegon County. Although park systems were being established, little land was acquired for county parks before 1920, and the first real progress in acquiring park property was made between 1921 and 1930. Table 10-1 gives a summary of the number of parks acquired before 1920.

During the 1920s, there was a marked interest in county parks in many sections of the United States. According to information received in 1925–1926 by the National Recreation Association (then called the Playground and Recreation Association of America), 31 counties had one or more county parks, and at the close of that decade the number of reported parks increased from 22 to 415 and the acreage expanded from 2169 to 108,485.[6] Furthermore, of the more than 3000 counties in the country, the number reporting park and recreation agencies in 1966 rose to 358, as compared to 257 in 1960. In 1966 these park and recreation agencies increased their total number of park areas to 4149, totaling 691,042 acres, as compared to 2610 such park areas in 1960 having 430,707 acres.[7] Hence we see that county park acreage has shown remarkable growth each decade during the century and will continue to do so in the 1980s and 1990s.

A number of counties have acquired large acreage under their supervision. These include Cook County, Illinois; Pima County, Arizona; Westchester County, New York; Union County, New Jersey; Wayne County, Michigan; Milwaukee County, Wisconsin; Los Angeles County, California; King

**Table 10.1 County Parks Acquired Before 1920**

| | *No. of Parks* | *Acres* |
|---|---|---|
| Before 1900 | 21 | 781 |
| 1900–1910 | 20 | 1396 |
| 1900–1920 | 22 | 2169 |

[5]National Recreation Association, *County Parks*, loc. cit.

[6]Ibid.

[7]*Recreation and Park Yearbook, 1966* (Washington, D.C., National Recreation and Park Association, 1966), and *Recreation and Park Yearbook, 1950–1961*, (New York: National Recreation Association).

County, Washington; Cuyahoga County, Ohio; Nassau County, New York; and many more.

### Expanding Use of County Parks

Park areas acquired by counties previous to the present century were used primarily as sites for county courthouses and provided no space for active recreational use. Many of them, however, were noted for their planted and shade areas and were equipped with benches, fountains, and statues. At times, these parks were used as meeting places but mainly served as passive "in town" parks, and no provision was made to use these areas for a program of activities.[8]

As county fairs developed, legislation was passed in many states authorizing counties to acquire lands for county fairgrounds, and in the development of these grounds, facilities were included for racetracks, grandstands, picnic grounds, baseball diamonds, and buildings suitable for recreational use. However, most of these facilities were not used as park or recreation centers and were open to the public only during the county fair season.[9]

The establishment of park systems and the acquisition of land for recreational use pointed the way for county authorities to develop areas according to their size, type, and location. Although earlier use of these parks was primarily passive, later developments called for expansion of facilities for more active recreation.

The services performed by county parks and recreation areas were to provide people with an opportunity "(1) to come in contact with the beauties and wonders of nature both in its natural state and as adapted and arranged by man, and (2) to participate in a variety of wholesome and enjoyable activities in attractive surroundings not too far distant from their homes."[10]

## CHANGING CONCEPT OF PARK RESPONSIBILITY

The dedication of new park areas by municipalities, counties, metropolitan authorities, and state governments has been increasing in recent years. The responsibility of city and state governments to establish park areas has become clear, but the responsibility of county governments has not been so clearly defined.

[8]National Recreation Association, *County Parks*, loc. cit., p. 2.
[9]Ibid.
[10]National Recreation Association, *County Parks*, loc. cit., p. 82.

## Municipal Responsibility

The principle that municipalities are responsible for the provision of recreation opportunities within their borders is clearly defined. This responsibility includes the acquisition, development, and operation of park and recreation areas that range in size and function from circles, squares, and ovals to neighborhood parks, playgrounds, playfields, and beaches. Some municipalities are even responsible for adding recreation parks outside their corporate limits which include mountain camps, reservations, and in a few cases, parkways. However, these extraterritorial facilities are principally for use by residents of the municipality.[11]

## State Responsibility

State parks have been established primarily to preserve unusual scenic or recreational resources within state boundaries. They generally comprise thousands rather than hundreds of acres and are located without reference to the population centers. In some instances those areas having historical or geological features are preserved as state parks even though they occupy relatively small space and acreage. The development of state parks usually provides for rehabilitation and protection of the property, as well as for limited activity. However, this activity is usually appropriate to the setting, such as picnicking, camping, sightseeing, hiking, or swimming.

## County Responsibility

The county responsibility for parks lies somewhere between that of the municipality and that of the state. County parks are generally larger than municipal parks and are developed for passive as well as for active recreational use. More than two out of three county parks are in unincorporated areas of the county. Furthermore, the size of county parks varies from one acre or less to many thousands of acres. A series of guides were given in 1963 by the National Association of Counties to assist county governments in identifying their role in making provision for parks and recreation. It stated:

> The special role of the county is to acquire, develop, and maintain parks and to administer public recreation programs that will serve the needs of communities broader than the local neighborhood or municipality, but less than statewide or national in scope. In addition, the county should plan and coordinate local neigh-

[11]National Recreation Association, *Memorandum on Park Planning in Erie County, New York* (New York: National Recreation Association, 1951), p. 5.

> borhood and community facilities with the cooperation of the cities, townships, and other intra-county units, and should itself cooperate in state and federal planning and coordinative activities. Where there is no existing unit of local government except the county to provide needed local neighborhood or municipal facilities and programs, the county should provide such facilities and programs, utilizing county service districts, local assessments, and other methods by which those benefited will pay the cost.[12]

It is also rather interesting that as far back as the 1920s and 1930s, statements were made and published as to the features that should identify county parks. For example, the following statement was made in a 1930 publication on this matter.

> County Parks should not be imitations of City Parks. They should provide attractive scenery but in addition they should have plenty of outdoor play room and places for assembly for large community gatherings as well as restful recreation for small groups and individuals. They should be plentifully supplied with benches, tables and open fireplaces where meals may be cooked and served; meadow room for games of all descriptions; hill and valley trails for hiking and horseback riding, and good water for boating and canoeing. Besides natural fitness and scenic advantages, County Parks should have a good supply of drinking water and plenty of sanitary conveniences. They should be made easily accessible to all users and the cost to acquire, develop and maintain them should be kept within reason. Historical or traditional associations naturally add to a park's attractiveness.[13]

## THE PURPOSE OF COUNTY PARKS

A county park and recreation system has as its prime purposes the giving of park and recreation service as well as the provision of recreation opportunities to people within the county. Although the function of most counties has been to meet the needs for regional park and recreation areas and facilities, a number of county authorities also provide a local recreation service to people in the unincorporated county territory and rural areas.

A look at current practices makes it apparent that one of the primary roles of counties is to provide a regional recreation service. In other words, the acquisition, development and maintenance of regional park and recreation areas and facilities are the proper responsibilities of the county. In no way does this mean that the county cannot conduct services of a local nature for the inhabitants in the unincorporated portions of the county. Actually, many counties

[12]National Association of Counties, "Policy for County Parks and Recreation," *Recreation* (June 1964), pp. 271–72.

[13]National Recreation Association, *County Parks*, loc. cit., p. 5.

provide such services. To clarify these two types, let us distinguish between a "local" recreation service and a "regional" one.

## Local Recreation Service

A local recreation service is designed to meet the needs of people living in close proximity in an urban environment. These services may be highly programmed and include provision for sports leagues and teams, classes, drama groups, music, arts and crafts, and the like, or for types of areas or facilities that are usually identified with a neighborhood or community, such as playgrounds, recreation centers, swimming pools, playfields, ball diamonds, tennis courts, or gymnasiums. It is evident that a county can provide this type of service, although many county authorities feel that this is the proper responsibility of the incorporated cities.

## Regional Recreation Service

A regional recreation service, on the other hand, affords people of an entire region an opportunity to participate in various kinds of recreation activities not generally possible or readily obtainable in the smaller community or city. Participation in these activities requires special or unique areas and facilities identified as "regional" in character and known as "regional" parks or facilities. This type of park lends itself to specialized activities that require either much space or special topographical features. These would include camping, boating, fishing, hunting, swimming, winter sports, and nature study. The park may also include space for such facilities as picnic sites, hiking trails, bridle paths, outdoors theaters, zoological or botanical gardens, observatories, museums, yacht harbors, marinas, or other special areas and structures.

Regional services are more closely identified with area and facility use than with organized activities or programs. In fact, many of the county parks and/or recreation departments have come to identify themselves primarily with the giving of this type of service. Indeed, it is quite common to see county park departments make use of special topographical features such as rugged coast, spectacular scenery, beautifully wooded areas, streams, lakes, waterfalls, and the like and provide picnic sites, camp grounds, trailer sites, launching ramps, trails, and aquatic areas.

Every county authority is concerned with the scope of park and recreation services. But regardless of the size of the county, when a park and recreation service is provided, it will consist of the two types of services just described, either singly or together.

Perhaps the greatest issue in connection with county provision of a local recreation service is that faced by the heavily populated county. The problem can be illuminated by closely examining the characteristics of the metropolitan county.

**The Metropolitan County.** Many counties have grown in population until the cities within them have become inextricably enmeshed in a complex pattern of relationships. This growth has not only given rise to incorporation of "urban clusters" but has also given rise to what is known today as the "urban fringe area." Moreover, many heavily populated counties have taken on the appearance of metropolitan communities. Schultz describes such an area, as follows:

> A metropolitan area is a densely populated region, predominantly urban in character, which has developed a substantial degree of unity in its social and economic life. Ordinarily by far the greater part of its population is assembled in an aggregate of distinctly urban communities which are related so closely to one another, geographically as well as economically and socially, that they constitute a complex super-community with a number of distinguishable centers of population.[14]

**Fringe Area Problems.** The outgrowth of decentralization and expansion of population into all parts of the county includes what are known as "fringe areas," which are clusters of people who reside in the unincorporated areas of the county. In an earlier period these areas were usually annexed to neighboring cities or when reaching a certain size, they sought incorporation. However, during the past 30 years there has been a trend away from incorporation and annexation. This has been due mainly to the lower tax rates that prevail and the ability of the people in these areas to depend on the county government to provide most of their needs for municipal-type services.

This pattern of local government is further complicated as population continues to expand in these unincorporated urban areas. The unprecedented growth of unincorporated areas and the giving of service by the county to these areas have resulted in a controversy between city and county taxpayers. It was claimed that municipal-type services that were provided to unincorporated areas from general county tax funds were given largely at the expense of the city taxpayer, and it was stated that people in the incorporated cities must pay county taxes and also provide their own municipal services at city expense. Thus it is claimed that county residents are given a "free ride" at the expense of the city taxpayers.

This controversy is of major concern to both urban and county authorities. The question arises: How can we resolve this troublesome issue?

[14]Ernest R. Schultz, *American City Government: Its Machinery and Processes* (New York: Stackpole & Heck, 1949), p. 179.

While the problem basically arises out of the inability of unincorporated communities to expand their municipal boundaries except through annexation elections that the inhabitants vote down because of higher taxes and other reasons, the issue can be resolved partially by (1) legislation creating special service areas and (2) establishment of metropolitan or regional park and recreation districts.

California experimented with service areas by passage in 1953 of the County Service Area Law. This law grew out of a controversy between city and county partisans over the giving of service by the county to heavily populated unincorporated areas. This controversy has been summed up as follows:

> The League of California Cities maintained that many municipal type services provided unincorporated areas from general county tax funds were so provided largely at the expense of the city taxpayer. The county resident thus was receiving a "free ride." With particular reference to parks, it was the contention of the League that cities taxed their own residents to maintain city parks, and the same city residents were then taxed by the county to maintain county parks. The same arguments were advanced with respect to certain types of police service—as radio patrol cars, structural fire protection, etc.[15]

**Operation of Service Area Law.** A service area is simply a means for providing a separate tax levy to pay for such services as may be specifically designated in the resolution of formation. All operations are under the control of the board of supervisors and are administered by a regular county department. There are no separate governmental hierarchies or governing bodies.[16]

A county service area may be established to provide any one or more of the following types of extended service within the area: (1) extended police protection, (2) structural fire protection, (3) local park, recreation, or parkway facilities and services, and (4) any other government services that the county is authorized by law to perform and that the county does not also perform on a county-wide basis within and without the city.[17]

The proceedings necessary to form a service area may be commenced under either one of two methods: (1) The board of supervisors can institute proceedings anytime a written request signed by two members of the board is formally filed with the board. (2) A petition may be filed with the board by not less than 10 percent of the registered voters residing within the proposed service area.

[15]"County Service Area Law as Applied to Parks" (Los Angeles: Chief Administrative Office, Hall of Records, 1953), p. 1. (Mimeo.)

[16]Ibid.

[17]California Government Code, Section 25210.4.

## MANAGEMENT OF COUNTY PARKS AND RECREATION

Many factors must be considered in establishing an effective administrative organization to provide for a county park and recreation system. These are:

1. *Managing Authority*. Should the service be provided under jurisdiction of a board or commission, or should it be under the direct authority of a county supervisor or manager or a roads or public works superintendent?
2. *Scope of Service*. Should the service include only acquisition and maintenance of properties or organized program services or both?
3. *Political Considerations*. Should the program of services or facility development be offered only to inhabitants in the unincorporated portion of the county, or should it be given to all people including those in the incorporated jurisdictions?
4. *Regional Park Authority*. Should the acquisition, development, and maintenance of regional parks and facilities be solely the responsibilities of the county or a regional authority, if one exists, or should these services be provided by municipalities within the county if they so desire? In the latter case the county would only supplement such areas or facilities that presumably exist under municipal auspices.

### Managing Authority

As the movement toward the establishment of county and regional park systems has progressed in recent decades, there has been an unmistakable trend toward the establishment of county park and recreation boards and commissions. The park (and recreation) board or commission is one of the predominant forms of managing authority in the country today, although the types of agencies carrying on park and recreation functions vary greatly from county to county.

The variety of management authority is influenced by state legislation that makes provision for this service, but in large measure the size and population of the county greatly influences the type and kind of managing authority. In large metropolitan counties a separate department may be established to provide this service. On the other hand, this service may be managed in a small rural county by the road and highway department, a conservation department, a public services agency, or even be assigned to one of the county supervisors. Regardless, the park and recreation board or commission has gained much favor as the managing authority and is used in both a policy and advisory capacity. Most of the large metropolitan counties have a department organization within the county structure. Exceptions, of course, are the special metropolitan park districts found in Chicago, Cleveland, Boston, Detroit, Oakland, and others that provide this service to the metropolitan community. Moreover, in many instances these special park districts overlap municipal and county boundaries.

## Scope of Service

Although state legislation influences the form of authority under which the park and recreation function operates, the internal organization of a department is geared to meet the recreation needs of county inhabitants. Therefore the scope of program services will depend on the size and complexity of the county operation. As was indicated earlier, the rural or sparsely populated county is usually concerned only with area and facility development. Little consideration is given to program. The more heavily populated county may design its department of parks and recreation to provide for a variety of services. These may be

1. To acquire, develop, and maintain regional park and recreation areas and facilities.
2. To provide an organized program of recreation services at county areas, parks, and certain school properties. This service may be of two general types: (a) countywide services such as athletics, music, arts, crafts, or drama and (b) services that meet the needs of the residents of urban and rural unincorporated areas.
3. To offer to all county inhabitants types of program services that are either too expensive or impractical for most cities to offer individually (e.g., organized camping).
4. To provide contractual services of a specialized nature to cities or special taxing districts in the county that these entities feel they cannot provide for themselves, such as dramatics, music, and other cultural activities.
5. To provide encouragement and special services to small communities, industries, or agencies through training sessions or consultation.
6. To encourage actively incorporated cities within the county to establish their own local public recreation services.

County departments of parks and/or recreation may provide all or only two of these types of services.

## Political Considerations

The question of what jurisdictions to serve is one of the perplexing questions faced by county recreation and park authorities. In many instances a program service is given only to inhabitants in the unincorporated areas of the county.

This is enlarged in other instances to include area and facility development in the same county territory. In still other cases the county may be the single managing authority for all city and county recreation and park services. In the final analysis this question must be resolved by each county authority. The complexity of county government, the overlapping of services by the many government jurisdictions, the scope of program services provided by the municipalities, and many other considerations will all be factors in determining the jurisdiction of service.

It is important to note that one method of rendering service by the county to incorporated cities within their boundaries is to negotiate special contracts for service between the county parks and recreation department and the incorporated cities. These contracts usually state that certain services will be provided by the department for a specified amount of money. Such services usually call for special talent and leadership, which are used to enrich special phases of a city's program. Examples of activities under such service include sports, music, drama, nature, arts and crafts, and the like.

An entire park and recreation service may be provided to an incorporated city by the county under this contractual agreement. The local corporation would pay to the county the cost of the service from which the former received direct benefit. Of course, participation would be on a voluntary basis.

## Regional Park Authority

The regional park and recreation needs of a county are not met by merely adding regional functions to each municipal recreation and park department or agency. To do the job requires that the county or regional authority be given this responsibility.

The duty for providing and administering regional park and recreation areas and facilities could reside with the county or with a regional park district or authority. Clearly, delegation of regional responsibilities to cities tends to dissipate and extinguish the singleness of authority, purpose, and policy that is needed by a single agency having this obligation.

To go even further, if a city within the county has acquired or developed a park or recreation facility that, by reason of its size, development, or unique features, has assumed regional significance and use, it should be permitted to transfer this property to the county for operation and maintenance. County authorities should not ask local communities to share in financing the operation or maintenance of these areas after transfer has taken place. Moreover, any transfer should be on a voluntary basis on the part of the city, and any property given to the county should continue to receive the same or increased quality of service and maintenance.

## RECOMMENDED SCOPE OF COUNTY OPERATIONS

A look at park and recreation administration in American counties shows a wide divergence of operational procedures and departmental objectives. However, in the final analysis each has the goal of serving the leisure-time needs of people. Therefore in looking at the patterns of service, one should be concerned with the following matters.

### Acquisition of Land

Most counties are deficient in parkland. The standard most often cited calls for a minimum of two acres of regional parklands for every 100 inhabitants. Therefore counties should make every effort to acquire regional recreation land and water resources before a population influx causes land values to reach a point where they are prohibitive in cost. Such acquisition should include property for multiple use. Beaches, mountains, valleys, rivers, and the like should be considered in the regional recreation plan.

### Development of Land and Facilities

Each county should establish a long-range recreation improvement and development plan for areas and facilities. Development of existing and newly acquired property should be based on considerations of

1. Erosion control and pollution abatement of our beaches and water areas.
2. Beach and access frontage along coasts, lakes, and rivers.
3. Development of regional reservations and land preserves for general use.
4. Use of reservoirs, flood-control works, dams, and other, similar public properties for recreation purposes.
5. Development of special facilities—roadside parks, bridle paths, parkways, yacht harbors, arboretums, botanical gardens, zoos, wildlife sanctuaries, museums, camps, and similar features.

### Operational Programs

All park and recreation programs at this level of government should have countywide significance. Local programs should be conducted for towns, municipalities, or other urban areas only through formulation of contracts for such service, unless the county is the single program agency. A program service at county schools and parks or other properties that serve the needs of a rural population would fall within the framework of a general county service.

### Coordinated Planning

It should be the function of the county to act as the coordinating body for all park and recreation agencies operating and maintaining facilities and areas within it.

### Governing Authority

The governing authority for county parks and recreation should rest with a county parks and recreation commission or board. The membership should consist of from five or seven members who represent the supervisory districts of the county, or they should be selected in any other representative manner. Members should serve for overlapping terms of at least three years, with no salary except expenses.

### Finance

Financing for the acquisition, development, and operation of countywide areas and facilities should be provided through a special tax levy on all of the county inhabitants. Bond issues should be used to bring the park system up to standard through land or property purchases.

## THE REGIONAL RECREATION CONCEPT

In recent years park and recreation planners as well as government authorities have become deeply concerned with the problems associated with urban growth and metropolitan "regionalism." Our great cities are no longer self-contained units, by reason of our exploding birthrate, technological advances, and continuing prosperity. Urban areas are overflowing with people. In fact, the population in our metropolitan areas may well double in a span of 30 to 40 years, and this situation raises an enormous complex of problems begging for solution. Making provision for parks and recreation is only one of a myriad of services that are needed in the framework of government functions. Indeed, planning for the metropolitan region calls for creative and imaginative leadership of the highest quality.

To understand the role of parks and recreation in the regional complex, one must consider the "region" as a planning unit, the place of recreation within metropolitan regionalism, and the organization of government to provide services compatible with sprawling urban expansion.

### The Region as a Planning Unit

To clarify many of the questions that relate to the acquisition of regional parks and facilities in a metropolitan area, we should answer the question "What is a *region*?'' A definition of this term will make for a better understanding of the characteristics of regional parks and facilities and will make it possible to establish standards and determine criteria for their development. There is no easy answer to this question, and the determination of a regional area will remain to some extent controversial. Some think of a region as a division of a nation, or even a state, while others think of a geographical unit such as a river valley or a plain as an essential regional characteristic. Moreover, climate, soil, topography, resources, production. and cultural expressions have also been regarded as determining factors.

Administrative boundaries between cities are man-made, changeable, and artificial in scope; hence it is hard to give them regional dimensions. However, it is generally agreed that a region is an interrelated part of an area—a part that is self-contained by reason of geographical factors, natural resources, climatic conditions, and transportation routes. Lewis Mumford defines a region as ''a complex of geographic, economic, and cultural elements.''[18] Within such a region, communities become interrelated and a part of the whole. Each community shares in the fortunes of the region.

## The Metropolitan Region

A region may be a specialized type such as a ''metropolitan region,'' which consists of a dominant ''core'' city and its surrounding communities. It is a web of relationships blended around the dominant center and its subordinate parts. Moreover, it is a cultural expression, which to a great extent, is the result of newer methods of transportation and communications, for it has only been by means of the private automobile, rapid transit, the newspaper, radio, television, and the telephone that satellite communities have become closely integrated.

Metropolitan areas develop a regionalism by the rapid growth and development of subordinate cities and urban unincorporated areas within the influence and orbit of a large ''core'' city. The orientation of these communities around a central city assumes a form of integrated unity. These areas, in turn, must function as a unit as they look to solutions of their communal problems.

A crisis exists for regional areas in political and social organization. Problems arise and are attacked in a fragmented manner. Provision for water, sewage, transportation, parks, housing, and the many other needed services calls for solution of various problems. All services need to fit into a general rational framework of unity. While there is no single solution to the problems of metropolitan regionalism, at least a general pattern of regional organization and services must emerge.

[18]Lewis Mumford, *The Culture of Cities* (New York: Harcourt, Brace & World, 1938), p. 367.

Difficulties arise out of fragmented government. Instead of an integrated unity, a welter of government jurisdictions become involved with their immediate problems, not realizing, of course, that these problems have regional significance. However, signs of cooperative planning are emerging. For example, coordinated action to solve common problems within the region has occurred in a number of instances, examples of which have been when cities have combined to form water districts or sewer districts, to provide utilities or transportation facilities, or to furnish parks beyond their political boundaries. Although communal problems have been legally recognized relative to economic situations, political and social organization has lagged behind in establishing precedents for unity.

## Regional Parks and Facilities

The whole concept of "metropolitan regionalism" is centered around a unit in which people have common problems (involving, for instance, living conditions) and similar interests, regardless of political boundaries. Therefore planning for parks and recreation calls for the establishment of areas and facilities that serve the entire metropolitan community. One obstacle or problem in planning for such a region has been the great number of conflicting and overlapping political and administrative jurisdictions into which the area is divided. While these boundaries have been a limiting factor in metropolitan park development, they must be appraised and recognized in a practical sense of planning.

Another major problem inherent in the development of metropolitan regions has been the serious lag that exists between park needs and existing areas and facilities. Most park and recreation properties have been acquired and operated within the confines of separate cities within a county or metropolitan region. Recreation needs do not stop at city limits, and little attention has been given to acquiring and developing special park properties and facilities to serve people of the entire region.

Regional park and recreation areas and facilities can be divided into two types: (1) those large regional parks that serve a section of the metropolitan area or the entire region and (2) those special recreation facilities or areas that primarily serve a particular or specialized recreation use but are still regional in scope. The former group includes certain types and characteristics that distinguish it from the latter group.

**Local and Regional Parks.** A relationship exists between local and regional parks and recreation areas. Local properties are adapted to local conditions and needs. The effective radius of service for such areas as well as their size will vary with each type of unit, but these areas are located and developed in accordance with their respective functions within reasonable travel distance of the people to be served. These areas are judged in terms of service to the neighborhood and the community. Although large recreation parks and reservations are provided by some cities, these areas have most of the characteristics of regional parks.

The regional class of park and recreation areas and facilities involves somewhat different problems from the local, although no hard and fast rule can be drawn between the two classes. Regional park properties include beaches, forests, lakes, mountains, or such other park areas and facilities as are impossible to duplicate in the smaller government units of the metropolitan areas. While such large or special areas and facilities may be made available by a local government, primarily for the use of those living within the city, and at local expense, experience shows that municipalities alone will not provide for an adequate system of regional parks for the entire population of a metropolitan area. For this, there is needed a larger government entity or public authority that is specifically charged with the responsibility of acquiring, developing, and maintaining a comprehensive and adequate system of regional parks and special recreation facilities.

**What Is a Regional Park?** Any attempt to define a *regional park* encounters certain difficulties. The factors of definition are relative, not absolute. Some people have escaped the difficulties of giving a concise and clear definition by stating what a regional park is not; others have bypassed the difficulties by pointing out only predominant characteristics and determining factors. Although these considerations help determine the function of a regional park, they fail to indicate what makes a park a *regional* park. Some of the distinctive characteristics of a regional park that have emerged and developed in large metropolitan areas provide a clue to a definition.

Many people in a metropolitan area come together in strategic places to participate in various kinds of recreation activities that are generally not possible or readily obtainable in the smaller community or city. Participation in these recreation activities requires special and unique park and recreation areas and facilities. Moreover, the unique property features lend themselves to specialized activities or programs that attract and serve people from all parts of the metropolitan area or region. *Therefore a regional park is a recreational area that, by its unusual development or unique features, gives people of an entire region an opportunity to enjoy certain types of recreation activities. It possesses natural features and is intended to give people a chance to get away from an urban environment, but its primary purpose is to provide pleasant surroundings for engaging in a variety of special recreation activities that lend themselves to the park's setting.*

A regional park is a recreation area permanently set aside for the use of the people, but it is also a particular kind of park. The factor that appears to distinguish a regional park more than any other is the kind of natural features and facilities that formed the basis for establishing the area.

This type of park usually comprises an area of at least 100 acres, although smaller properties having unusual historical, scientific, or scenic interest may serve this purpose. Because of its size, this area is usually located near or outside the boundaries of urban communities and, in some cases, some distance from the cities it serves. The availability of property usually determines location and size, but in closely built-up metropolitan areas, acreage of less than the desired standard is sometimes acquired. Every major section of a large metropolitan area needs access to one of these properties.

One of the values of the regional park lies in the effective use of its natural state, and a large portion of the area is usually devoted to wooded areas, meadows, and other natural features. Furthermore, this type of park makes possible participation in activities that require either large space or special outdoor features. These activities may include camping, swimming, boating, water skiing, fishing, winter sports, and nature study. Accessible portions of the park may also include space for such facilities as boat-launching ramps, picnic sites, beach fronts, hiking and bridle trails, outdoor theaters, zoological and botanical gardens, arboretums, observatories, museums, and other special features.

**Regional Properties: Special Recreation Areas and Facilities.** In addition to regional parks, other areas have special regional significance or use and serve the population of a number of cities or the entire metropolitan area. These regional properties are characterized by the development of specific or special recreation areas and facilities or the preservation of some natural, scenic, historical, or scientific feature. Areas and facilities of this type may be established throughout a region, and their acreage and size depend on use and service.

A county or regional park and recreation system usually has several types of recreation areas and facilities that may be classed as "regional." In general, these properties may be classified into the following types: (1) sports areas, (2) cultural, historical, and scientific areas, (3) scenic and vista areas, (4) hiking, riding, and nature trails, (5) natural and passive park areas, and (6) water and shoreline areas.

***Sports Areas.*** A sports area, which may serve a number of communities or the entire metropolitan region, is primarily designed to meet specific athletic needs or provide facilities for a highly specialized sport or game. A number of specific sports areas should be strategically located to serve the needs of a metropolitan area, in order to provide space for baseball, softball, soccer, track, tennis, handball, swimming, and various other games and sports.

Such areas, if they are to meet the particular need for which they are established, should be within reasonable proximity of a large number of young people and adults. A regional sports area may include any one or a combination of the following: a combined game and athletic field, a rugby or soccer field, a golf course, an archery range, a pitch-and-putt course, a driving range, a group of baseball and softball diamonds, shuffleboard courts, tennis courts, bowling greens, handball courts, casting pools, roque courts, skeet and pistol range, a controlled model airplane area, or a swimming pool. Any of these sports areas may be developed in sufficient number to serve the needs of a metropolitan region or county. Furthermore, to develop each of these areas or facilities in sufficient number to satisfy the needs of a number of communities is clearly a regional function.

Although each of these sports areas or facilities may be developed to serve regional needs, it is the "sports center" that requires special planning and consideration, because this facility provides for multiple sports activities.

A regional sports center is a highly developed area that provides a variety of athletic facilities, primarily for the use of young people and adults; it serves the needs of those from a number of communities or a region. It makes possible the development of a complete athletic facility, which is not possible in the smaller neighborhood playground or playfield. Furthermore, the center is used for multiple purposes, in that it provides different athletic facilities in one area. Twenty acres would be considered a minimum size, with 30 to 50 acres more desirable.

The sports center serves a number of communities and should be as central and accessible as possible to the people it serves. It is recommended that the effective service radius of the center be not less than 5 miles and preferably 8.

***Cultural, Historical, and Scientific Areas.*** Within the confines of most metropolitan areas there exist certain cultural, historical, and scientific sites that are attractions to the people of the region. Some of these areas are endowed with historical lore and have played a significant role in the history of the area, whereas others have proved to be attractions from a natural or scientific point of view. There are few communities that are not endowed with historical properties, which would include Indian trails, missions, cabins of early settlers, and early burial grounds—all of which are part of the tradition of the region. Many of these sites are worthy of restoration or preservation.

Cultural centers provide people with an opportunity for the enjoyment of educational activities. These would include art, music, dance, drama, and crafts, or special facilities such as museums, zoological gardens, arboretums, or garden centers.

Standards do not readily apply to these types of areas, for unlike developed parks, they are not provided in every section of the metropolitan area. The location of these areas or facilities is determined by the nature of the property available. However, cultural centers should be located in areas where they seem to be needed.

***Scenic and Vista Areas.*** A recreation resource that has regional implications is that of vista areas that give panoramic views of mountains, coastal areas, rivers, canyons, or deserts. The location and size of this type of regional facility will vary widely, and usually, it is developed in conjunction with the highway and freeway system that radiates throughout the metropolitan area. Roadside rest areas and picnic sites are usually located here.

***Hiking, Riding, and Nature Trails.*** This type of regional facility is often incorporated in the large recreation park or the reservation. It is considered a separate facility only when trail sites are acquired that are continuous for many miles. Such trails are generally established as part of a complete trail system that traverses the region or county and links the various parts of the regional park system. Moreover, lands for such trails usually follow washes, riverbeds, canyons, or tidelands. To make long trips possible, a trail system will occasionally have adequate stopover sites for picnicking and overnight camping.

***Natural and Passive Park Areas.*** A natural and passive park area is one that has lakes, forests, streams, or other natural features and is used mainly for camping, swimming, picnicking, fishing, or nature lore. Because of topography, for example, rugged terrain, or other natural features including the unspoiled beauty of the out of doors, this natural or passive area caters to people wanting a day's outing in a unique outdoor setting. Properties of this kind will range in size from very few to several hundred acres. Moreover, this type of park only differs from the large recreation park in that it usually has smaller acreage developed around a unique natural feature.

***Water and Shoreline Areas.*** The use of water and shorelines makes possible many recreation pursuits, and the value people attach to such areas is evidenced by the high price of this type of property. Although shoreline areas have a number of recreation possibilities, the greatest demand is for a regional public beach that makes possible swimming, strolling, or enjoyment of the coolness and refreshment of the water or breezes. Shorelines along rivers, lakes, or ocean fronts also make possible the development of yacht harbors, marinas, camp areas, fire rings, picnic sites, boat ramps, and moorage areas.

## THE REGIONAL PARK SYSTEM

For many years, county and regional park authorities have been searching for solutions to problems arising from the growth of urban areas. Since the turn of the century many recommendations have been made, some good, some bad, by various commissions, citizen study groups, taxpayers' associations, and government research bureaus as to the solution of these problems. All these groups agree that an adequate recreation authority is needed.

It is clear that regional recreation needs cannot be met by units of city government. Experience has shown that a larger unit of government is needed to provide this function. Hence the best possible solution to these problems seems to be the establishment of a metropolitan park and recreation district, or performance of this function by the county. Since a discussion of county organization was provided, consideration will be given to metropolitan park and recreation districts.

## Metropolitan Park and Recreation Districts

The first metropolitan park district was established in 1893 around Boston. The success of this operation spread to other large cities, and it became a model for metropolitan park organizations.

The term *park district* as used here applies to a local park entity. It is a government unit with power to tax, borrow, and expend funds. With few exceptions, park and recreation matters are its sole concern, and it is responsible for its own policy and administrative action.

Metropolitan park and recreation districts may vary in their governmental organization. In many instances the governing board is elected; in others, appointed. In any case the board or commission selects the park superintendent, manager, or director to administer the operation of the district.

State statutes that give the legal framework for district organization are rather specific concerning the organizational framework of the governing body. The law usually specifies the manner of selection of officers, committee organization, and adoption of rules for the conduct of business and executive organization. In most instances the number of board members will vary from 3 to 15, with 5 being most frequent. Terms are overlapping. Appointments may range from three to seven years. Most laws specify that members serve without compensation.

Resort is often had to bodies of this kind to provide services that transcend the boundaries of government jurisdictions. For example, some bodies will serve one county, whereas others will serve a part of a county or even extend beyond political boundaries, but regardless, all may be grouped as follows:

1. Those serving the people of one metropolitan county. (An example would be the Forest Preserve Districts of Illinois.)
2. Those serving the people that live in parts of several counties. (Examples would be the park districts of Boston, Cleveland, and Washington, D.C.)
3. Those serving the people who live in parts of one county. (Examples would be the Metropolitan Park District of Tacoma, Washington,

or the Willamalane Park and Recreation District in Springfield, Oregon.)

Districts as units of general government are similar in organization and operation in the various states. Examples of district organization follow.

In Ohio the state constitution provides that laws may be passed for the conservation of the natural resources of the state, and enabling legislation has been enacted that permits any county or portion thereof to create a park district, which may or may not extend beyond the boundaries of the county. The governing body of the district consists of a park board of three members appointed by the county probate judge, and the county treasurers and the county auditor serve as ex officio members. Under terms of this legislation, Cleveland established a metropolitan park district in 1917, and a park board was appointed to act as the governing authority. Fifty-six municipalities are included in the district, and its operation is supported by general and special tax levies. Akron also established a metropolitan park district in 1924, under a governing board of three members, and this system is supported by general tax levies. Both Toledo and Hamilton County have also established park districts.

County park and recreation systems in Illinois operate under terms of a County Forest Preserve Act that was passed by the state legislature in 1913. This act permits any county or portion of a county in the state to create forest preserve districts, and the governing body of each district is composed of five members who are appointed by the president of the board of county commissioners. A forest preserve district may be established by petition and a majority vote of the electors within the district boundaries. Although these districts may cover only a portion of the county, all districts so far established cover the entire county.

The authority of a district board is extensive. This body has general power of taxation for corporate purposes; it may dispose of property; and it may issue bonds in an amount not to exceed 0.6 percent of the assessed value of the taxable property in the county.

Before 1933 the city of Chicago and several of the communities in the Chicago metropolitan area each had a levy for local park purposes. Because of the number of independent districts having separate responsibility for developing their own park and recreation system within the city, these governing units were consolidated into a single Chicago Park District through passage of the Park Consolidation Act of 1933. This new district was given extensive authority including the power to collect taxes, issue bonds, and acquire property. Moreover, it is entirely independent of the city government.

Metropolitan park districts in Washington State have been established in both Tacoma and Yakima under state law, which calls for a governing board of five members. The park district and the school district jointly administer the recreation and park work of metropolitan Tacoma, and two members of the metropolitan park board also serve on the Tacoma Advisory Recreation Commission.

The Yakima Metropolitan Park District, which was approved by vote of the people in 1943, can operate in and out of the city and own property in its own name and can levy a tax for park purposes. Other very large regional park and recreation systems are the Maryland Park System; the East-Bay System in Oakland; the Huron-Clinton Metro Parks System in Detroit, Michigan; and others too numerous to list.

### Questions for Discussion

1. Explain the purpose of a county as a unit of government. How does a county recreation area differ from that of a city (if at all)?
2. Trace the historical development of county parks and recreation. How has it paralleled the course of municipal development?
3. What is the recreation role of the county? Why can the municipality not take over this role?
4. Distinguish between the recreation role of a recreation and park district and that of a county park and recreation department.
5. How does the "fringe area" problem relate to a county recreation operation? To a district operation?
6. What type of managing authority would you establish to conduct a county recreation service? What factors must you consider?
7. What is the scope of service of a county recreation department?
8. Can you make a case for separating parks from recreation at the county level? Explain.
9. How do county parks differ from state parks? From city parks?
10. How do you account for the growth of metropolitan regions? Why has this created problems for recreation?
11. How would you classify regional parks and recreation areas?
12. What is a "regional park"? A regional recreation facility?
13. Should cities operate any kind of regional area or facility? Why or why not?
14. What is the difference between a reservation and a regional park? Could they be the same? Different?
15. Give the advantages and disadvantages of establishing a regional park and recreation district.
16. Can you distinguish a difference between a metropolitan park district and a regional park district? Explain.
17. Give the arguments for and against a municipality giving its regional recreation areas to a county for operation.

## SELECTED REFERENCES

Butler, George D., *Recreation Areas: Their Design and Equipment* (2d. ed.). Ronald, New York, 1958, pp. 1–5; 131–145.

Butler, George D., *Introduction to Community Recreation*. McGraw-Hill, New York, 1959, pp. 159–176.

Gabrielsen, Alexander M., and Miles, Caswell M., *Sports and Recreation Facilities: For School and Community*. Prentice-Hall, Englewood Cliffs, N.J., 1958, pp. 1–16.

Hjelte, George, and Shivers, Jay S., *Public Administration of Recreation Services*. Lea and Febiger, Philadelphia, 1972. Chap. 7.

Kraus, Richard, *Recreation and Leisure in Modern Society*. Prentice-Hall, Englewood Cliffs, N.J., 1971. Chap. 3.

Scott, Harry A., and Westkaemper, Richard B., *From Program to Facilities in Physical Education*. Harper & Bros., New York, 1958, pp. 1–48.

Sessoms, H. Douglas; Meyer, Harold D; and Brightbill, Charles K., *Leisure Services*. Prentice-Hall, Englewood Cliffs, N.J., 1975, pp. 131–137.

# CHAPTER 11

# Program Administration

The recreation programs provided by a public authority can be classified into three general categories.

1. The provision of recreation areas, buildings, and facilities for recreational enjoyment.
2. The furnishing of professional leadership for organized individual or group activities.
3. The giving of assistance to individuals or self-determining groups that wish to enjoy recreational pursuits or opportunities through their own continuing membership.

Therefore the purpose of public recreation is to serve people of all ages throughout the year with a variety of recreation opportunities for enriching their lives through the use of areas and facilities, leadership resources, and special guidance or assistance services for individuals and groups. These opportunities may serve merely to awaken new interests, or they may be opportunities for self-expression, creative effort, physical activity, or merely for relaxation in a wholesome environment. These are not all the possibilities, but they do suggest the scope of programs and the kinds of experiences that constitute what is called "recreation." Moreover, the range of recreation program is unlimited. Actually, programs include all the means by which a recreation authority works to achieve its purpose for being. They take into account the leisure-time needs of people, the resources of the community, and the growth of persons as individuals and in groups. In short, "recreation programs" may be defined as all services and activities, as well as other experiences in which people engage, under the auspices of the recreation system. They are the end products of administration—the mediums for the achievement of recreation objectives.

## OBJECTIVES OF PROGRAM

All program is oriented toward reaching the objectives of the recreation authority. While most systems formulate a statement of their more general objectives based on community needs, all to some extent embrace the aim of developing the individual or group through participation in some phase of program. A concise statement of objectives was given in Chapter 1, but they are restated here as follows:

1. *Emotional and Physical Health.* To develop a sound body and mind through wholesome, vigorous, and creative activities.
2. *Character Development.* To build character through rich, satisfying, and creative leisure-living patterns focused toward the attainment of socially desirable attitudes, habits, and values.
3. *Widening Interests.* To open new interests that provide satisfying outlets for individual development.
4. *Citizenship.* To develop through recreational associations of people a respect for the worth and dignity of individuals and faith in democratic action.
5. *Skills.* To develop skills in the arts of leisure-time living that raise the level of the refinement, culture, and happiness of people.
6. *Social living.* To develop and strengthen social relationships within the family and the community through close group associations and activity participation.
7. *Economic Value.* To strengthen the morale and economic efficiency of the community through expanding leisure-time interests and improving social living conditions.
8. *Community Stability.* To develop community stability by providing an environment that is conducive to wholesome family living and community life.
9. *Long-range Enjoyment.* To enjoy an activity that can be used for a lifetime.

## PROGRAM LAG

Unfortunately, many recreation programs lag behind the needs of the community. This lag is due to the changing social and economic patterns of American life, the lack of understanding of the true objectives of community recreation, the lack of adequate recreation resources, and the lack of training of the leadership staff in the full scope of program services.

Although there are a number of exceptions to this statement, the deficiencies are all too apparent when close scrutiny is given to the program of many recreation agencies. On the other hand, the lack of financial support for expanding services has been a major cause of program lag. This is a reasonable excuse. But surely, this should only focus further attention on the need for public understanding of the full scope of public recreation.

Organized recreation programs originally centered around play for children. Little thought was given to unorganized services financed by public funds, except for the development of parks. However, time brought changes. Program concepts expanded to meet the social needs of the 1930s, 1940s, and 1950s. In other words, there was expansion from a play movement to a recreation movement. The recreation orbit will continue to expand and adapt to a changing society. Ideally, recreation programs should be in the forefront of social progress. But are they? It is unfortunate that the answer is no, except for a number of special instances. Programs have not been conceived with imagination and launched with courage. Too often, the programs are dull and monotonous and the scope of activities and services is limited. A new age is on us, and programs, whether social, physical, cultural, or intellectual, must adjust to this age and embrace a new concept—that a great cultural renaissance is here. Simple time-killing activities are no longer acceptable; there is need for programs of the highest quality. Recreation must strive consistently and skillfully to satisfy the creative and intellectual hungers of people through wholesome and socially sound leisure programs.

## PROGRAM PLANNING

The area of program planning, as was stated earlier, involves "all services and activities, as well as other experiences in which people engage, under the auspices of the recreation system." Therefore in the development of a recreation program, definite procedures or steps are recommended to be taken, and basic principles of program planning should be followed. Actually, the aim of program planning is to meet the expanding leisure needs of people in a dynamic society. Moreover, all program should be focused toward worthwhile objectives.

## Steps in Program Building

There is general agreement that the following basic procedural steps should be taken in constructing the recreation program.

1. Formulate a clear statement of program objectives.
2. Study the needs of individuals by various age categories.
3. Study the basic social and psychological wishes of all people.
4. Study the social and environmental conditions and trends of the community.
5. Survey the existing recreation areas and facilities.
6. Select a program of activities on the basis of the findings of steps 1 through 5, and have it conform to guiding principles of program planning.

**Formulating Objectives.** Every recreation authority needs to formulate a clear and concise statement of its program objectives. Consideration should be given in such formulation to individual and societal factors as well as to the philosophy that underlies the basic assumptions of the recreation authority. Although some of the more general objectives were stated earlier in this chapter, each recreation system must draft its own statement of objectives.

Most program objectives of recreation systems are either long range or immediate. The long-range objectives are associated with the overall goals or broad areas and purposes of the agency. They involve general purposes such as "worthy use of leisure time," character development, good citizenship, and the like. On the other hand, immediate objectives are more practical and denote the attainment of an end. They are measurable and more visible. The development of a recreational skill, good health, courtesy, emotional control, safety practices, practical efficiencies, and the like are examples of this category.

**The Needs of Individuals by Age Levels.** A study of the characteristics of individuals at various age levels provides the basis for the selection of activities suited to their needs and capacities. It is important to adapt a program to each stage of development of the individual. Too often, programs are formulated without consideration given to these important factors. However, a word of caution is in order here, for no rigid application of age, personality, or activity characteristics should be made for all individuals. Many persons will deviate from the average, and the human variable must be considered in all instances. Furthermore, no age period is isolated from the preceding or the following one. The characteristics of one overlap those of the other, and age categories are given only because the characteristics listed are more apt to predominate or be exhibited. There are eight age and activity classifications.

## Period I Age 4–6 (Kindergarten Age)

*Age Characteristics*

Intensive self-interest. Wants companionship. Takes initiative. Active imagination. Beginning of social relations. Attention span short. Needs routine. Wants to please leader. Selfish. Quarrelsome. Responsive to suggestions.

*Activity Characteristics*

Tires easily. Eye–hand coordination incomplete. Full of activity. Likes to play in informal groups. Loves motion. Interested in activity rather than results. Loves dramatic make-believe, storytelling, singing games, rhythmics, balancing stunts, clay modeling, beadwork, paints, crayons, picture books, blocks, and collecting. Believes in Santa Claus and fairytales.

## Period II Age 6–9 ("Big Injun" Age)

Know-it-all age. Self-assertive. Impatient. Impulsive and restless. Very inquisitive and noisy. Imitates teachers and leaders. Easy to impress. Sensitive to failure. Period of questioning. Overestimates ability. Expanding interests. Attention span increasing. Needs adult approval. Sex antagonism developing.

Small-muscle coordination developing. Posture may be poor. Small heart. Poor endurance. Arm and shoulder girdle strength lacking. Bones soft. Exceedingly active and wants physical activity, team games, climbing, hunting, and chasing. More group participation. Interest in games of low organization, stunts, wrestling, and games involving rivalry. Likes rhythmics, singing, nature, pets, and collecting.

## Period III Age 9–12 (Preadolescent Age)

Start of gregarious spirit. Extremely restless. Fatigues readily. Period of codes, passwords, and cliques. Ill at ease. Peer group important. Beginning of hero worship. Tends to be moody at times. Desire for adventure. Power of concentration developing. Time of self-consciousness. Loses temper quickly. Developing individuality. Growing independence. Wants to make own choices.

Endurance improved. Excellent health. Weak in upper extremities. Coordination becoming automatic. Wants to play well. Developing interest in highly organized activity. More competitive. Concerned with own personal performance. Likes self-testing activities and modified sports. Girls like rhythm. Team sports, swimming, rhythmics, group games, relays, and individual athletic events are popular. Also likes camping, collecting, music, and reading.

### Period IV Age 12–15 ("Gang" and Early Adolescent Age)

*Age Characteristics*

Age of puberty. Rapid growth. Poor posture. Awkwardness. Hero worship at its peak. Strong group loyalty. Sensitive to others. Gossip-and-giggle period. Overenthusiastic. Rebellious against authority. Wants adventure and excitement. Craves understanding. Seldom represses anger. Cliques strong.

*Activity Characteristics*

Greater strength after puberty begins. Separation of sexes in body contact sports. Need for team games and variety of activities. Likes outing activities, hobbies, crafts, and all active games and sports. Interest in mechanics. Reading, scouting, and parties popular.

### Period V Age 15–18 (High School Age)

Sex stress. Seeks status and recognition. Extreme loyalty. Imitation strong. Hero worship. Idealism. Daydreaming. Critical attitude. Development of self-confidence. Nervous and overactive. Good looks important. Wishes to excel. Wants group approval. Attitude of revolt against parents. Easily embarrassed. Enlarged mental activity. Senses very acute.

Strength increases. Improved coordination. Age of team games. Likes reading, group activity, social dancing, parties, and other corecreational activities. Interest in physical prowess. Music, books, dramatics, and hiking popular. Interest high in mechanics and varsity-type sports for boys and soccer and field hockey for girls. Individual sports, volleyball, tennis, badminton, and bowling also popular.

### Period VI Age 18–30 (Young Adult Age)

Maturity complete except in lower age level. Habits, attitudes, values becoming developed. Thinking of marriage and family. Interest in vocation. Striving for success. Developing permanent interests. High vitality. Motor mastery. Feeling of responsibility. Emotional balance. Interest in intellectual activity.

Activities of youth carry into early adult life. Interest still high in sports. Likes hiking, dancing, swimming, and social activities. Interest in individual sports, clubs, lodges, music, drama, hobbies, civic activities, and passive and family activities.

### Period VII Age 30–65 (Adult Age)

*Age Characteristics*
Full maturity. Recognizes responsibilities. Interest in vocation. High personal pride in achievements. Interest in youth work. Adjusts to work environment. Physiological body changes.

*Activity Characteristics*
Likes family activities, social recreation, civic clubs, lodges, and business affairs. Interest high in less vigorous activities including tennis, golf, bowling, fishing, swimming, hobbies, clubs, classes, music, art, drama, and hiking.

### Period VIII Age 65 and over (Retirement Age)

Mature judgment. Less physical vigor and stamina. Concern for others. Needs more rest. Interest in civic affairs. Slower tempo in living. Wants to feel needed and a part of the stream of life.

Needs modification of activities. Wants outlets for unused time. Likes social participation, club activities, and hobbies. Interested in intellectual pursuits, community service, games, gardening, arts and crafts, walking, golf, dancing, and travel.

**Basic Human Needs and Wishes.** Individuals and groups often express their basic wishes through recreation without recognizing the force that impels them. Butler states that "one characteristic of all forms of recreation is that each provides an outlet for some basic urge or need."[1]

One of the basic elements of recreation is that it brings satisfaction to the individual participant. But satisfactions are related to the wants of people. The basic problem of recreation systems—in fact, of all social groups—is to create an environment in which the basic needs of people find satisfaction. Therefore program planners must be aware of the major drives that explain the direction of individual wants. Much of an individual's recreation is based on these deep-rooted, universal human drives.

Since everyone, child and adult, has the same basic wishes and seeks the same general goals, how do people reach these goals through recreation? First, let us identify these basic motivations and relate them to recreation activities. In our culture, people wish recognition, achievement, affection, security, social approval, new experiences, and beauty and harmony.

[1]George D. Butler, *Introduction to Community Recreation* (New York: McGraw-Hill, 1949), p. 207.

***Recognition.*** Early in life, the individual struggles for recognition and status. This struggle to achieve distinction, receive praise, stand out as an individual, gain prestige, and be a worthy person is a lifelong one. It represents a basic goal in American culture, one to which the child readily responds. Every individual wishes ego support and social approval, which give him or her assurance and confidence. Children and youths are naturally hungry for social approval. This craving is often satisfied through participation in recreation activities. To receive recognition in sports or other forms of athletics, drama, music, dance, club work, and the like opens the door to all kinds of success patterns. In fact, programs should enlarge in scope to the point where everyone may develop a talent for performance that is enjoyable and worthy of group approval.

***Achievement.*** Achievement is related to recognition, for, indeed, achievement brings recognition. To achieve is also to succeed, and success brings prestige and ego satisfaction.

Our society has made a fetish of achievement, which is basic to our American philosophy and is dominant in our educational patterns as well as in our leisure. Hence, to recognize it, and to provide situations in which individuals can succeed and feel the thrill of supremacy, adds a new dimension to recreation.

The desire for achievement is satisfied in part by an individual selection of activities, competitive or noncompetitive, in which there is personal enjoyment and a degree of success. Participation in dramatics, music, club work, sports, and other recreation programs provides a medium for achievement. Also, creative arts and crafts, drama, and hobby activities provide for self-expression. To have the satisfaction of a task well done, to develop a talent or skill that brings group approval, to attain honors through a competitive situation, or to feel power over others regardless of the significance of the event provides an individual with a feeling of dignity and personal worth. Needless to say, achievement provides a motive force so strong that it is a compulsion in our value structure. Therefore a broad and varied recreation program in which every individual can experience personal satisfactions and a degree of achievement through success and creativeness should be the goal of every recreation system.

***Affection.*** Affection is related to love and the feeling of being accepted and wanted. All individuals yearn for affection and similar emotional relationships with others. In fact, individual personality is formed in large measure through affection and love. The feeling of being wanted provides the base for emotional security and the feeling of belonging. On the other hand, the rejected individual is likely to feel insecure, anxious, and out of things. Painfully, he looks for evidence of response and continually seeks approval and attention.

The longing for affection and the longing to feel needed are deep rooted and universal. Fulfillment provides a sense of self-confidence and well-being. When unfulfilled, these longings produce loneliness, frustration, and despair. Provision of an environment that tempers the emotional shock of rejection and satisfies this basic need is a challenge to recreation leadership. For, surely, through a broad and varied program of activities focused on the needs and abilities of individuals, recreation experiences can restore or reinforce the feeling of belonging. Indeed, they can bring satisfaction, contentment, and close personal relationships.

***Security.*** People are creatures of habit. Early in life, individuals establish routines and ways of doing things that become habitual. Strange situations and environments requiring behavioral patterns that are different present problems of adjustment to most individuals. Every person wants to feel secure, both physically and socially. But freedom from fear of physical harm and security for loved ones are not enough. To feel secure in society requires an ability to adjust to new situations and appraise new stimuli in a changing environment.

All individuals feel to some extent insecure in strange situations and seek means to adjust to them. Unconsciously, one fears looking ridiculous in a new social environment. The changing world in which we live has created a high degree of mobility of people. In fact, constant adjustment to new groups has become almost a universal experience.

The role of recreation in adjustment to working with new groups, in making people feel they belong in the neighborhood and community, and in providing a sympathetic environment for children and youths presents an expanding concept for program planning. Individuals are constantly coming to grips with threats to their security. Tension at home, economic insecurity, illness, unemployment, peer-group rejection, teacher hostility, and other factors too numerous to mention are examples of such threats. Sympathetic leadership and understanding can do much to make people feel secure. Recreation programs must adjust so that every child has a chance to develop feelings of security and freedom from fear.

***Social Approval.*** Social approval is related to respect, group acceptance, fellowship, and the esteem in which one is held by others. The wish for it is the social urge manifested by the desire to belong and be a part of the social group to which one feels related. To be ridiculed and scorned is among the most painful of experiences. In fact, to be isolated from one's peers and social groups and to be laughed at and humiliated are to be driven to the limit of endurance. Suicide in our society is considered by many to be largely a result of rejection or isolation. Every person, child or adult, wants to belong and to be a sharing member of a group. Indeed, the deepest satisfactions in life come from having intimate group ties, deep friendships, and a feeling of kinship with others.

Participation in individual and group activities should present situations where a person is encouraged to find his or her role in the group, to feel that he or she belongs. It is not difficult to plan recreation programs to meet this need, if leaders recognize it. As we might expect, recreation leaders have great influence on children and youths, and on their images of themselves. Camping, club work, social activities, athletics, debating, drama, and the like give strength to fellowship and close personal ties.

***New Experiences.*** The desire for excitement, adventure, and new experience is a basic wish for all people. In fact, it is considered a fundamental aspect of personality development. To express interests or do something different adds "spice" to life and a zest for living. Certainly it is a part of our culture pattern to seek thrills, variety, excitement, and adventure. One only has to look at the commercial exploitation of this need to see its place in American tradition.

The quest for new situations can be carried out through hiking, camping, mountain climbing, travel, and also in creative activities such as hobbies, painting and other kinds of art work, and music.

***Beauty and Harmony.*** The desire for beauty and harmony in our environment is a universal one. The beauty of stars and sunset; the tranquility of a wooded glen; the comfort of lakes, forests, and slow-moving streams; or the grandeur of breathtaking scenery provides an emotional experience of deep magnitude.

Opportunities should exist in recreation programs for one not only to appreciate beauty but also to create it. To camp, fish, take scenic trips, or travel is to experience beauty and the harmony of our national environment. But opportunities should also exist to create beauty. Work with paints, wood, clay, or cloth to create articles of beauty, or expression of beauty through music, dance, poetry, or drama, provides a means of self-expression and satisfies a deep-seated human need.

**Social and Environmental Factors.** There is a need to study the social, environmental, and governmental factors that affect recreation needs in developing programs. Knowledge of the people in a community, their distribution and density, ethnic and cultural backgrounds, age and sex composition, incomes, and educational levels provides reliable data for understanding the specific and general recreation wants of people as they distribute themselves throughout a city or metropolitan area. Moreover, governmental essentials and citizen participation are also factors that influence the success of the program. These factors can be summed up as follows.

***Population Distribution and Density.*** The growing density and distribution of population of an area point to the need for expanding recreation programs and facility development. What is the pattern of density? What kinds of dwelling units house families in various parts of the community? Are certain neighborhoods changing in character? Is population decreasing? All of these and other questions need to be answered in program planning. Density is related to the service radius of recreation facilities; changing neighborhood patterns affect facility development; and type of dwelling units is related to space requirements.

***Ethnic and Cultural Backgrounds.*** Many communities have great contrasts in the ethnic and cultural backgrounds of their people. This presents problems in providing recreation programs and facilities that meet the needs of groups that have deep-rooted traditions. Who are the minority groups? Where do they live? What kind of housing is available to them? Are the various ethnic groups isolated, or are they assimilated by other groups? Answers to these questions are needed in program planning.

***Age and Sex Composition.*** The percentage and distribution of people by age groups and sex assist planners in determining the type of programs that meet their needs. Fast-growing new communities have a high proportion of children and youths; older neighborhoods may have a high percentage of older adults. To identify people by age classifications, and know their activity and personality characteristics, is to gain invaluable information for program planning. Indeed, programs can only have meaning if they are planned to meet the needs of specific age groups.

In program planning, consideration was always given in previous years to sex differences. There was much discussion about the handicaps of endurance, strength, and speed on the part of women, and it was generally desirable to have a different program for boys and girls after the fourth grade. Research has indicated that the assumed handicaps of endurance, strength, and speed are not necessarily different between boys and girls, and therefore programs should be planned for both of them similarly. It may be desirable, of course, to have special programs for girls because they may simply want them, rather than because of sex differences. The very fact that girls are now engaging in many kinds of activities that were almost exclusively programmed for boys means that plans need to be developed for increasing these areas and facilities. Girls now play basketball, handball, racquetball, tennis, baseball, and the like. Certainly this will call for the need for additional facilities to take care of this expanded use.

In sum, physical, social, cultural, and mental activities lend themselves to co-recreational participation, and most of the previous social taboos no longer bind girls and women's participation in vigorous sports and other kinds of activities.

***Income and Educational Levels.*** Knowledge of the income and educational levels of people in a community assists program planners in determining the kinds of services that meet the needs of the various subcultures comprising the community. Surely, low-income families have kinds of program needs different from those of upper-middle-class families. Their recreation opportunities are different; their skill levels vary, and program objectives have a different focus.

***Governmental Essentials.*** Every community needs basic legislation or legal authorization to acquire, develop, and maintain recreation areas and to develop thereon a program with well-qualified recreation leadership and a dependable income source.

***Citizen Participation.*** Every effort should be made to secure citizen participation in program planning. Citizen participation is vital if worthwhile programs are to be attained. Actually, this is the only way of discovering the potential interests, skills, talents, and inner feelings of the people to be served.

**Survey the Existing Areas and Facilities.** As program needs are identified, it is important to survey all existing areas, buildings, and facilities to determine their adequacy and availability. Too often unused space, as well as unused buildings, are overlooked when they could be available if requests were made for their use. Old fire stations, libraries, auditoriums, coliseums, abandoned railroad right-of-ways, lodge buildings, old school structures, and the like could be added to the inventory of available space and buildings of the recreation and park organization. It is assumed, of course, that old and unused structures would be renovated or remodeled to meet fire and safety requirements.

**Principles of Program Planning.** The selection of activities is the most difficult phase of program planning. Program objectives, individual needs, and environmental characteristics as previously outlined must be evaluated and crystallized into program content that is adapted to the individual. This calls for the highest recreational statesmanship. Consideration should also be given to guiding principles that can act as criteria for program.[2] These are:

1. *Program planning should involve consideration of diversified interests and include a wide variety of offerings.*

If community recreation is to endeavor to provide for the varied interests of human beings, it must be prepared to offer a rich and diversified program. Such a program must also, however, have a concern for what happens to the individual through the participation.

[2]Principles used are taken by permisssion from Athletic Institute, *Recreation for Community Living* (Chicago: The Institute, 1952), pp. 140–45.

Program fields generally include such things as sports and games, arts, crafts, music and rhythmics, the dance, social activities, dramatics, nature, outing, and camping; hobbies and collecting; and service activities. It is to be noted that these program fields overlap and that there are many varieties and levels of participation. The concept of program must be broad enough to include contemplation and enjoyment of stars, sunsets, majestic scenery, great art, and music. These, too, are recreation—and recreation with great spiritual significance. Consideration should be given to community customs, nationality interests, traditional folkways, and community attitudes.

A process of selection both in the offerings of the program and on the part of the participants goes on continually and influences the type of program available at any particular time.

**2.** *Community planning should take into account all ages, all economic and racial groups, all creeds, and both sexes.*

Community recreation does not endeavor to meet all recreation needs, but only those that can be best provided and financed by society. It must not, however, exclude consideration of any group in the community because of social, racial, religious, or economic considerations. Because of limitations of finances, leadership, and facilities, it is generally necessary to work out priority schedules and give first consideration where need is greatest and where there is least opportunity to find other ways of providing desirable recreation experiences.

A part of the community recreation program is concerned with the stimulation of interests that individuals and groups continue to pursue on their own time with a minimum of assistance.

Program offerings should be provided indoors and outdoors on a year-round basis. They should include varied opportunities for participant and spectator for group activities and for the pursuit of individual interests.

**3.** *Program planning for community recreation agencies should be a joint venture whereby the community at large, program participants, the professional staff, and governing bodies assume joint responsibility for both planning and execution of activities.*

In a democratic society, it is assumed that the will of the people is the determining factor in what the recreation program shall include. The duly elected representatives of the people, members of boards, committee, and members of recreation councils all have a part in the determination of what is to be included. If community recreation is to maintain support and function most effectively, with all agencies complementing one another and recognizing that each has its own contribution to make, the machinery for group planning must be provided and kept functioning.

The sharing of program participants in the planning process, within the limits of their ability, is also important. The community recreation program is often a self-starter for many interest groups that later become self-sufficient in whole or in part. Hobby groups, little theater clubs, and music groups are illustrative of this type of organization. Oldsters want the opportunity for self-direction; teenagers need a planning participation role if their interest is to be maintained; and members of sports associations should make decisions relative to eligibility rules. Children's groups, especially in such situations as clubs and camping, should have opportunities to participate in the planning process. Planning by the participant is in itself one of the finest of recreation experiences and may be one of the best types of training for citizenship in a democracy.

**4.** *Program planning should be related to the physical, mental, social, and emotional characteristics of individuals at various age levels.*

Program needs arise out of the very nature of the human organism. The varied needs, interests, and appropriate activities of all age groups are [to be] considered. . . .

Individual and group activities should be of a type and character to satisfy the desire for activity; for adventure and new experiences; for fellowship; for self-expression; for the urge to find and enjoy beauty, to satisfy curiosity, to receive recognition in the group, and to create and to serve. To the extent that program offerings satisfy these and other desires, will they find participants and achieve those social and individual values expected of them.

**5.** *Program planning should provide an opportunity for participation at varying levels of proficiency.*

As a result of education, environment, physical and mental capacities, and other factors, individuals have a wide variety of skills, tastes, and knowledges. It is an obligation of program planners to provide opportunity for the individual at whatever level he or she desires or is able to participate. Extending this opportunity may require the teaching of performance skills or providing an opportunity to use already highly developed skills. Provision for different degrees of skill and for helping individuals become more proficient in the activity is essential, since there is high correlation between skill and satisfaction. Quality is important at every level of skill.

**6.** *In program planning, utilization should be made of standards developed by national agencies with such modifications as may be expedient to meet local conditions.*

The promulgation of program standards implies the education of the public to an understanding of them. Over a period of time many organizations and agencies have developed standards in the fields of activity included in the recreation program. Among them are the American National Red Cross, National Section on Women's Athletics, Amateur Athletic Union, United States Lawn Tennis Association, American Camping Association, and numerous others. Not only have these organizations developed standards, but they have also adopted codes which govern participation. These standards have been developed as the result of considerable experience and study and afford, in most cases, a sound basis for operation.

**7.** *Long-range planning for program is a prerequisite to planning for community organization, finance, leadership, areas, facilities, and legislation.*

It is a fundamental concept that the services to be rendered through a program should be the basis for determining the kinds of facilities and the type of leadership and finances to be needed over a period of years. Only as needs are known, can we plan intelligently for the other aspects of community recreation.

**8.** *Planning should enlist all those community resources that can provide variety and enrich the program.*

The extent to which program(s) may expand is not governed solely by operating budget and the abilities of paid leaders. Enlistment of community resources to provide for greater program volume and enrichment is desirable as a program policy as well as for public relations. In every community there exist numerous self-organized and directed special interest groups. They include music federations, theater groups, athletic organizations, and arts and science clubs—to mention a few. These organizations come into being and are held together because of special interests which the members pursue diligently. Assistance from these groups should be enlisted to the greatest extent possible. Not only are they invaluable in planning, but they represent an excellent source of volunteer leadership for program expansion. . . .[A] great number of civic, fraternal, and patriotic organizations found in each community . . . often undertake recreation service projects. Program planning should take cognizance of all these projects with a view to offering professional guidance and direction and also with a view to avoiding conflicts in activity.

Organized recreation agencies assist to the extent possible in helping to establish policies and practices in the conduct of these events that are consistent with sound recreation policy. Exploitation of participants, expensive rewards, and a "win at any cost" policy are undesirable factors that by planning and direction, professional leaders can help eliminate.

Program planning cannot adequately take place unless the planners keep vigilant and alert to the community resources that exist and continually develop. Added assistance in program planning, expansion in program content and volume, interpretation of basic recreation principles, and a united community effort in providing recreation service become dividends to the community.

**9.** *Continuous evaluation and measurement should be provided for in program planning.*

A continuous evaluation of program content and quality must go on. Recreation leaders, members of boards and committees, and participants should have a part in the evaluation process.

The following are some of the bases for evaluation.

(a) Do participants receive basic satisfactions from the recreation experience?

(b) Does it serve basic human needs??

(c) Is the response to the program sufficient to justify its continuance?

(d) Is the cost excessive in terms of the values received?

Accurate program records and valid methods of measuring and evaluating the results of the program are essential.

## CLASSIFICATION OF ACTIVITIES

One of the most comprehensive listings of recreation activities was made by a National Workshop on Recreation in 1954, under the sponsorship of the Athletic Institute. Its eleven major program classifications are presented here.

## I Arts and crafts

*Elementary*

Paper bag crafts
Puppets
Plaster carving
Carving—soap, wood
Drawing
Finger painting
Clay modeling
Copper foil
Vegetable printing
Stenciling
Basketry
Gimp craft
Finger weaving
Card weaving
Hooking
Spatter printing
Paper-mache
Candle making
Mobiles
Whittling
Wood stamping
Shell craft
Tie dyeing

*Advanced*

Leather Carving
Jewelry enameling
Painting
(Advanced levels of activities as listed)

*Intermediate*

Games—checkers, puzzles, bean bag, box hockey
Leather—modeling and tooling
Simple metal jewelry
Wood carving
Woodworking (hand)
Basic ceramics
Painting and sketching
Block printing
Dry paint etching
Hand and simple loom weaving
Tin can craft
Art metal craft
Plastics

## II Dance

*Folk Dance*

American Folk Dances
- Square dance
- Round or couple dance
- Longways dance
- Circle dance
- Solo dance

*Creative Rhythms for Children*

Free rhythms
Identification rhythms
Dramatic rhythms
Rhythms games
- Singing games
- Simple folk dances

Folk Dance of Other Lands
  Ethnic groups
Social Dance
Dance Mixers
Ballet

Modern Dance
  Conditioning and free exercise
  Art of movement
  Concert dance
Tap, clog, character dance

## III Drama

Blackouts
Ceremonials
Charades
Children's Theater
Choral Speech
Community Theater
Creative Drama
Demonstrations
Dramatization
Festivals
Formal drama
Grand opera
Imagination plays
Impersonations
Light opera
Marionettes
Monodrama
Monologue

Musical Comedy
Observances
Operettas
Pageants
Pantomime
Peep Box
Plays
Puppetry
Script-in-hand
Shadow plays
Shows
Skits
Story reading
Storytelling
Stunts
Symphonic drama
Tableaux
Theater in the round

## IV Games—Sports—Athletics

***Informal Games and Activities***

Ball games—Dodge-ball, Circle Pass Ball, etc.
Circle Games
Goal Games
Hitting or Striking Games
Net Games
Running Games
Tag Games
Stunts
Relays
Apparatus Play

*Individual and Dual Sports*

Individual—archery, bicycling, boating, bowling, fishing, etc.
Dual—badminton, fencing, horseshoes, table tennis, etc.
Team Sports
Combative Sports
Women's and Girls' Sports
Co-recreation Sports

## V Hobbies

*Collecting*

Stamp[s]
Antiques
Firearms
Models
Books
China
Paintings
Others

*Educational*

Reading
Ornithology
Astronomy
Horticulture
Entomology
Zoology
Sciences
Others

*Creating*

Writing
Woodworking
Sculpture
Painting
Photography
Cooking
Gardening
Others

*Performing* (Use of body skills)

Hiking
Swimming
Hunting
Fishing
Magic
Camping
Bowling
Chess
Canoeing
Tennis
Others

## VI Music

*Singing*

Informal singing
Community sings
Choruses
Quartets
Ensembles
Glee clubs
Solos

*Playing*

Rhythm instruments
Melody instruments
Harmony instruments
Fretted instruments
Bands
Orchestras
Chamber music groups

*Listening*
- Home music
- Records
- Radio
- Television
- Concerts

*Creating*
- Song making
- Other music making

*Rhythmic Movement*
- Purely rhythmic
- Simple interpret[at]ion
- Singing games
- Folk dances
- Play party games

*Combined Activities*
- Folk dancing
- Musical charades
- Shadow plays
- Pageants
- Talent and variety shows
- Others

## VII Outdoor Recreation

*Nature Activities*
- Scenery and observation
- Collecting
- Nature experiments
- Gathering wild food
- Nature trails
- Nature talks and exhibits
- Nature games

*Outdoor Living*
- Informal fun
- Basic campcraft skills
- Camping

*Outdoor Sports*
- Winter sports
- Water sports
- Riding and packing
- Hunting and fishing
- Others

*Outdoor Arts and Crafts*
- Projects (Use of natural materials)
  - Wood
  - Bark
  - Fibre
  - Minerals

*Trips and Outings*
- Tours and travel
- Informal outings

*Plant Culture*
- Gardening
- Forestry
- Landscaping
- Animal care
- Others

## Miscellaneous

- Museums
- Zoos
- Arboretums
- Parks and forest preserves
- Gardens

## VIII Reading, Writing, and Speaking

*Reading*
- Great Books program
- Book review clubs
- Reading classes
- New books club
- Reading the classics
- Mystery story club

*Writing*
- Business and social writing club
- Writing for fun
- Technical writing
- Creative writing
- Newspaper writing
- Contesting

## IX Social Recreation

*Activities*
- Games
- Informal drama
- Music
- Dance
- Co-recreation sports
- Relays
- Arts and crafts
- Novelty events

*Events*
- Parties
- Banquets
- Informal drama
- Outings
- Dances
- Snow and ice sports
- Progressive games program
- Family recreation
- Teas and coffee hours

## X Special Events

*Exhibits of Objects*
- Hobby show
- Science fair
- Arts and crafts exhibit
- Flower show
- Doll show

*Performances*
- Circus
- Talent show
- Concert
- Dance exhibit

*Mass Activities*
- Folk dance festival
- Winter carnival
- Celebrations

*Social Occasions*
- Veterans' Day parade
- Cherry blossom festival
- Others

### Skill Contests

Marble tournament
Jacks tournament
Fishing derby
Bait and fly casting contest
Men's baking contest

### XI Voluntary Service

*Administration*
- Board members
- Committee workers
- Fund raisers
- Facility planning
- Decorations

*Program*
- Club leaders
- Activity leaders
- Camp counselors
- Special event assistants

### Service

Clerical
Libraries
Legal
Maintenance
Public relations[3]

## PROGRAM ESSENTIALS

Earlier in this chapter reference was made to the three general categories of recreation programs: (1) use of areas and facilities for recreational enjoyment, (2) use of leadership resources for organized individual or group activities, and (3) assistance services to individuals or self-determining groups. These will now be discussed in more detail.

### Use of Areas and Facilities

Too often, reference is made only to organized recreation programs and services when actually, for many, use of recreation lands or facilities is all that is needed. The whole concept of unorganized programs that provide essential services for people without supervised leadership needs exploration. People are mobile in today's society and want to go places and do things on their own. To open new program horizons so that individuals and families can enjoy the recreation resources of the community and region represents an aspect of program planning that has many times been overlooked. Beaches; swimming pools; golf courses; lakes for fishing, boating, and water sports; camp grounds; scenic drives; bicycles and nature trails; winter sports areas; fly-casting pools; libraries; museums; zoological gardens; picnic areas; and the like are examples of facilities of this kind.

It cannot be too strongly emphasized that in development of recreation resources for people, consideration must be given to their rising standard of liv-

[3]Athletic Institute, *The Recreation Program* (Chicago: The Institute, 1954).

ing, higher educational achievements, improved physical health, and ever increasing mobility. The types of recreation programs and facilities of yesterday are not good enough for the people of today. New vision and imagination are needed.

Abundant outdoor recreation resources are available inside or outside the community if competent administrative direction and skillful leadership are provided in planning them. Such programs provide real opportunities for service to all vigorous age groups, and special opportunity exists here for the ill and the handicapped. Camping, nature programs, exploration, collecting, gathering of wild foods, horseback riding, trips on land and water, fishing, hunting, archery, and visits to scenic, historical, or scientific sites open a wide field of program possibilities. Furthermore, enjoyment of outdoor recreational pursuits requires skill in planning special facilities, areas, and structures. And what of the indoor possibilities for expanding unsupervised recreation services? An entire new horizon is open to planning in this direction. It is too late to worry over jurisdictional battles about who does what in providing these services. Whether they are educational or recreational, active or passive programs under the supervision of park departments, schools, public recreation agencies, youth organizations, cultural commissions, or outdoor recreation authorities are still recreation that people engage in for the satisfaction it brings them. Close study of the individual, the family, and the group living patterns of dynamic America discloses unlimited program possibilities. Beauty, high standards, and a focus on needs compatible with the goals of education and democracy should be the guiding stars for the future.

## Leadership and Organized Programs

People want opportunity to participate in activities or programs that are "organized," those that by their nature require leadership and equipment and could not be implemented without them. Someone must provide the leadership and the facilities for musical programs, shows, team play, dances, nature lore, dramatics, social activities, crafts, and the like. In the main, these activities also require a variety of areas, buildings, or facilities.

Mention was just made of the wide scope of activities and services possible under public recreation and park auspices. Although many of these activities are self-directed or are provided at all times to individuals and groups without supervision or leadership, a large number of them do require organization, direction, or guidance. In other words, they need to be organized.

All organized recreation programs are conducted either indoors or outdoors. Expanded indoor programs are usually related to the operation of playgrounds, playfields, or recreation facilities. Both indoor and outdoor programs are conducted throughout the year. While organized recreation programs are sometimes classified by seasons and special events, the activities are similar except for the different emphasis given to seasonal activities and the greater responsibility placed on programming for children and youths during the summer months. Generally, climate, customs, school vacations, and the traditional seasonal play patterns of people influence people's choice of activities.

Organized recreation programs will be discussed here under the following classifications: (1) indoor activities, (2) outdoor activities involving playfields, special areas, and recreation facilities, (3) citywide special events, and (4) playground programs.

**Indoor Activities.** Indoor programs involve the use of school buildings, recreation centers, or specialized structures where activities can be provided throughout the year. Programs of a social, cultural, and educational nature require meeting rooms, classrooms, auditoriums, game rooms, or social rooms, and special activity space for art, music, sports, drama, and the like. Also gymnasiums and swimming pools are often an integral part of the community or the recreation center building.

The number and kinds of indoor recreation facilities indicate the scope of program services that can be organized and provided within the community. For example, the general recreation building may be a multipurpose or all-inclusive structure in which a multitude of activities can be organized. The public schools may also be used extensively in this manner. In addition to these centers, indoor facilities that are more specialized in use may be available in the community. Separate auditoriums, gymnasiums, skating rinks, indoor swimming pools, museums, theaters, art studios, clubhouses, teen centers, dance halls, community houses, armories, garden centers, camp lodges, and the like fall within this category. Most common, of course, in program planning is the use of the all-inclusive recreation building or schools. Here the facilities to provide a varied recreation program are found. Only the resourcefulness of the leadership and the diversity of the facilities limit the kinds of activities. Although programs are focused primarily on the needs of youths and adults, children are not excluded. Actually, many programs are planned for families. However, activities in the recreation center tend to be more specialized, since rooms and facilities are used for specific purposes.

Activities within the recreation center can generally be classified into two categories: (1) sports and games and (2) social, cultural, and educational activities. Many of the activities are provided through special classes that have limited enrollment. Activities are usually listed as to time and place, and the leadership may consist of part-time workers. People come to participate in specific activities; there is little of the "drop in" feeling that characterizes informal programming. However, many make use of facilities that do not require lead-

**Table 11.1 Fall and Winter Program Mt. Scott Community Center Portland, Oregon**

| *Time* | *Monday* | *Tuesday* | *Wednesday* | *Thursday* | *Friday* |
|---|---|---|---|---|---|
| 8:30–12:00 | Eye openers Basketball—open | Mixed volleyball | Eye openers Basketball—open | League basketball games | Open basketball |
| 9:00–12:00 | WWI Veterans Club | | Tole painting | Garden club | Overeaters anonymous |
| 10:00–11:30 | Kermit & Co. (Preschool for 3 years) | Preschool B 4–5 | | Preschool B 4–5 years | Open basketball |
| 12:00– 1:00 | Women's slimnastics | Preschool woodshop | Women's slimnastics | Preschool trampoline A | Open basketball |
| 12:00– 9:00 | Weight-lifting | Weight-lifting | Weight-lifting | Weight-lifting | Weight-lifting |
| 12:00– 3:30 | | Golden Age Dance Club | | | |
| 1:00– 2:30 | Preschool A 4–5 years | Art & movement 3–4 years | Preschool A 4–5 years | | Open basketball |
| 2:00– 3:00 | Preschool tumbling 4–5 years | Fall tennis | Preschool trampoline B | Staff meeting | Fall tennis |
| 3:00– 4:00 | Preschool ballet A 3–4–5 years<br>Open skate grades 1–3<br>Open gym grades 1–3 | Kids cooking grades 1–5<br>Sports fitness grades 1–3<br>Moms' & tots' skating<br>Open gym grades 1–3 | Floor hockey<br>Open skate grades 1–3<br>Open gym grades 1–3<br>Makeup & monsters grades 1–5 (October only) | Holiday crafts grades 1–5<br>Sports fitness grades 1–3<br>Open gym grades 1–3 | Preschool ballet B 3–4–5 years<br>Open skate grades 1–3<br>Friday surprise-gym grades 1–3 |
| 3:00– 5:00 | *Holiday Workshops* - one day only - adult<br>Hallowe'en costumes, bread dough ornaments, Santa candy dishes, tree skirts, crocheted bells & snowflakes, gingerbread houses, barnyard, pine needle baskets, straw brooms | | | | |

| | | | | | |
|---|---|---|---|---|---|
| 4:00– 5:00 | Open gym grades 4–8 | Sports fitness grades 4–8 | Basketball Goldenball team practice | Sports fitness grades 4–8 | Friday surprise grades 4–8 |
| | Soccer grades 4–8 | Beginning volleyball grades 5–8 | Soccer grades 4–8 | Open gym grades 4–8 | Woodshop grades 4–8 |
| | Beginning tumbling A | Beginning boxing grades 1–6 | Beginning tumbling B | Advanced boxing grades 1–6 | Advanced boxing grades 1–6 |
| | Advanced boxing grades 1–6 | Beginning ballet grades 1–6 | Beginning boxing grades 1–6 | Beginning ballet grades 1–6 | |
| | | | Disco dance grades 4–8 | | |
| 4:45– 6:00 | Open skate grades 4–8 | Roller hockey grades 4–8 | Open skate grades 4–8 | Open skate grades 4–8 | Open skate grades 4–8 |
| 5:00– 6:00 | Intermediate gymnastics grades 3–8 | Advanced gymnastics grades 3–8 | Intermediate gymnastics grades 3–8 | Advanced gymnastics grades 3–8 | |
| | | Intermediate ballet grades 1–6 | | Intermediate ballet grades 1–6 | |
| 5:00– 9:00 | Boxing—adult and high school | Boxing—adult and high school | Boxing—adult and high school | Boxing—adult and high school | Boxing—adult and high school |
| | Open gym Basketball | | Open gym Basketball | Open gym Basketball | Open gym Basketball |
| 6:00– 9:00 | | Tole painting | Open trampoline grade 5–adult | | |
| 6:00– 7:00 | Beginning fencing grade 7–adult | Women's slimnastics | Beginning fencing grade 7–adult | Women's slimnastics | Ki-Aikido grades 1–8 |
| | Business persons' conditioning | | Business persons' conditioning | | |
| 6:30– 7:30 | Gymnastic team grade 3–high school | Ladies' weight training | Gymnastic team grade 3–high school | Ladies' weight training | |

**Table 11.1 Fall and Winter Program Mt. Scott Community Center Portland, Oregon (Continued)**

| *Time* | *Monday* | *Tuesday* | *Wednesday* | *Thursday* | *Friday* |
|---|---|---|---|---|---|
| 6:30– 9:00 | Adult skating | Archery club | Skate lessons Beg.-Int.-Adv. | Family nite skating | Open skate grades 7–12 |
| 7:00– 8:00 | Intermediate fencing grade 7–adult | Karate grades 7–8 | Ki-Aikido high school–adult | Tap dance high school–adult | |
| 8:00–10:00 | Karate high school–adult<br>Camera club | Karate high school–adult<br>Mixed volleyball recreational and power | | Chess club | Trailer club |
| 8:00– 9:00 | | | | Dance disco | |
| 9:00–10:00 | | | | Partner disco | |

Term Special Events: *Hallowe'en Haunted House*, Oct. 31, 6–9 P.M., all ages, free
*AAU Boxing Tournament*, Nov. 10 & Dec. 15, all ages, free
*Volleyball Tournament*, Nov. 3, all ages, free
*Obukan State Judo Tournament*, Nov. 17, all ages
*Cerebral Palsy Dance*, Nov. 7, High School, free
*Christmas Open House*, Dec. 12, all ages, free

ership. An example of a center program is that of the Mt. Scott Recreation Center of Portland, Oregon, shown in Table 11-1.

**Outdoor Activities.** Many organized programs are conducted out of doors and require special areas or facilities. Most common are organized active sports or other athletics. Also baseball, basketball, tennis, golf, archery, track, swimming, and the like are part of most outdoor recreation programs. Team sports are often organized on a citywide basis, while individual sports are organized on a more informal basis, although contests and tournaments are arranged in many instances.

On the other hand, mention should be made of the many outdoor recreation opportunities other than those in the traditional categories of athletics and team sports. The following list is suggestive of the scope of such activities.

| | | |
|---|---|---|
| Fishing | Camping | Tennis |
| Hunting | Mountain climbing | Skiing |
| Hiking | Boating | Skin diving |
| Swimming | Golf | Sailing |
| Scuba diving | Skating | Jogging |

While many individuals will participate in these activities without supervision or leadership, it is not uncommon to see these recreation opportunities sponsored by clubs or special interest groups. Indeed, stimulation of such kinds of organization will often provide the spark to awaken new interest in them.

**Citywide Special Events.** Special events may be conducted either indoors or outdoors, but they are of such significance that attention should be focused on them. They take many forms. Art festivals, talent shows, holiday celebrations, and many other events are included in this category. Such programs depart from the daily routine and are highlights of program planning. It is not unusual to see a special events the culminating feature of many weeks or months of daily planning and activity participation at a playground, community center, or playfield. Not only do such events sustain interest, discover talent, and develop leadership, but they also provide a means for fostering community unity through integration of effort on a specific project. Furthermore, citywide special events have appeal throughout the community and provide "spice" in program planning. The eyes of the community focus on the recreation agency on these occasions.

While there are many kinds of special events, the interests and customs of the public determine its choices. Community pride in environment, products produced, or historical lore provides popular themes for many such events.

Roundups and pioneer days in the West, cherry festivals in Michigan, tulip festivals in Washington State, Indian festivals in Arizona, and many other events are examples of this category. However, possibilities for special events are boundless. Additional suggestions are:

Flower show
Science fair
Arts and crafts exhibit
Fourth of July celebration
Halloween party or parade
Easter-egg hunt
Winter carnival
Rock concert
Community picnic
Boat regatta
Christmas pageant
Hobby show
Fishing derby
Folk dance festival
Circus
Country music festival

Playgrounds or community centers will often have special programs such as doll shows, exhibits, parties, costume parades, and the like. But in most instances they are limited to the neighborhood, and hence do not fall under the "citywide" classification. Nevertheless, such events do meet neighborhood needs and add much to local programs.

**Playground Programs.** Programming for playgrounds, like that for community or recreation centers, must involve consideration of the factors of interests, the times when activities should be scheduled, the number of leaders, the availability of facilities, the sizes of the groups to be served, the skill of the participants, and the age and sex differences. However, it should be noted that playgrounds serve children of elementary school age at the neighborhood level—although some playgrounds are integral parts of larger areas such as community parks and encompass programs for older boys and girls. Regardless of this, playground programs are focused on the interests of children and consist of daily, weekly, seasonal, and citywide events and activities.

Summer playgrounds are of major concern to all recreation departments—large and small. Many communities do not have playgrounds except during the summer months, and in these instances the playgrounds are the major feature of the community recreation service. Actually, this is one of the most important phases of program planning for any or all public recreation organizations.

Child interest determines playground program content. Hence a study of the needs, interests, and play habits of children is important in organizing activities and developing daily, weekly, and seasonal programs.

The daily playground schedule gives the events or activities that are provided for a particular day. In the making out of this schedule the following guides should be considered.

1. Use the first period for getting the area ready for the daily schedule of events, inspection of apparatus, posting of announcements, flag raising, and other routine matters.
2. Use the midmorning hours for active games and sports for the age groups that predominate.
3. Schedule quiet activities or games for the period prior to the lunch hour.
4. Leave the lunch hour free from any scheduled activity unless it be a picnic, a club luncheon, or a wiener roast.
5. Have the first period after lunch devoted to quiet activities. Apparatus play, music, storytelling, club meetings, nature lore, dramatics, and the like are good for this time of the day.
6. Use the middle of the afternoon for special features, tournaments, athletic contests, and team games. This is the time when attendance is usually the largest.
7. Plan the late afternoon hours around individualistic activities, club meetings, arts and crafts, quiet games, dramatics, and the like.
8. Use the early evening hours for twilight league games, family nights, informal play, or special events, if the playground area is adequate to handle such activities.
9. Develop playground special events to be held one day per week to highlight the program.

Planning of the daily schedule of activities calls for good judgment and the ability to improvise. The schedule should be flexible and subject to change when necessary. Furthermore, activities or classes should be scheduled when children or members can be present. No hard and fast rules that standardize procedures or programs should exist. Also, it should be noted that a number of classes or activities may be conducted at the same time if space and leadership are available. In such instances programs may even be developed according to age divisions (children under 8, children 8–11, children over 11).

The weekly playground program provides for a diversity of activities. Instead of having the same schedule of events each day of the week, a richer program is possible by having a variety of activities scheduled for one, two, or three times per week. Dramatics or nature activities may be held once a week; music, twice; folk dancing, three times; team sports, each day of the week. Such diversity adds much to the program and provides an opportunity to use the special talents of the leadership staff. In addition, it provides for many more activities than otherwise would be possible. An example of a weekly program is shown in Table 11-2.

The seasonal playground program highlights the activities for a particular season and provides diversity, interest, balance, and progression of activities.

**Table 11.2 Weekly Forecast Mt. Scott Playground Portland, Oregon**

| *Time* | *Monday* | *Tuesday* | *Wednesday* | *Thursday* | *Friday* |
|---|---|---|---|---|---|
| 8:30– 9:30 | Lap swim, adults | Lap swim, adults | Lap swim, adults | Lap swim, adults | Lap swim, adults |
| 8:30A.M.– 9:30P.M | Open basketball | Volleyball | Open basketball | Open basketball | Open basketball |
| 10:00–11:00 | Mixed tennis | | Mixed tennis | Incredible edibles ages 4–6 | |
| 10:00– 1:00 | | | Tole painting | | |
| 10:00–12:00 | Swim lessons | Swim lessons | Swim lessons | Swim lessons | Swim lessons |
| 10:00– 3:00 | Boxing, grades 1–6 | Boxing, grades 1–6 | Boxing, grades 1–6 | Boxing, grades 1–6 | Boxing, grades 1–6 |
| 10:00– 3:00 | Open skating | | Open skating | Open skating | Open skating |
| 10:30–12:30 | Weight lifting | Weight lifting | Weight lifting | Weight lifting | Weight lifting |
| 10:30–12:30 | | Parenting group | | Parenting group | |
| 10:30–12:00 | | Moms' & tots' skating | | | |
| 10:30–11:30 | Crazy crafts grades 1–3 | Gimp craft | Crazy crafts grades 4–8 | Gimp craft | |
| 11:00–12:00 | | Mixed tennis | | Mixed tennis | |
| 12:00– 1:00 | Federal lunches | Federal lunches | Federal lunches | Federal lunches | Federal lunches |
| 12:00– 2:00 | | | Women's Tennis League | | Camp cooking grades 1–6 |
| 12:45– 1:45 | Women's exercise | | Women's exercise | | |
| 12:30– 8:00 | Open swim | Open swim | Open swim | Open swim | Open swim |
| 12:30– 3:30 | | | | Trips on Tri-Met (bus trips) grades 1–4 | |

| | | | | | |
|---|---|---|---|---|---|
| 1:00– 2:00 | Art & movement ages 3–4 | Ladies' tennis<br><br>Golf lessons | | Disco dance grades 7–9 | Ladies' tennis |
| 2:00– 3:00 | Art & movement ages 5–6 | Tot tumbling ages 4–5 | Fun hour games | Fun hour games | |
| 2:00– 5:00 | | | | One-day craft workshops<br>Barnyard, crochet, fabric painting, etc. | Friday surprise special event (pet shows, etc.) |
| 3:00– 4:00 | Softball | Preschool ballet A ages 4–5 | Softball | | |
| 4:00– 5:00 | Advanced arts and crafts | Preschool ballet B ages 4–5 | Floor hockey<br><br>Arts and crafts | Trampoline grades 1–8 | |
| 4:00– 9:00 | Adult boxing<br>Weight lifting | Adult boxing<br>Weight lifting | Adult boxing<br>Weight lifting | Adult boxing<br>Weight lifting | Adult boxing<br>Weight lifting |
| 5:00– 6:00 | Dinner<br>Roller hockey | Dinner | Dinner | Dinner | Dinner |
| 5:30– 6:30 | Business persons' conditioning | | Business persons' conditioning | | |
| 6:00– 7:00 | Beginning fencing | Mixed tennis | | Mixed tennis | |
| 6:30– 7:30 | | | | Dance disco | |
| 6:30– 9:30 | | Adult skating | Golf lessons | Family nite skating | |
| 7:00– 8:00 | Intermediate fencing | | | | |
| 7:45– 8:45 | | | | Beginning disco | |
| 8:00–10:00 | Karate | Karate | Karate | | |
| 9:00–10:00 | | | | Partner disco | |

While programs are made for each season or combination of seasons, the summer season receives major attention. And the techniques or methods used in structuring a summer playground program can be used for any season.

It is common practice for many recreation systems to schedule activities around a weekly theme such as "music week," "circus week," "nature week," or "hobby week." In some instances all playground activities are centered around one basic theme or are focused toward one or more special events or even a combination of events or themes. However, any method used to break monotony, drabness, or narrowness of program and to provide for rich, varied, creative, and challenging experiences throughout the season is a worthwhile goal.

Planning of playground programs for the other seasons is much the same as planning for the summer, except that activity scheduling, in most instances, is limited to after-school hours, weekends, or holidays. The use of playground shelters or buildings is highly desirable when inclement weather forces the program indoors. Because of the limited playtime after school, the program is devoted largely to active games and sports. However, there are some notable exceptions to this pattern in cities that have excellent indoor facilities and buildings and make use of outdoor lighting. In these instances full-time operation is planned for from 3:00 P.M. until the closing hour of 9:00 P.M., or later, or the program may be scheduled for morning, afternoon, and evening hours. The indoor facilities supplement the outdoor ones, and the program is varied and challenging.

With the expanding use of schools and the increase in the number of indoor playground buildings, varied and richer programs are being offered for the fall, winter, and spring seasons. New developments in outdoor lighting, building construction, and facility design are opening new possibilities for future playground operation.

Some recreation and park agencies conduct summer day camp programs instead of the playground programs. In actuality, however, these programs are very similar, and the planning and operation at the day camp program is carried out in the same manner as those previously discussed.

## Program for Self-Determining Groups

An area of recreation programming that should not be overlooked is that of providing assistance to individuals, clubs, groups, organizations, industries, teams, committees, and self-interest associations that want the satisfactions of recreation but wish to maintain their identity as a group. Such units may be identified as a "teen-age" club, a boy scout troop, a business people's organization, a church club, a garden club, a local industry, an athletic or sports team, a civil club, a hobby group, and the like. Much has been done in this respect with regard to sports organizations or teams; little has been done in widening opportunities for other interest groups. Each community has literally hundreds of clubs, organizations, and self-interest groups or organizations. To provide the impetus in widening the recreation opportunities for them is a challenge to the recreation agency. Furthermore, such involvement provides a medium for securing excellent volunteer leaders and enlargement of participation.

## CREATIVE ELEMENTS IN RECREATION PROGRAMMING[4]

The city of San Mateo Department of Parks and Recreation developed an innovative and imaginative plan for classifying their types and kinds of recreation services and the priority that they should be given because of the large number of recreation opportunities offered by the department and the seasonal characteristics of some of the activities.

### Program Determinants

Recognition was given to the premise that certain program determinants are necessary for a department to function in any fashion. Because of the large number of recreational opportunities offered and the seasonal characteristics of some of the activities, it was imperative that a system be devised to determine where and when moneys are to be expended for these efforts.

**Basic Fundamentals.** Primary to our method of operation is the recognition that certain fundamentals—facilities, staff, and equipment—are necessary to initiate any fundamental recreation program. These would include items such as communications, business service and clerical needs, buildings, pools, fields and playgrounds, as well as basic equipment such as typewriters, playground apparatus, kilns and sporting equipment, and other overhead items.

**Basic Level of Service.** Besides these fundamental elements, the next item for expenditures by the department is the provision for a "basic level of service." This refers to those services that are totally funded from city revenues. Examples would include a drop-in basketball program where a facility and supervision are provided; perhaps a Halloween costume parade; or merely leadership for a senior citizen meeting.

**Core Program.** As we expand from basic levels of service, the next step of activity would be the core program. These are the activities that are an integral part

[4]This section on creative elements was prepared and used with the permission of Willard Shumard, Director of Parks and Recreation, San Mateo, California, September 4, 1979.

of a sound recreation program and would include items that may be partially or fully sustained but are essential for citizen recreation. Examples of this would be recreational and instructional swimming, children's little theater, folk dance, day camping, oil painting, and numbers of others that give the city a broad base of leisure-time interests.

**Optional Activities.** Beyond core programs are the optional activities. These are partially or fully sustained by revenue producing activities that, while very important, are available only after all other needs are met. Examples of this would be adult swimming competition, ski trips, gardening, and optional use of facilities.

Within the framework just described there must be a determination within each level of the priority of each activity relating one to another. For example, is boating safety more important and does it have a higher priority than a track meet. The priority system must also relate one group of activity to another so that internal priorities may be determined, for instance, establishing the relationship between ceramic classes and a teenage dance. This concept is illustrated schematically in Figure 11-1.

## Service Categories

Informational input used to determine into which level specific services are categorized include: goals and objectives, philosophy of service responsibility, population and demographic data, citizen surveys, requests and opinions, numbers served and cost per participant.

Those services categorized to be included in the basic level of service completely subsidized by the department are:

1. To provide for the opening and operation of municipal facilities at minimum levels including recreation centers; swimming pools; a fine arts studio; and utilization of athletic areas, beaches, playgrounds; and other outdoor facilities. This provides basic information and referral services, a limited number of flexible, informal, self-determined (drop-in) activities, and assistance to permit and self-sustaining groups. Levels of hours of operation will be determined on a year-to-year basis.
2. To provide space for basic community service—that is, charitable and educational programs, including governmental and quasi-governmental meetings and functions such as councils, commissions, and voting. Incorporated in this classification are agencies that provide the department reciprocal services and facilities such as the three public school districts and to a lesser degree private schools and agencies.

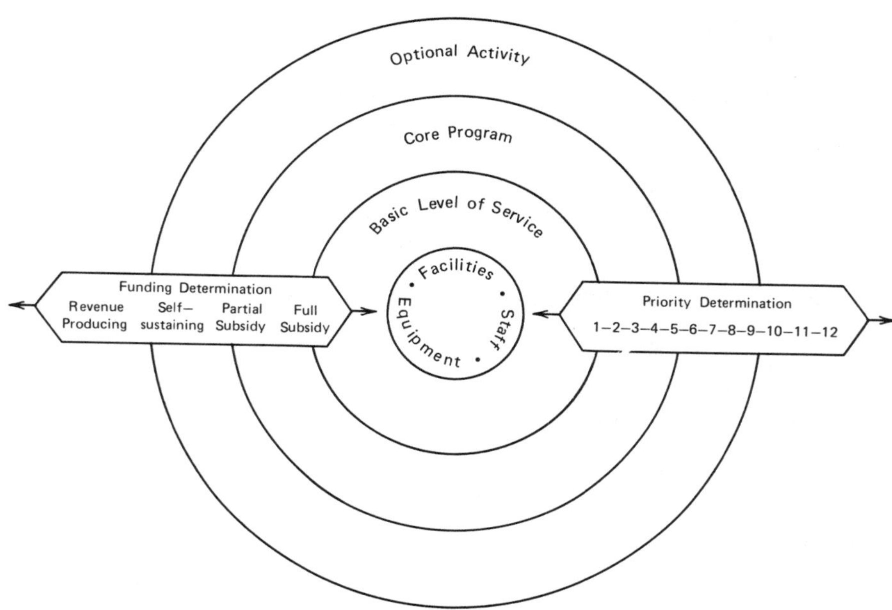


**Figure 11-1.** Service concepts.

3. To provide for elementary school age youth, junior and senior teens, and senior citizens, predetermined levels of
   (a.) Supervised but flexible, self-determined activities such as active and passive games, gym play, recreation swim, and social opportunities.
   (b.) Formalized social and interest opportunities such as clubs.
   (c.) Structured and supervised activities such as Saturday specials, open houses, elementary tournaments, holiday and seasonal special activities, fundamental individual and team sports and exposure to cultural activities, special events, trips and workshops.
   (d.) Special opportunities in (a)–(c) for minorities, including the physically handicapped and disadvantaged youth.
4. To provide for a year-round basic playground program conducted on school and municipal facilities in the summer and on municipal facilities adjacent to recreation centers in the fall, winter, and spring.
5. To provide for community service and training opportunities for adults, senior citizens, and teens through a volunteer program. In addition to providing program participation, this program offers the department a significant resource for low-cost leadership.

6. To provide art exhibits, theater productions, dance programs, and special performing groups that provide expression for participants and entertainment, information and a goodwill contribution to civic groups and agencies as well as to the public at large.
7. To provide for citizens' exposure to basic physical fitness, water safety, basic camping skills, and survival programs.
8. To provide professional and technical guidance to city government, other community agencies, schools and the community at large.

Those services determined to be in the core program category of partially subsidized level of service are those expanded, enlarged and more specialized versions of services in the core program such as intermediate camping, additional recreation swim opportunities, adventure clubs, slim-trim, cooking classes, instructional swimming, volleyball, some permit groups, some preschool programs, the teen musical, music camp, special activities, trips and excursions, and special productions.

Activities determined to be in the optional or self-sustaining category are highly specialized activities such as diving, judo, preschool programs, permit groups of the lowest priority, and most of the highly specialized cultural arts' programs for adults. There must be an overriding public need without the private enterprise available to meet the demand.

The final criteria consideration must be a determination of the level of expenditures by the department. Each activity, both within the major divisions and within the department, must be evaluated in view of its fiscal expectation. Is it paid for completely by budget? Is it partially sustained? Is it fully subsidized? Or does it produce revenue?

### Criteria to be Used for Categorization

The criteria for determining what should be placed in basic categories of service, partially subsidized, self-sustaining, and revenue producing are:

1. Basic levels of service are those budgeted activities, facilities and services that are a primary recreational need of the community. These are items that are very fundamental to broad, general, leisure interests of the citizens.
2. Partially subsidized activities are those supplemented by the budget and are essential to a core program but are not necessarily an expectation of all citizens. These are still in the fundamental range interest but have some specialization involved.
3. Self-sustaining activities may or may not be a part of the core program but are of such a nature as to require a higher degree of

specialization, including skilled leadership or particular types of equipment or facilities. Some of these activities, however, meet a current community demand.

4. Revenue producing criteria are those items that will provide a source of funds beyond the cost of the activity. Usually these are very specialized areas of interest that require expertise beyond that readily available within the department.

## Program Analysis

To measure the success or failure of the recreation dollar, one must find a method to weigh properly the many factors that affect expenditures. Some factors include the number of participants; the number of participants' hours involved in the activity; the number of spectator recreation hours, in the case of a spectator-oriented event; as well as the number of hours for professional personnel to plan and supervise an activity.

The National Recreation and Park Association and other technical resources have found that the criteria measuring the character building and preventative aspects of recreation is not readily identifiable. We continue to use the subjective judgment of the department staff to determine these outcomes.

## Additional Elements in Recreation Programming

Mention was made of the three general categories of recreation programs. Additional special features should not be overlooked to take care of changing needs as we project into the 1980s and 1990s. Most programs still follow traditional patterns of the past, but more and more changes are taking place that call for innovative program planning. For example, more and more playground programs are being replaced by specific classes and activities and the use of special themes, day camps, and other organized activities.

It is also important that we give consideration to the special needs of ethnic minorities. More and more of our cities are made up of such minorities, and representatives of these ethnic groups should have a place in program planning. It is not unusual, of course, to see these ethnic minorities reflect interest and pride in their heritage, which is reflected in their programs.

New developments are also taking place in establishing programs with private industry, expansion of services to the handicapped, special programs for youth and young adults, programs for senior citizens, and programs that have just recently developed, which make use of special equipment or facilities. For example, some cities have developed special facilities to take care of the growing interest in skateboards, trampolines, artificial fish ponds, roller skating, and bicycling. Also, it is not unusual to see a tremendous expansion in the entire area of cultural programs that include music, drama, dance, fine arts, and the like.

Programs are also being developed around the social needs of people, and more should be done in this area. Social centers are badly needed in our small and large communities to provide opportunity for people to become better acquainted with one another and to develop social relationships of long standing.

The energy crisis in America and the possibility of gasoline rationing with increased cost of transportation, will have its affect on mobility of people and their search for leisure activity. A rebirth of family and neighborhood recreation programming is growing, and the development of organized programs that involve more use of county and regional parks is facing us now and in the future.

### Questions for Discussion

1. What are the general categories of service provided by a community recreation department?
2. What is meant by "program"? How is it related to administration?
3. What is meant by "program lag"? Do all departments have such a lag?
4. Why are organizational objectives important to program planning?
5. Explain the steps in program building.
6. Why should a professional recreation leader know the needs of individuals by age periods? What are they?
7. Explain the basic wishes and needs of individuals. How are they related to program planning?
8. How do social and environmental factors influence program? What are they?
9. What is meant by "principles of program planning"? Give five such principles.
10. How would you classify recreation activities into basic groups?
11. How would you present priorities for your recreation services to the public?

## SELECTED REFERENCES

Ball, Edith L., and Cipriano, Robert E., *Leisure Services Preparation*. Prentice-Hall, Englewood Cliffs, N.J., 1978, pp. 29–95.

Butler, George D., *Introduction to Community Recreation*. McGraw-Hill, New York, 1976, p. 231.

Hjelte, George, and Shivers, Jay S., *Public Administration of Recreation Services*. Lea and Febiger, Philadelphia, 1972. Chap. 18.

Kraus, Richard G., *Recreation and Leisure in Modern Society*. Prentice-Hall, Englewood Cliffs, N.J., 1971, Chaps. 15, 16.

Kraus, Richard G., *Recreation Today*. Goodyear, Santa Monica, Calif., 1977. Chap. 1.

Kraus, Richard G., and Curtis, Joseph E., *Creative Administration in Recreation and Parks*. Mosby, St. Louis, Mo., 1977. Chap. 6.

Lutzin, Sidney J., Ed., and Storey, Edward H., Assoc. Ed., *Managing Municipal Leisure Services*. International City Managers Association, Washington, D.C., 1973. Chap. 7.

Sessoms, H. Douglas; Meyer, Harold D.; and Brightbill, Charles K., *Leisure Services*. Prentice-Hall, Englewood Cliffs, N.J., 1975. Chap. 11.

Tillman, Albert, *The Program Book for Recreation Professionals*. Mayfield, Palo Alto, Calif., 1973. Chaps. 1–10.

# CHAPTER 12

# Recreation and Park Finance

The extent to which public authorities acquire and develop recreation and park properties and provide programs and services on them depends in large measure on the amount of money made available for these purposes. The very foundation of recreation service is contingent on a dependable source of income, as well as on the willingness of public authorities to spend it for essential services. In short, the financing of public recreation programs and services is concerned primarily with (1) the securing of income and (2) the expending of this income. Although there are some differences of opinion concerning finance practices, every recreation and park agency is vitally concerned with its income source. Without financial support, there would be no recreation service. If funds are not wisely expended, recreation and park agencies will lose public trust and will not realize their full potential.

## NEED FOR REVENUE

Everything that is needed to operate a recreation and park agency requires money. Personnel, supplies, equipment, and facilities are expensive. Actually, most problems facing the recreation and park executive today involve, in one way or another, financial matters. While some people feel that there is little need for expanding public recreation support, they fail to see the changing public conceptions of recreation. The public has come to embrace the belief that recreation is a public function and should be tax supported. Although the controversy over taxation for recreation and park services varies in intensity in various jurisdictions, the battle has largely been won in most states. Furthermore, a changing conception of the roles of the county, the state, and the nation in provision of recreation and park resources is rising. In early days, the major responsibility for public recreation rested principally with municipal authorities. Today all levels of government are expanding in this direction. In fact, additional millions of dollars of public monies are expended annually for recreation services.

### Increasing Costs

Public recreation costs have increased markedly during the past two decades. These increases are due to a number of factors, not the least of which is the increasing regard of our people for wholesome recreational opportunities.

Like every other public venture, recreation must "stand on its own feet" and compete for public funds. While recreation and park expenditures have shown remarkable increases, they are still generally below a respectable standard. Whereas some cities will spend more than $20 per capita for recreation, others spend less than one-third of this figure. This inequality of expenditures for recreation by the various government jurisdictions is due to:

1. The lack of income resources to provide adequately for government functions including recreation and parks.
2. The lack of imaginative leadership in awakening a public consciousness of the need for creative use of leisure time.
3. The differences in community structure and recreation needs of people.
4. The lack of interest of some government authorities in park and recreation services.
5. The type of government agency and its size and population density.

It would not be fair to state that increases in recreation and park expenditures have been large without giving an explanation for these increasing costs. In the first place, the quality and quantity of recreation service have increased with each passing decade. Better leadership, a greater variety of services, and expansion of present, and addition of new, areas and facilities have been provided. Where once public recreation was considered as hardly encompassing more than the operation of a few playgrounds for children, today recreation services include parks, recreation centers, museums, zoological gardens, beaches, multipurpose sports centers, performing arts facilities, and the like. Actually, every phase of recreation service has been improved and enlarged. Second, recreation expenditures have increased with the rise of population. For, indeed, an expanding population calls for more facilities, areas, leadership, and conveniences. Third, public recreation costs have increased as the dollar has depreciated in purchasing power. Salaries have tripled since the 1950s, and each fiscal period sees the need for additional expenses to merely maintain the status quo. Hence the rising cost of goods and services accounts for a large proportion of any increase in spending.

## PUBLIC RECREATION REVENUE

How does a recreation department secure funds to finance its current service and responsibilities? Money for the support of public recreation comes from a number of sources. For the most part, these programs are supported by tax funds, but additional sources of income may be fees and charges, grants, gifts, concessions, interest from investments, or special money-raising events. By far the largest source of income is tax funds. However, in recent years the percent of revenue received from fees and charges has grown rapidly ranging from 20 to 50 percent in many communities. This trend has developed with the passage of several state and local tax limitation laws following the "Proposition 13" issue in California. The approval of the "Proposition 13" referendum placed a taxing limitation on property drastically cutting the revenue sources of local governments. The amount of revenue received from federal and state grants has also increased considerably in the last two decades with the advent of matching fund grants for land acquisition and development and a variety of federal government financed job programs.

The revenue problem that is faced by all government departments continues to be a big one for recreation and park agencies. Although recreation and park officials should be concerned with getting larger appropriations from the legislative body, they should not overlook internal economics whenever possible and should consider basic means of getting the most out of their tax dollar; (1) the elimination of waste and (2) the development of staff efficiency.

There is no excuse for waste in a recreation and park system. To eliminate it is one of the jobs of management. When staff is impressed with the importance of making small savings and following thriftful practices that in no way affect the operation of program, income is stretched that much further.

Much can be said for getting more out of the tax dollar by increasing the efficiency of staff, for more than half of all recreation income is expended for personnel or leadership services. Provision of the means and the direction that increase staff efficiency truly enlarges the scope of operation. In seeing that the department works at a maximum of efficiency, recreation and park officials must provide a climate for encouraging staff to develop new ideas and to practice modern techniques of operation. Specifically, it is the job of management to give the employee a feeling of belonging, to consider his or her feelings and need to be a part of the organization. With good supervision, the employee will want to develop his or her greatest potential in reaching department goals. Too often, the recreation executive overlooks this most important human element in expanding the tax dollar. What better way is there to enlarge services than through more efficient staff practices and operations?

In addition to getting the most out of the tax dollar, the recreation and park agency must continually promote and justify expenditures for public parks and recreation. An economic equivalency index developed by Robert Wilder, administrator of the Washington State Interagency for Outdoor Recreation, can be very helpful in justifying park and recreation expenditures. Information about the economic equivalency index may be found in Appendix I.[1]

## Taxation

Early in the recreation and park movement, financial support came from private sources; government bodies were reluctant to include recreation services among their responsibilities. However, as the value of community recreation came to be accepted by the general public, these services were added to the family of government functions and tax funds were appropriated in ever increasing amounts for their support.

The property tax is a chief source of income at the local level. Although there continue to be many arguments pro and con as to whether this tax is a fair tax representing the true wealth of a community, it nevertheless bears the brunt as the major local tax source. Therefore it is important to consider briefly an example of how the tax is computed. A tax is levied on the assessed value of property as determined by the assessor. The tax levy is shown in terms of mills (tenths of a cent) on the dollar, cents on the hundred dollars, or dollars on the thousand dollars of assessed value. Thus when a park district refers to its millage tax rate, it means that this millage is the amount of tax taken from each dollar of assessed property value. A tax bill is computed by multiplying a property's assessed value by the millage rate. For instance, if property is assessed at 25 percent of market value, and a home has a value of $30,000, the assessed value of the property would be $7,500. Then if the tax rate is 50 mills, one takes this figure (0.050) and multiplies it by the assessed value (0.050 × $7,500). The tax in this case would be $375. If the tax rate is 5 mills, the figure of 5 mills (0.005) is multiplied by $7,500, making the tax bill $37.50.

Reference is made in some areas to a "combined" millage rate. This rate includes the millage of all taxing districts within which the property is located. For example, it will include the rates for the city, the county, and the school district.

[1]Provided through the courtesy of the National Recreation and Park Association, Arlington, Va. and Robert Wilder, October 1979.

**Tax Limitations.** All states have established maximum local tax limitations for the protection of the taxpayer. However, the ratio at which property is assessed is usually left to local option. Therefore the assessment ratio can be as important as, or even more important than, the tax rate. Constant lowering of property assessment ratios makes it necessary to increase the tax rate to raise the same amount of revenue. Only by having some kind of state control of our assessment procedures and ratios can this problem be eliminated.

Tax limitation laws take the form of placing a maximum on the millage of assessed value of taxable property that may be assessed, or on the millage of true cash value of taxable property. They are usually expressed in terms of mills per dollar or cents per hundred dollars. The reason behind such limitations is to establish a ceiling, or a limit beyond which local authorities cannot go in levying property taxes without approval from the people.

An undesirable feature of tax limitation expressed in terms of maximum millage on assessed value is that while tax rates have a ceiling, assessment ratios may vary from year to year. The ratio of assessed values to true values may fluctuate from one political jurisdiction to another. For example, a tax of 150 mills in one city with an assessed value of 25 percent is the same as one of 75 mills in another city having an assessed value ratio of 50 percent. Hence tax limitations only have validity if they are based on true property value or non-fluctuating property assessment ratios.

**Allocating Tax Revenue.** Although taxation is the primary source of recreation revenue, the means by which this income reaches the recreation and park agency is important. Public moneys are allocated to the recreation and park authority or department in three ways: (1) appropriations from the general fund, (2) special millage tax levies, and (3) special taxes.

***General Fund Appropriations.*** In most cities and counties, current operating costs of recreation and park departments are financed through appropriations made by the legislative body to the recreation and park department on the basis of the approved departmental budget. There has been much discussion by recreation and park authorities as to the advisability of having funds allocated from the city general fund as against having a special levy for recreation. The structure of the government usually dictates the finance pattern with most recreation and park departments being financed through general fund appropriations. The major exceptions are those special recreation and park districts that operate as a separate governmental unit appropriating and levying their own taxes.

Financing recreation and park services from the general fund has many advantages.

1. It gives the chief executive officer for the city flexibility in determining the budget for overall community needs. Each department thus stands on its own record and projected work programs.

2. It makes the recreation and park executive demonstrate and prove the value of the service being rendered. The executive must constantly be alert to interpret the program and the quality of the jobs being done.
3. It does not hamper the recreation and park department in securing funds if the need can be shown.
4. It does not subject the recreation and park authority to fluctuating assessed property values in determining its operating budget.

On the other hand, the chief disadvantage of securing funds in this manner rather than by a special tax levy is that the recreation service may be vulnerable in times of crisis. It is the first service at which a city council or county commission will look for possible budget cuts. Hence the recreation and park authority is under constant pressure to protect its income source. But still if a department maintains a good rapport with the public and does an outstanding job, the validity of this point may be dissipated.

An explanation should be given as to what constitutes the general fund. Actually, it is the unrestricted fund into which most government income is placed. Each government body such as the municipality, the county, and the state has a general fund. Moneys from this fund are used to finance general government services. This is contrasted to the special, or earmarked, funds restricted to specific services. For example, when a tax levy for park purposes can be expanded only for parks, or a highway levy only for roads and highways, or a library levy only for libraries, it becomes an earmarked fund. Thus expenditures can only be made from this fund for the function specified. Furthermore, if a city earmarks any of its revenue source to a designated service, that source becomes a special fund. A county or a city may have a number of such special funds in addition to the general fund. This situation creates many problems inasmuch as the budget that represents the general functions presents a warped picture of government services and fiscal control.

***Special Millage Tax Levies.*** Some recreation and park authorities receive their principal source of income through a special tax levy. In instances when the levy is used, a millage tax on the assessed property value is designated, the income from which is earmarked for recreation. A number of states permit communities to make use of the tax levy.

The chief argument for the special tax levy for financing recreation services and facilities is that it provides a dependable income source free from political control. Much of the earmarking of revenue for recreation grew out of the belief of legislative bodies that if the public was willing to tax itself for this function, the tax would provide a source of revenue without tapping the general fund. This belief is still held widely today. In many instances community recreation was initiated only through a referendum, which allowed the public to vote for this service and its supporting special tax.

***Special Taxes.*** While tax resources for recreation and parks are generally provided through appropriations made from the general fund or through special tax levies, there are occasions when communities levy a special tax, the income from which is used to finance public recreation and parks. A tax on liquor, amusements, general merchandise, or any other product or service where some or all of the moneys collected are earmarked for recreation and parks would constitute this source.

## FEES AND CHARGES

Fee charging by public recreation and park agencies is not a new practice. In a publication on fees and charges the Heritage Conservation and Recreation Service stated that, "At least 50 percent of public park and recreation providers charge user fees."[2] Actually, few systems exist that do not involve charging for some service, and over the last two decades there has been quite a growth in the acceptance of fees and charges as a valid and important source of income.

Although the use of fees and charges is generally accepted and widely used by recreation and park agencies, there is still a concern of their use by many public agencies. Some of the reasons suggested as to why public recreation should be free are:

1. Charging restricts participation by the group most in need of the activity.
2. Charging tends to influence the development of services on the basis of income rather than the needs of the people.
3. Charging means double taxation.
4. Charging does not put public recreation on the same basis as education, as public interest demands.
5. Charging makes it more difficult to secure appropriations from governing bodies for recreation and parks.

The charging of fees can present a restriction that would result in some persons not being able to participate in a program. However, it should be pointed out that this is not the intent of charging fees and various means can be used to ensure that those persons who cannot afford to pay a fee may participate. Some recreation and park agencies provide scholarships and others provide for different fee rates according to the economic conditions in the community. One solution that seems to work well for programs serving youth

[2]U.S. Department of the Interior, Heritage Conservation and Recreation Service, *Fees and Charges Handbook*, U.S. Department of the Interior, Heritage Conservation and Recreation Service, Washington, D.C.: 1979.

is to allow the participant to earn his or her way by helping the recreation leader prepare materials for the program or by assisting in the cleanup at the end of the program.

Care must be exercised by the public recreation and park agency to ensure that program services are not developed on the basis of income as opposed to the needs of the people. Generally, however, recreation and park staff members are "people oriented" seeking to serve people and therefore will not tend to set up programs on the basis of a profit motive.

A question asked from time to time by participants is: "If I pay taxes, why should I also have to pay a fee to participate? Isn't this double taxation?"

With the increased costs of offering recreation services and the restrictions on increased tax support, it becomes impossible to continue to provide all of the services that the public demands. Therefore there is a need to charge fees if the service is to be offered and those who want that service can pay for the additional cost. This is not double taxation.

While it is sometimes stated that public recreation and park services should be provided on the same basis as public education, there are some important differences in the two programs that could well open this point of view to question. Participation in a recreation activity is a personal matter encompassing a wide range of activities. Many of these activities are not wholly essential to the recreation system and are provided only as an extra service or convenience to participants. Indeed, charges for these extra services can well be made without endangering a policy of free service for those activities in which the public as a whole is interested.

A concern of some recreation and park executives is that an increase in revenue from fees and charges will result in reduced appropriations for the governing body. When a fee system is implemented, every effort should be made to obtain an agreement that the increased income from fees will not result in the appropriations to the department being reduced. It should be pointed out that the fees will allow the recreation and park department to improve and increase services necessary to meet the public demand and these fees should not be used to replace existing funds.

Another factor for the recreation and park managers to consider is the effect that "Proposition 13"[3] in California and other tax restriction laws have had on government operation, especially recreation and park services. In many cases the effect has been to cut the funds for recreation and parks. Therefore

[3]Proposition 13 was a referendum passed by the voters in California in 1978. It limited the amount of property tax on real property to 1 percent of the value of that property, placed a freeze on property valuation and restricted the annual increase in assessed valuation to 29 percent and required a two-thirds vote of the state legislature to increase state taxes and a two-thirds vote of the local electorate to increase local taxes.

recreation and park executives might do well to consider the early establishment of other sources of income to protect against this future possibility.

## Legal Authorization

A study of fees and charges must take into consideration the legal authorization under which a public jurisdiction performs it recreation and park function. The framework within which local public authorities can perform their official function can be found in city charters and ordinances, state enabling acts, or other legal enactments. This legal sanction to conduct recreation programs and services is given to every local jurisdiction by the state government. Hence the authority given a department of local government to assess fees, dispose of moneys collected, or carry over moneys from one fiscal year to another is determined by the legal framework of the governing jurisdiction. For example, in Illinois the state statutes clearly spell out the power of park districts to collect fees when it is stated that park districts have the power "to establish fees for the use of facilities of the districts. Fees charged non-residents of such district need not be the same as fees charged to residents of the district."[4] With this authority, park districts may make and regulate fees and charges as well as determine the kinds of services warranting a charge.

## Liability and Fees and Charges

The effect of charging has many ramifications for the recreation and park agency, but one that has been of great concern to public recreation and park agencies is the question of the effect of charging on the liability of a municipality or county for injury to patrons resulting from use of facilities or service. Initially the court decisions tended to relate to whether the recreation program was provided by a government agency (public function) or by a private agency (proprietary function).[5] If operated by the government and the fees were not used to make a profit, the court decisions reflected little or no liability. However, when the recreation program was operated by a private agency or where higher fees were charged that resulted in profit, the court decisions reflected greater liability.[6]

In recent years the courts have tended to rule that governments can be held liable for damages and the doctrine of governmental immunity is no longer valid. At the same time state legislators recognizing this change have

[4]The Illinois Association of Park Districts, *The Park District Code* (Springfield: The Association, 1976), p. 51.

[5]Charles E. Reed, *Charges and Fees* (New York: National Recreation and Park Association, 1932).

[6]George D. Butler, *Introduction to Community Recreation* (New York: McGraw-Hill, 1949).

passed laws allowing governments to obtain liability insurance, and in many cases the local governments have been authorized to levy special taxes for this purpose.

This trend in the courts and the legislature bodies is emphasized by Dr. Betty Van Der Smissen:[7]

> In keeping with the social philosophy of the early 1970s, the rights of the individual for redress are being upheld more and more, yet at the same time efforts are being made to protect the governmental entity through permissive legislation for liability insurance and through technical restrictions, such as notice of claim time limitations and maximum amounts of insurance.

It is imperative that all public recreation and park authorities be alert to the effect of charging on liability and that liability insurance be carried to safeguard the agency against damage suits.

## Policies Governing Charging of Fees

As pointed out earlier, the present trend is for more and more public recreation and park agencies to use fees and charges as an important source of income. Probably the most important reason for this trend is that today with rapidly increasing costs and the restrictions being placed on tax income, park and recreation, agencies have had to find new sources of income in order to continue to provide the facilities and program services demanded by the public. In many cities any expansion of services can only be done if additional funds are found to cover the basic costs. Thus new programs are instituted only if the revenue from fees will cover the cost of the program. Fees are also used quite extensively to support special recreation programs where there is an excessively high cost due to the need for special facilities or leadership, and fees are also used to control use of special facilities such as golf courses, tennis centers, and the like. At the same time, there are many concerns about fees and charges as discussed earlier. With the increase in use of fees and the expressed concerns about their use, it is very important for each recreation and park agency to develop a clear policy to guide them in the use of fees and charges.

All planning for a recreation system begins with agreement on fundamental beliefs. Such beliefs are known as "principles" and are used as general guides for the determination of policies and procedures in finance planning. Once principles have been established, policies flow out of them. Every recreation and park agency must have sound policies as a basis for operation. Poli-

[7]Betty Van Der Smissen, *Legal Liability of Cities and Schools for Injuries in Recreation and Parks, 1975 Supplement*, (Cincinnati: W. H. Anderson, 1975), p. 5.

cies, not to be confused with objectives, provide the guidelines for reaching the agencies' goals.

Policies relating to fees and charges vary widely among recreation agencies. The two extreme points of view are: (1) that all services should be provided without charge and (2) that all services should be made self-supporting insofar as possible.

It is necessary to find the dividing line between the two extreme points of view if a sound, definite policy on this matter is to evolve. Before attempting to formulate a base for policy, one must consider some of the existing charge patterns.

## Existing Charge Patterns

The majority of recreation and park agencies only charge for use of specialized facilities, special accommodations, and program services that entail expenses above the normal cost of general operation. Charges in general are made for the following categories of service.

1. *Special features and facilities*. Including use of golf courses, swimming pools, boating facilities, bathhouses, tennis courts, picnic accommodations, game areas, bowling greens, and the like.
2. *Exclusive occupancy (rental)*. Including exclusive use by individuals or groups of clubrooms, auditoriums, gymnasiums, swimming pools, stadia, and the like.
3. *Rental of Specialized Equipment*. Including rental or service charge for use of costumes, athletic equipment, radios or phonographs, projectors and screens, kitchen and banquet equipment, boats, horses, and the like.
4. *Materials and Supplies*. Including those items that are used and consumed in special activities such as drama, music, handicrafts, social activities, and the like.
5. *Special Instructors*. Including only those giving highly specialized instruction that does not fall within the general services of the program staff.
6. *Special Accommodations*. Including such accommodations and items of a special nature as parking for special occasions, fuel for heating, special meetings, and special services.
7. *Athletic Leagues or Similar Programs*. Where participation is restricted to members of the league or association.

## Policy Guidelines

As an initial basis of support in the development of a fee policy, the recreation and park agency might refer to or quote from the "Fiscal Position Statement" of the National Recreation and Park Association, which relates to the use of fees and charges:

> *Fees and charges will be utilized to continue and to expand leisure opportunities:* Fees and charges are an effective and justifiable means of providing fiscal resources to continue and to expand leisure opportunities that require special equipment, materials, leadership costs, and/or special use of facilities. It is essential that some of this special cost be shared by the participants unless the economic means of the participants dictate differently.[8]

The following general principles are suggested as policy guides for fee charging by any recreation and park system.

1. All fees and charges for recreation services should be in conformity with the long-term program policy of the recreation and park system.
2. The fees and charges should supplement tax appropriations as a source of recreation revenue and should not be the primary source of funds for the agency operation.
3. The extension of recreation services should be based on need and not the income-producing value of the services.
4. The kinds of services entailing a fee or charge should be reviewed periodically in terms of the aims and objectives of the department.
5. The collection and disbursement of all fees collected should be governed by sound business procedures and administrative controls.
6. The concession operation or lease agreements of a departmental facility should be governed by the policymaking body that determines the fee policy.
7. Recreation areas and facilities used primarily for general community recreation purposes under departmental auspices should be made available at no charge to participants, unless special costs accrue to the department.
8. The recreation areas and facilities, when not in use for departmental activities and programs, should be made available at no charge (unless special costs result to the department) to a nonprofit and nonrestricted membership group for events and activities consistent with the aims and standards of the recreation department. In this category are:
   (a) Recognized community groups engaged in recreation programs and activities.
   (b) Recreation programs and services under the sponsorship of school authorities or other government institutions.

[8]Robert F. Toalson, "Fiscal Resources" *Parks and Recreation*, IX(2) (February 1974): 40.

9. The use of public recreation areas and facilities by private and closed membership groups should be discouraged by a recreation and park system by charging fees comparable to commercial rates. In no instance should such private use hamper the ongoing departmental programs.
10. The recreation services and programs of the recreation and park organization should be provided to the public without charge, but special services or exclusive temporary privileges should entail a charge. Since *special services* can be an ambiguous term, the following services are listed as constituting this charge category.
(a) *Expendable Materials.* Special materials and supplies furnished by the department, used in such program features as arts and crafts, and retained by the participant. (Such expendable materials entailing a charge consist of special items not generally provided free to participants in the ordinary programs conducted.)
(b) *Consumable Materials.* Fuel, food, or other consumable items furnished by the department.
(c) *Specialized Instruction.* Specialized leadership or instruction offered to individuals or groups involving services too specialized to be part of the basic recreation program.
(d) *High-Cost Facility.* A facility requiring a relatively large capital expenditure for construction and involving more than average maintenance and operating expense (swimming pools, golf courses, etc.).
(e) *Use of Equipment.* Rental items including game equipment, kitchen supplies, tools, musical instruments, costumes, boating equipment, and the like.
(f) *Protection of Property.* Checking of clothing and valuables, special parking accomodations for participants in programs, or police surveillance of an area.
(g) *Exclusive Occupancy.* The exclusive temporary use of recreation centers, swimming pools, auditoriums, picnic grounds, or other recreation areas or facilities.
(h) *Admissions.* Admission to special departmental events where profits are used to extend the activity or cover the cost of the event.
(i) *Nonresident Use.* Use of facilities, services, or programs by nonresidents of the community.

## MISCELLANEOUS REVENUES

Revenue sources for the operation of public recreation and park agencies, in addition to the primary sources previously indicated, are moneys secured through (1) grants, (2) gifts, (3) concessions, and (4) special money-raising events.

## Grants

In a number of instances moneys are secured through subventions from federal, state, and county governments to support local recreation and parks. Federal matching grants have been a prime resource for local agencies to use when purchasing and developing land for park and recreation use. Federally funded work programs for training and development of workers have also been used by recreation and park agencies to provide operating personnel. This is especially true of the departments in larger cities. The work program grants can be very helpful but serious problems may develop when the grant expires and there are no funds to continue to pay the staff hired under that program.

State and county governments generally find themselves hard pressed to generate enough income to meet their own expenditures; however, many states do provide some special aid programs. For example, the state of New York through its division for youth provides grants for recreation programs for the elderly.[9] The state of Illinois has provided matching grants to local agencies for land acquisition and development and the state of Delaware provides local assistance grants for recreation as do many other states. Generally, however, these grants are very limited.

As tax funds are becoming increasingly tight, some recreation and park agencies have obtained financial support in the form of grants from private foundations. The majority of these grants have been obtained to initiate or support special recreation programs for the handicapped and senior citizens, and the arts.

## Gifts

Gifts represent a small but growing portion of financial support for the operation of public recreation. Gifts of land, buildings, and structures have been very important to many recreation and park departments in the expansion of their physical facilities. In recent years many agencies have made special efforts to obtain gifts to support current operations. The majority of these gifts are for the sponsorship of a particular program or to provide scholarships for participants in those programs with a registration fee. Publicity received by local business and service groups sponsoring a recreation program is a positive factor that should not be overlooked by recreation and park agencies seeking financial support. Several recreation and park agencies prepare special catalogs of possible gifts that may be donated and distribute them to local businesses, foundations, estate planners, and other interested persons or groups.[10]

[9]New York State Office of Parks and Recreation, *Federal and State Aid for Parks*, (Albany: Recreation and Historic Preservation, New York State Office of Parks and Recreation, 1977).

[10]A special publication, *Gifts Catalog Handbook*, was prepared by the U.S. Department of Interior, Heritage Conservation and Recreation Service in 1978 to provide assistance to local recreation and park agencies in the preparation of a gifts handbook.

### Concessions

Many recreation authorities secure revenue by granting the privilege to a private party to sell services or commodities to the public on or within the recreation and park property or structures. A contractual agreement is usually drawn up between the city and the concessionaire, specifying the conditions of operation. While there are advantages in allowing a private party to provide a concession service to the public, recreation and park officials should scrutinize closely the policies on concession operation. Rules and regulations should be drawn that protect the public and establish a high standard of operational practices. In most instances concession contracts are open to competitive bids based on an established sum or a percentage of gross receipts. It should not be overlooked that some recreation and park agencies operate their own concessions.

### Special Money-raising Events

Not to be overlooked in the financing of recreation service is the method used in many small communities of raising money through fund drives or special events. Income secured in this manner does not provide a dependable source of revenue and should not be encouraged as a means of financing public recreation and park services. Yet in the small village or community it has financed many a special recreation project during the summer months. Money received from community barbecues and other special events, street corner donations, and sale of candy or confections has financed camp programs and athletic leagues, and in some instances has provided leadership for playgrounds.

## PUBLIC RECREATION AND PARK EXPENDITURES

Many millions of dollars are spent annually by public recreation and park authorities in America and the total continues to grow year by year. The spending and accounting of this money provides a challenge for recreation and park officials, for unless the best provisions are made for expenditure control and efficient operation, a weakening of confidence in public recreation and parks will inevitably result.

Once funds have been approved and appropriated to the recreation and park department or agency by the legislative body, they are expended according to a work program or financial plan. This fiscal plan is known as the "budget." The budget is reviewed and analyzed in Chapter 13. Expenditures can be classified into two groupings: (1) current expenditures and (2) capital expenditures.

## Current Expenditures

*Current expenditures* are the annual operating expenses for the recreation and park department. They comprise the day-to-day spending to carry on the ongoing program of activities and services. Since recreation is basically a service, the largest item of current operating expense in any department is that of recreation personnel. In fact, personnel accounts for almost two-thirds of the spending program of most recreation and park departments. This is not unusual when we consider the recreation function as basically focused on leadership and personal services. Indeed, only when maintenance, plant operation, and general incidental costs overshadow leadership services need we be alarmed.

**Standard for Current Expenditures.** Since cities and counties differ widely in their need to provide recreation and parks, it is difficult to prescribe a definite standard as to how much money should be spent annually for this service. Cities vary in their capacity to finance a program. The size of the community or taxing jurisdiction, the proximity of municipalities to state and regional parks, the variety of social and economic conditions, the tax limitations, and many other factors tend to influence the amount of money that will be expended for recreation by local authorities.

A standard is merely a guide or rating device by which a department or agency can compare its operating methods and procedures with an acceptable measure. It is a safeguard or an ideal arising from professional practice. Not meant to be rigid, it is only a general indication of what is required in fiscal planning.

The standard that was advocated by the National Recreation Association recommended an annual public expenditure for local parks and recreation at $6 per capita[11] from government sources. This amount was divided into $3 per capita for recreation programs and $3 per capita for maintaining general park areas. This standard was generally used in the 1950s but is no longer valid.

Today the National Recreation and Park Association recommends a minimum expenditure of $20 per capita. Expenditures by recreation and park agencies vary considerably, ranging from as low as $5 per capita to over $40 per capita. In information obtained from the Sports Foundation Inc., the average per capita expenditure for all of the recreation and park agencies entering the 1979 Gold Medal Awards program was $40.35. However, there were seven agencies listing very high per capita expenditures of over $100. If these seven agencies were removed from the sample, the average per capita expenditure for

[11]National Recreation Association, "How Much Should a City Spend for Recreation?" Bulletin No. 38, (New York: The National Recreation Association).

the remaining agencies would be $24.85.[12] In any case the per capita figure used today should be $20 or more and not the old standard of $6 per capita.

A second reason making it hard to determine an accepted standard of expenditures is, the difficulty in comparing cities on the basis of per capita expenditures. For example, in some cities the care and maintenance of street trees add greatly to the expense of the department. In other instances departmental expenditures may include the cost of maintaining a yacht harbor, a municipal football stadium, a big league ball park, or even cemeteries. Furthermore, different accounting methods are used by municipalities and counties in charging costs against various funds and accounts. In short, per capita costs cannot be used as a scientific basis for making fiscal comparisons between communities.

## Capital Expenditures

*Capital expenditures* are outlays for items that result in the acquisition or addition of fixed assets such as land, new buildings, remodeled buildings, equipment, additions to structures, and the like. Known also as "capital outlay," this category of expenditure can be financed by recreation authorities through three methods: (1) current taxes, (2) special assessments, and (3) issuance of bonds.

**Current Taxes.** Some public authorities advocate the payment of capital outlay from current taxes. They hold that it is undesirable to go into debt, particularly if the government unit is in good financial condition. Proponents of this plan feel that a tax should be sufficiently large to pay for improvements, construction, or land acquisitions, or if this is not feasible, that moneys should be placed into a building fund each year to take care of improvements anticipated in the long-range plan.

The advantage of financing capital improvements or additions out of the current income lies in the saving of money through the elimination of interest charges. Without question, the total cost of improvements is much less when financed on a "pay as you go" basis. In fact, as much as 50 percent of the cost of an improvement can be expended for interest on a bond issue maturing over a 20- or 30-year period. However, most municipal, county, and independent recreation or park authorities do not have the current fiscal resources to permit extensive use of this plan. Tax limitations and the constant pressure to extend services out of current income leave little money available for major improvement or site acquisition projects.

[12]Information obtained from 1979 Gold Medal Award Entries, The Sports Foundation, Inc., Chicago, Ill.

An ideal situation would be if capital outlays could be financed out of current income. This method should not be dismissed lightly as completely impractical. While most recreation and park systems must go into debt to finance capital improvements, no system should overlook the possibility of using some current revenue for improvements or construction. A recreation and park system is in constant need of improvement. To rely wholly on bond issues when such debt is incurred at five- or ten-year intervals is not being realistic. Hence if some current funds could be used or set aside each year for capital outlay, expansion needs would not reach critical proportions. Obviously, therefore the ideal is to have a combination plan of financing capital inprovements partly through current income and partly through long-term loans.

**Special Assessments.** Special assessments provide a means by which a government body can defray the cost of a specified public improvement project. Based on the "value of service" theory that "an individual should share in the costs of government in proportion to the value of the benefits he or she derives,"[13] a special assessment is a special tax levied on property for a public improvement that adds to the value of the property. In other words, special assessments are justified on the basis that special benefits accrue to the property in the form of appreciated values, and hence owners of the property benefited should defray the cost of the improvement.

If properly used, special assessments can be a useful device for financing needed projects. While much has been written pro and con concerning the fairness of this taxing method, it nevertheless is used by many American communities. Used mainly for streets, sewers, water mains, lights, paving, and the like, special assessments could just as well be used to finance neighborhood playgrounds, parks, or recreation buildings.

**Issuance of Bonds.** The time comes when recreation and park authorities cannot finance a program of capital expenditures from current fiscal resources and must therefore go into debt. Going into debt enables recreation and park authorities to acquire property or to build or improve structures for present use, but it permits payments to be spread over a period of years. Actually, it is similar to installment plan buying; "you get now what you pay for later." Of course, such buying is expensive, for any borrowing involves an interest charge. However, when it is necessary, long-term borrowing can be a sound fiscal policy. Borrowed funds should be amortized in as short a period as possible, and in no circumstances should the repayment period extend beyond the life of the improvement. Actually, only capital improvements having a long period of usefulness should be financed through long-term debt.

[13]W. J. Shultz and C. L. Harriss, *American Public Finance* (Englewood Cliffs, N.J.: Prentice-Hall, 1959), p. 181.

Most recreation and park authorities today issue bonds to finance their capital improvement programs. In fact, the issuance of bonds is the most common means by which a public body involves itself in long-term indebtedness. A bond is merely a legal contract or promise by a government body that it will repay the money it borrowed at a specified time with a fixed rate of interest.

The chief argument in favor of bonding is that it spreads the cost of payment for capital improvements over a period of years and thus requires future generations to help pay for an improvement they will use. This eliminates any undue hardship on present taxpayers.

## TYPES OF BONDS AND BONDING PROCEDURES

All states provide for local governing authorities to issue bonds for capital outlay. Bonds may be classified in terms of their method of retirement, that is, as (1) term bonds, (2) serial bonds, and (3) callable bonds. These, in turn, may be further classified as to the procedure used in the payment of interest and principal, that is, as (1) general obligation bonds, (2) revenue bonds, and (3) special assessment bonds.

### Term Bonds

A *term bond* is one where the entire amount of the issue, or the principal, comes due at the end of a specified period of time. The *straight-term bond*, where the interest on the debt is paid at stated intervals for the life of the issue, represents the poorest and most expensive type of public debt financing. However, it is rarely used today. A modification of straight-term bonds, *sinking fund bonds*, is used more often.

Sinking fund bonds are merely term bonds wherein the government body sets aside annually sums sufficient to meet the long-term issue. In theory, the fund accumulates each year until it reaches the amount of the principal; the bonds are then retired. It would seem that with good financial management, the money placed into a sinking fund would make for fiscal flexibility. However, this is not the case. In actual practice, it is difficult for government bodies to administer sinking funds. They find it convenient to tap this fund for emergency uses or to borrow from it to pay for current operations without bothering to repay, with the result that the money is not there when due to retire the issue. The history of sinking funds has seen such incompetent management that many states have banned their use. Here is a case where good theory has not worked in practice. Consequently, most public bodies use serial bonds to pay off long-term indebtedness.

## Serial Bonds

The *serial bond* plan for financing capital outlays provides for payment each year of a specified amount of the principal and interest of the total debt for the life of the bond issue. This plan for debt payment is similar to that of a person paying off a loan on a home or an automobile. A certain amount of the debt is reduced through regular payments of approximately equal amounts.

There are a number of advantages to use of the serial bond plan. First, as bonds are retired at regular intervals, the total interest costs are reduced. Second, it provides a systematic plan for the liquidation of the entire debt by the end of a fixed period of time in the future. Third, obligation for debt service can be stabilized and thus more easily budgeted. Fourth, investors prefer serial bonds over straight bonds. Fifth, it provides for greater fiscal flexibility, when new debts are added to existing indebtedness.

The retirement schedule for serial bonds may be adjusted to the needs of the public recreation and park authority, provided, of course, that a portion of the debt is liquidated each year. For example, the public body may use *straight serials* or *annuity serials.* A *straight serial* is one where payments of equal amounts are made on the principal for the life of the issue. This reduces the amount of interest charges as the issue approaches maturity; thus the total annual payments for debt service decrease each year. This annually decreasing debt service cost provides some financial leeway in budget planning.

An *annuity serial* is one where payments of equal amounts (principal plus interest) are made throughout the life of the issue. During the first years, interest charges are very high in proportion to principal, but as the issue grows older, the reverse is true. Nearly all long-term home financing uses this method, since its advantage for a homeowner lies in having the same fixed payment each year for the term of the debt.

*Straight serials* are perhaps to be preferred, since they provide for a faster reduction of principal and hence a saving on interest. They are also flexible enough to provide for larger installments on principal as the interest costs are reduced. Furthermore, it is better to have a decreasing debt service as new issues are planned for the future.

## Callable Bonds

*Callable bonds* are merely issues having a callable feature. In other words, the issuer has the option of calling in bonds for payment at any time or at a specified time before maturity. If bonds were sold at high interest rates, recreation and park authorities may make substantial savings by calling in the bonds and reissuing them at lower interest rates. Obviously, bonds having a callable feature are not so attractive to investors as those that are noncallable. Hence the interest rate is slightly higher, unless special provisions are made that protect the investor. For example, bonds may be callable only after five or ten years from time of issue. This makes the issue more acceptable to investment firms. However, the market for callable bonds is more limited than would be the case without the callable feature.

Issuing callable bonds may be a good practice, if prevailing interest rates are high or if the governing body has the necessary funds to pay off all or a part of the issue before maturity. However, bonds should not be refunded merely to extend the term of payment, except under unusual circumstances.

There should not be confusion between callable bonds and serial bonds. Actually, a callable bond merely provides a refunding feature before the full maturity of the issue. A serial bond issue may have the callable feature and still not be called before the scheduled maturity date. In other words, the decision rests with the issuing body as to whether there would be an advantage in refunding the issue.

### Payment of Interest and Principal

Bonds may differ in their negotiability. If the bond has coupons attached that specify the amount of interest payable at specified times, the bond is called a "coupon bond." The coupon is negotiable, and the interest payment is made to the person surrendering the coupon. The principal of the bond itself is also negotiable if due and presented for payment.

However, if the bonds are registered in the names of the owners, they may only be redeemed by them. These bonds are called "registered bonds." A register of bond ownership is kept in the office of the issuing agent, and all transfers or sales are recorded. Such bonds may be registered as to principal, as to interest, or as to both principal and interest.

### Bond Payment

It cannot be assumed that all bonds are payable from the general revenues of the issuing body. This is usually the case, and paying for debt service in this way is most prevalent. Bonds payable in this manner are called "general obligation bonds." If, however, payment is made from special assessments, the bonds are called "special assessment bonds." If payment is made from revenue or other earnings of a self-supporting facility such as a stadium, auditorium, or golf course, the bonds are called "revenue bonds."

General obligation bonds have the security of the governing unit issuing them, through claim against the tax resources of this body. The tax rate is fixed to service the retirement of the bonds. These bonds are most attractive to the bond market if the credit rating of the issuing body is good.

Few recreation and park agencies use special assessment bonds, but revenue bonds are more widely used. Some recreation facilities lend themselves to financing in this manner. Particularly, ball parks, marinas, stadia, tennis and racquetball centers, and structures catering to needs that bring in a large revenue source fall in this category. Before proceeding to use revenue bonds, one should make a careful study of the income potential. Insufficient revenue received from the facility may require that general revenues of the agency be used to pay off the bonds.

## Lease-Rental Agreements

Where legal tax and debt limitations prevent a recreation and park authority from financing necessary recreation projects above a specified ceiling, a device has been used by school districts to circumvent this restriction through the establishment of "building authorities." For example, in some states, school building authorities have been established under state law for the purpose of constructing school buildings, the cost of which is financed in the name of the authority through issuance of revenue bonds secured through receipts received from rental of the facility to the local board of education or similar body. Under this arrangement the board eventually gets ownership of the building.

Some states have passed legislation authorizing all local governments to utilize a "building authority" for capital improvements. A problem with this method is that it may sometimes be used to circumvent the public, who might vote down the issue on a referendum vote. Actually, it would be better to change debt-limitation requirements to allow public bodies to provide for capital outlays that are needed rather than to evade the issue through "building authorities."

## Bond Limitations

All states have a legal provision restricting the amount of debt a public authority may incur. Such restrictions are usually expressed in terms of a percentage of the assessed value. Bond limits may range from 2 percent of assessed value to 15 or more percent. In other words, bond limits will vary with the various states. In Oregon, for example, the limit for park and recreation districts is 2.5 percent of the true cash value of taxable property located within the district. Municipalities in that state have a limit of 3 percent. As was pointed out earlier in the chapter, attention must again be called to assessment ratios, for a true picture of debt limitations cannot be given until one considers the ratio at which assessed property value is related to true value.

All bond limitations have the purpose of protecting the taxpayer against abnormal debt service charges. When too great a portion of the annual recreation and park budget is drawn off for debt payment, the recreation program suffers. What is a safe limit? Is not the legal limit an adequate safeguard? Not necessarily, for the various states vary in their statutory debt limitations. Ellinwood attempts to provide a guide in this regard:

> Overall debt should not exceed 10 percent of the tax base. Preferably, overall debt should not exceed 10 percent of the assessed valuation. To help meet problems raised by the varying needs and functions of different levels of the government and in different areas, a second rule of thumb is offered: direct debt should not exceed about three and a half times total annual expenditures.[14]

It has also been suggested by some fiscal authorities that the tax rate to pay debt service should not be greater than 25 percent of the tax rate for current expenditures. Others have proposed that bond limits be placed at a percentage of true value of property. Here a rule of thumb might be that bonds could be issued up to a limit of 4 percent of true property value without an election and up to 10 percent with one.[15]

There is no hard and fast rule governing the amount of bonds that should be issued. Borrowing needs vary in the various governing jurisdictions. Despite this, however, each public body should make a maximum effort to meet its capital outlay needs and still maintain a sound debt policy and a stable tax structure.

### Term of Bonds

The average life of most bond issues is 20 years. However, issues may run from 10 to 30 or more years. There is no reason that the term of indebtedness should be for an extended period if an issue can be retired in a shorter period. A "rule of thumb" standard states that the life of the debt should not be greater than the life of the capital improvement project. Actually, this standard is too liberal if it means that the term of debt is adjusted to the life of the improvement. Buildings and recreation facilities may be used for 50 or more years. A better rule of thumb would be that one-fourth of each bond issued be liquidated within a five-year period or that the issue not extend beyond 20 years. Again, the financial ability of the park and recreation authority to pay off its indebtedness in a long or short period should be the criterion for the length of term of the issue.

[14]David M. Ellinwood, "Guide Posts to Success in Debt Management," *Municipal Finance*, XXIX (February 1957): 115–21.

[15]R. L. Johns and E. L. Morphet, *Financing the Public School* (Englewood Cliffs, N.J.: Prentice-Hall, 1960), p. 218.

## Approval of the Issue

Bond issues in nearly all instances require approval by vote of the people. This requirement, however, will vary in the various states. In some instances a majority vote of 60 percent or more is required to pass an issue; in other cases only the owners of real estate (freeholders) are allowed to vote. Other restrictive provisions are also sometimes imposed. Actually, these restrictions are not in the best interest of the community. No useful purpose is served. A provision calling for a simple majority of those voting to approve an issue is consistent with the principles of democracy, and it accords with the maxim that the will of the majority shall rule.

## Sale of Bonds

The procedures governing the issuance of bonds are regulated by state law. Therefore before undertaking the marketing of bonds, a recreation and park agency should secure the most competent financial counsel or bond attorney available. The consultant should be responsible for all the legal details concerning the issuance and sale of the bonds. This involves drawing up the necessary resolutions, the notice and time of sale, and the legality of the sale, which should be done by securing certification from a nationally known firm of bonding attorneys, attesting to the legality of the issue or that statutory requirements have been met. If possible, the consultant should also supervise the details involved in the calling of the bond election.

The financial consultant should also send information to a bond rating company to have the bonds rated. The bond rating system provides investors with a guide to the relative investment quality of the bonds. Ratings are given in three categories: A, B, or C, with three levels in each category: Aaa, Aa, and A. A rating of Aaa denotes the highest investment quality and a C rating is given to those bonds with the greatest investment risk. A high rating will increase the market value of the bonds and provide the agency with more bids.

After due advertising, the bonds are sold by competitive bidding, with the firm offering the most favorable interest rate being given the issue. The bond prospectus should provide the information needed by the buyer. An example of a bond prospectus is that of the Willamalane Park and Recreation District, in Appendix H.[16] It can be seen from this example that the buyer needs to know the amount of the issue, the place where the bids will be received as well as opened, the basis of the award, the tax rate, the method of payment, and any other pertinent information that makes the bonds attractive to the purchaser.

[16]Provided through the courtesy of the Willamalane Park and Recreation District, Springfield, Oregon, January 26, 1978.

## Bond Elections

Much could be written on the organizational procedures needed to put over a bond election. Such details are not necessary in a chapter on finance. However, a brief discussion on the steps needed to win public approval of a major capital outlay program is in order.

A successful bond issue campaign should be planned well in advance of the bond election. In fact, a year or more should be spent in systematically gathering data on recreation building and facility needs, which, in turn, should be presented in a manner that is easily understood by the public. In general, the steps necessary to put over a successful bond issue can be sketched as follows.

**Delineating the Need.** The first step involves the development of a long-range plan that shows the capital needs of the recreation and park authority. This plan not only outlines the specific projects but also provides supporting data on how the capital improvement program achieves the objective of meeting the needs of the community.

**Organizing the Action.** The second step is to develop a plan of community organization, for the purpose of informing the public of the need for additional recreation areas or facilities. This involves establishing a citizens' committee that represents a cross section of community life. For example, this committee should represent the many "publics" that make up a community's economic, social, and political structure. Labor, women's groups, civic clubs, religious organizations, lodges, parent–teacher associations, management, professional organizations, and the like should be represented.

**Establishing Priorities and Amount of Issue.** Once steps 1 and 2 have been carefully prepared, the path is clear for planning the amount of the issue and the priorities for development. The citizens' committee should study the proposed capital improvement program prepared by the recreation authority and determine if all or a part of the projects should be financed through the proposed bond issue. If it appears that the improvement program will be greater than the committee feels can be financed at this time, or if it would not be expedient to ask for a sum that may tax the fiscal resources of the community, the committee should establish a schedule of priorities. This should be done regardless, inasmuch as the urgencies of the improvement program will vary with the various neighborhoods or communities within the service radius of the recreation authority.

Once the development plan has been translated into fiscal terms, the committee is ready to educate the public on the urgency of financing the capital improvement program through passage of the bond issue.

**Educating the Public.** The final step involves winning at the polls the public's approval of the proposed improvement program. This is an education task. For in order to engender a favorable feeling toward financing additional recreation facilities or areas, one must make the need crystal clear to the taxpaying public. The citizens' committee should plan its strategy so as to present the facts in such a way that they not only reach the public but also are readily understandable to it. Specific methods are:

1. Plan for open meetings in the various neighborhoods, to discuss the issues at hand with interested citizens.
2. Provide a speakers' bureau so that talks concerning the problem can be given to civic groups.
3. Organize a subcommittee on publicity. Print a brochure giving the facts, and provide for its wide distribution. Secure the support of news media including the local newspaper, radio, and television.
4. Plan a telephone campaign to take place just a day or two prior to the election. Call people and remind them to be sure to vote.
5. Develop a slogan that succinctly points out the need. Constantly repeat the slogan throughout the campaign.

**Planning Election Follow-up.** Once the election has been won, inform the public of your progress in carrying out the improvement program, and keep faith with the voters by doing the work you said you would do in the bond campaign.

## Investment of Bond Proceeds

The proceeds from the sale of the bonds will not be expended immediately. In fact, it may be two to five years or more before all of the funds are expended. Therefore it is important to invest these funds with the investments maturing as needed to meet the expenses of bond projects. The interest received from these investments can provide opportunities to proceed with additional capital improvements.

### Summary

Financing of public recreation requires sound planning of present and future recreation programs and facilities. Plans also need to be flexible to take care of contingencies and expanding needs. The job at hand is concerned with the securing of needed revenue and the spending of it for essential services. The recreation and park authority has a public trust to see that the moneys allocated to it are expended in a businesslike manner. Actually, the objective of fiscal management is to attain economy and still provide for the utmost efficiency in carrying out the departmental work program.

## Questions for Discussion

1. Why is there an increasing need each year for additional funds for recreation service?
2. What are the revenue sources for public recreation? Can each of these sources be increased each year, or have we reached our maximum?
3. What is meant by the term assessed value? If you have this value and a tax rate, how do you establish the amount of tax?
4. What is the purpose of having a tax limitation?
5. Public moneys may be given to a recreation system in a number of ways. Explain each.
6. What is meant by a general fund appropriation? Millage tax levy? Special tax?
7. Why should a recreation and park system be concerned with charging for services?
8. If a service is tax supported, can you make a case for charging?
9. Explain the guidelines that could be used in establishing a policy for fee charging.
10. List some grants that are available. Are there any problems with grants.
11. Distinguish between a current expenditure and a capital expenditure.
12. What are special assessments? Should they be used to finance capital improvement projects? Why or why not?
13. Distinguish between a term bond and a serial bond, between a general obligation and a revenue bond.
14. What is meant by a callable bond?
15. Explain the purpose of bond limitations.
16. Explain why it is important to have bonds rated.
17. Describe the steps that are needed to put over a successful bond election.
18. What would you consider a good per capita expenditure for a recreation and park agency? Explain.

## SELECTED REFERENCES

Aronson, J. Richard, and Schwartz, Eli, *Management Policies in Local Government*, International City Management Association, Washington, D.C., 1975.

Buck, A. E., *Public Budgeting*. Harper & Bros., New York, 1929.

Butler, George D., *Introduction to Community Recreation*, (5th ed.). McGraw-Hill, New York, 1976.

Edginton, Christopher R., and Williams, John G., *Productive Management of Leisure Service Organizations: A Behavioral Approach*. Wiley, New York, 1978.

Hines, Thomas J., *Budgeting for Public Parks and Recreation, Management Aids Bulletin No. 46*. National Recreation and Park Association, Arlington, Va., 1974.

Hjelte, George, and Shivers, Jay S., *Public Administration of Recreational Services* (2d ed.). Lea and Febiger, Philadelphia, 1978.

Johns, Roe L., and Morphet, Edgar L., *Financing the Public Schools*. Prentice-Hall, Englewood Cliffs, N.J., 1960.

Kraus, Richard G., and Curtis, Joseph E., *Creative Administration in Recreation and Parks*. Mosby, St. Louis, Mo., 1973.

Lutzin, Sidney, and Storey, Edward H., *Managing Municipal Leisure Services*. International City Management Association, Washington, D.C., 1973.
Lynn, Edward S., and Freeman, Robert J., *Fund Accounting Theory and Practice*. Prentice-Hall, Englewood Cliffs, N.J., 1974.
Moak, Lennox L., and Millhouse, Albert M., *Concepts and Practices in Local Government Finance*. Municipal Finance Officers Association of the United States and Canada, Chicago, 1975.
Schnitzer, Martin, and Chen, Yung-Ping, *Public Finance and Public Policy Issues*. Intext Educational Publishers, Scranton, 1972.
Schultz, William J., and Harriss, Chowell, *American Public Finance*. Prentice-Hall, Englewood Cliffs, N.J., 1959.
Snyder, James C., *Fiscal Management and Planning in Local Government*, Lexington, Heath, Lexington, Mass., 1977.
U.S. Department of the Interior, Heritage Conservation and Recreation Service, *Fees and Charges Handbook*. U.S. Department of the Interior, Heritage Conservation and Recreation Service, Washington, D.C., 1979.
Van Der Smissen, Betty, *Legal Liability of Cities and Schools for Injuries in Recreation and Parks, 1975 Supplement*. Anderson, Cincinnati, 1975.

# CHAPTER 13

# Budget Administration

Preparation and control of the department budget are two of the most important tasks confronting the recreation and park executive. The budget prepared by the executive staff and approved by the board is one of the most important single documents of the recreation and park agency. Hence a knowledge of the principles and procedures of budget preparation and control is of the utmost importance in recreation and park management.

The word "budget" is of English origin, which originally meant the public purse or money bag containing the revenue and expenditures of the state.[1] Today a budget may be defined as a financial plan used by management that forecasts the estimated income and expenditures of an operating organization for a given period of time, usually one calendar year.

Strictly speaking, the budget is a planned work program expressed in terms of dollars and cents. It sets forth in advance the monetary requirements for a public or private enterprise; it is a plan of action for the future.

## PURPOSE OF BUDGETING

Basically, three types of enterprises use budgets. These are: (1) the tax-supported government unit, (2) the nonprofit agency that must rely on voluntary

[1]Henry C. Adams, *The Science of Finance*, New York: Holt, Rinehart and Winston, 1898.

contributions and other sources of revenue, and (3) the profit-making firm. While each of these organizations may vary its budget procedures, each must see the value of long-range fiscal planning and control, and each must make use of sound budgetary principles and practices.

The practical advantages of budgeting for any organization are many. First, it substitutes planning for chance in fiscal operation. It forces the administrator to think in terms of work to be accomplished in terms of its cost, to foresee expenditure needs, to follow a work schedule, and to organize the staff to accomplish established goals. Second, it requires the executive to view the entire operation in terms of funds available and the needs of all divisions or subunits. This gives a complete picture of the financial requirements of the organization and tends to prevent budget padding and overemphasis on pet projects. Third, it promotes standardization and simplification of operation, by establishing objectives to be accomplished through elimination of inefficient practices. Fourth, it provides guidelines for staff to follow by having a convenient reference source when questions of work goals arise. Fifth, it provides members of the legislative body with factual data for intelligently evaluating the fiscal policy and practices of the organization or agency. For example, it supplies the facts on rather technical matters and thus helps legislators or policymaking bodies to comprehend the ramifications of work to be accomplished in terms of cost. Furthermore, it sheds light on the many parts that make up the whole and thus gives intelligent information for establishing fiscal policy. Sixth, it helps the taxpaying public to know how money is secured and the purposes for which it is expended. In fact, it becomes a valuable source of public information. Finally, it acts as an instrument for fiscal control. This is achieved by evaluation of expenditures in relation to work objectives and work accomplished. Such data assist the executive in modifying and adjusting work priorities as needs and emergencies arise. With all these advantages that provide for administrative control, coordination, and flexibility in the giving of needed public service, budgeting has proved to be an invaluable tool of management and, when properly used, a basic means of achieving the ends for which an enterprise is established.

## LEGAL ASPECTS OF BUDGETING

The general regulations that govern public budget procedures and practices may be found in state statutes or city charters. Such provisions are supplemented by local ordinances and administrative rules and regulations. It is highly desirable for the legal provisions to provide only the broad skeletal outline of the budgetary system, leaving details of regulations and procedures to local officials. In some cities the only legal requirement is to have an itemized appropriation bill passed by the city council; however, other elements are frequently required by law. There may be items relating to classification of expenditures, a time calendar for preparing the estimates, a date for submitting the budget document to the legislative body, how it should be published, the sources and limits of funds, and the conducting of a public hearing.

At the municipal level the authority designated to formulate the budget depends on the form of local government. In the council–manager form this responsibility rests with the city manager; the mayor–council form usually designates the mayor; and the commission-governed cities give this authority to the commission members, who exercise both executive and legislative powers. In other, older and less clearly defined forms of city government lacking integrated patterns of organization, the authority to formulate the budget may be delegated to a body or committee consisting of both administrative and legislative officials. In some instances laws provide for a special staff to assist in budget preparation.

## TYPES OF BUDGETS

The classification of budgets can be very confusing to casual observers, the beginning recreation and park administrator, and the members of agency boards. One reason for this confusion is the misuse or, more specifically, the interchangeable use of terminology by many authorities. A second reason is that budgets, particularly governmental budgets do not come in neat specific packages and may include combinations of the different types of budget processes or approaches.

The two types of budgets that the recreation and park administration will normally work with are the current operating budget and the capital budget. In addition, the administrator should have a working knowledge of the following budget processes: object expenditure, performance, and program.

### Current Operating and Capital Budgets

The current operating budget includes the estimated income to be received for a year and the proposed expenditures from that income. When adopted by the legislative authority the current operating budget becomes effective for the fiscal year.

The capital budget is the proposed plan of capital outlays and development, also for the current fiscal year. The capital budget is usually based on a capital program, which is a plan of long-range expenditures—usually five to ten years—for capital development and purchases. The capital budget therefore is the first year's expenditures of that program. In practice, the capital budget is normally shown as a part of the current budget.

## Object-Expenditure Budget Process

The object of expenditure format often referred to as a line item budget is the most simple form of budgeting and is still widely used by many local governments. This type of budget is prepared and presented to the agency board in terms of the types of goods and services to be purchased during the fiscal year. The usual listings are personnel, contractual expenses, supplies, and capital items.

The chief advantages of the object-expenditure budget process are that it is simple, is easy to prepare and understand, and provides for good control by the board or legislative body, because they can ensure that the money was spent for whatever it was appropriated. There are two major disadvantages of the object-expenditure process: (1) It lacks planning and evaluation in the budget process. In many cases the budget is prepared by using a percentage factor to increase the budget sums over the previous year's amount without any evaluation or planning. (2) It encourages spending rather than economizing. Managers fear that if they don't spend all funds allocated, their budgets may be reduced next year.

## Performance Budget Process

The performance approach to budgeting is a system where the expenditures are broken into detailed subunits in which measurements of performance can be instituted. This type of budgeting requires the organization to develop a means to evaluate and measure programs so that maximum efficiency can be obtained. From this, cost standards can be developed and succeeding budgets may be prepared by multiplying the unit cost standards by the number of units expected in the next fiscal year. For example, in an athletic league a cost may be determined for each game played. The budget for the next year can then be developed by multiplying the number of games projected by this standard.

The performance budget places the emphasis on the activities for which the funds will be spent rather than simply stating how much will be spent. The main advantages of this budget are the emphasis on measuring program results as well as expenditures, a more detailed description of each activity is required that gives the managing authority more information on which to make a funding decision and the development of an efficiency awareness by all staff as they consider expenditures in relation to unit costs.

A problem with performance budgeting is the fact that many governmental services do not readily lend themselves to good measurement in unit cost terms. Also, this type of budgeting requires more time and work to determine units of measurement and performance cost analysis. Most recreation and park agencies cannot provide sufficient staff to do the job properly; however, many agencies do use the performance budgeting process to determine unit costs for selected programs. This is especially true where fees are charged for participation in the program activity.

## Program Budget Process

The program budget process became popular in the 1950s and early 1960s and in the late 1960s developed further into the Planning-Programming-Budgeting System often referred to as PPBS. A program budget emphasizes the planning for an activity or program and assigning costs for that program. The budget is prepared and summarized in terms of the program rather than an object-expenditure presentation.

In PPBS budgeting the activities of the organization are grouped into programs with common objectives to be considered together, with an emphasis on the planning process. The planning period extends beyond one fiscal year usually involving fiscal planning for five years. In this process it is important that goals and objectives be set so that the various courses of action available to meet these goals and objectives can be evaluated as to cost and benefits of each. Management then has the information to make a decision on which course of action to follow.

The main advantage of PPBS is the planning process that involves the making of program policy decisions followed by the allocation of resources in the budget. The two areas that cause problems with this system are the difficulty in developing an acceptable statement of goals and objectives and the development of adequate methods to measure the success of the program.

**Zero-Base Budgeting.** The theory of planning-programming-budgeting is that all programs will be evaluated annually, so weak programs can be dropped and new ones added. This approach has been further developed and popularized as zero-base budgeting. As described by Peter Pyhrr, zero-base budgeting is:

> An operating, planning and budgeting process which requires each manager to justify his entire budget request in detail from scratch (hence zero-base) and shifts the burden of proof to each manager to justify why he should spend any money at all. This approach requires that all activities be identified in "decision packages" which will be evaluated by systematic analysis and ranked in order of importance.[2]

Once the programs' "decision packages" have been ranked, the resources can be allocated to those programs considered the most important. Zero-base

[2]Logan M. Cheek, *Zero-Base Budgeting Comes of Age*, New York: AMACOM, 1977.

budgeting is not a separate budget system but is an approach that can be used with all types of budgets.

### Selection of Budget Approach

The selection of the type of budget and the budget process may or may not be an option of the recreation and park manager. In many instances the manager must follow the directives of the city administration. However, even then the recreation and park manager may have some freedom to institute an internal budgeting system. It is possible to combine the processes of object-expenditure, performance, and program budgeting into one budget. Many organizations, for example, prepare their budgets using a program process but adopt a performance approach for certain programs where the unit cost standards can be helpful in making budget decisions. The budget information thus developed may then be translated to a line item or object-expenditure format for presentation to the legislative body. In any case the budget system should be designed to provide for adequate planning and control and evaluation, and to fit the unique features of that agency.

## THE BUDGET PROCESS

For discussion of the budget process, it is again necessary to make clear that recreation and park agencies or departments may be structured with a policy-making board that acts as the legislative body for final action on budget matters, while others may be under an advisory body that has no final legislative authority. In the latter instance final legislative approval on fiscal matters relating to the department is given by a city council, county board of supervisors, state legislature, or other legally constituted policymaking authority.

The recreation and park budgetary process passes through three stages, or phases: (1) preparation, (2) authorization or adoption, and (3) execution. The function of the executive is to control the first and third stages, the legislative body controls the second stage, with an independent check on the third. Hence cooperation between the executive and the legislative body is of the utmost importance. The fact that the executive and his or her staff prepare the budget document does not give them the authority to determine the budget. That authority rests firmly with the legislative body. Therefore the executive prepares the budget in tentative form in order for the legislative body to examine it critically, revise it within established limits, and provide the income for its implementation and execution.

In discussing the three stages of budgeting, one should emphasize that the budget process is continuous. It is cyclic in operation, new budgets being prepared as old ones are being executed. Actually, the executive begins planning more than a year before the budget year. Thus agreements are reached and continuous planning is achieved.

## Budget Responsibilities

Budgeting is a managerial function and therefore the recreation and park executive must be assigned the responsibility for budget preparation and execution. The term recreation and park executive, as used here, refers to the head of a recreation, park, or recreation and park department. In an autonomous department in which the executive is directly responsible to a policymaking board or commission, the recreation executive is at the top of the administrative hierarchy. In departments of municipal, county, or state government where recreation policy bodies do not exist, the recreation executive is responsible to the mayor, city or county manager, governor, or some other administrative authority. In these instances the recreation executive clears matters through the chief executive of the governing body. Regardless, however, the recreation executive is responsible for preparing and executing the departmental budget.

Although much of the work of budget preparation and execution will be delegated to other staff, it is, in the final instance, the recreation and park executive who bears the responsibility of the department budget report to the chief executive or, in other instances already indicated, the policy-making board or commission. From an organizational standpoint, each major division or subunit in the organization takes responsibility for its respective budget requests to the executive and for the execution of its budget after it is approved, assuming, however, that the department is large enough to do this. Hence budget preparation and execution is a team task. In many of the smaller departments the recreation and park executive performs more of this work but there must be continuous consultation with all staff. All key department personnel should know fiscal policy and the framework of budget procedure, for in the larger department it is only through the sifting of budget requests from the operating level to the top level of management that adequate information is secured on the fiscal needs of the department.

## The Budget Calendar

A budget calendar is merely an orderly chronological timetable showing the dates when necessary action must be taken in preparing the budget. Every governing body should establish a calendar that indicates deadlines for taking action on various phases of the budget program. In turn, each department within the government structure should develop its own budget calendar to dovetail with the master schedule.

The budget calendar should include a schedule showing the responsibilities of various officials and groups, the time when information is due, the time for completion of the various steps of preparing the budget document, and the time for presenting the tentative budget to the chief executive and legislative body. The calendar should also give excerpts of legal and administrative rules governing budget formulation and execution, for the convenience of those preparing the budget report. The budget calendar should be prepared at the beginning of the fiscal year. The fiscal year could be, but is not necessarily, the same as the calendar year. In many instances the fiscal year extends from July 1 to June 30; in other cases the fiscal year may be from April 1 to March 31, or it may be for some other annual period. In any case it is important that the budget calendar specify dates and deadlines as well as legal and administrative rules governing budget formulation. A budget calendar has been used by Oregon communities for many years as shown in Table 13-1.

## Budget Preparation

As already mentioned, budget preparation is a team effort. Therefore the staff should know explicitly the form of the budget document, which will vary greatly in the various governmental jurisdictions. However, there are basic considerations in presenting the financial plan with supporting data that are applicable to practically all units of government.

Budgetary experience indicates that the budget document should give a clear and concise picture of an organization's financial needs. Hence it would seem practical for this document to be as complete as possible and to be prepared in four basic sections.

The first part should provide the budget message in which the principal budget items are explained and proposed major changes are highlighted. A presentation of the fiscal status of the organization and recommendations for future programs and fiscal operations are usually included.

The second part should provide a balance sheet and summary of what the organization plans to spend and how it will secure its funds. This summary will give the recreation board the overall picture of what the department plans to do and will act as a guide to the study of details. Actually, this is the only part of the budget that citizens or taxpayers see, and in many instances this summary is printed separately and in pamphlet form for public consumption.

The third part of the budget document should provide the necessary detailed information relative to the second part. This would include revenue and expenditure estimates with comparative data for prior years, cost figures, explanatory statements, and work programs. This section should include all the detailed data and supporting facts needed to critically evaluate and justify the financial plan.

The fourth part should contain the proposed methods of raising revenue, for, by provision of comparative revenue data of preceding years with proposals for the coming period, the work of the recreation board or legislative

**Table 13.1 Timetable for Completing Municipal Budgets**

| *When*[a] | *Who* | *What* |
|---|---|---|
| Not later than | | |
| 1. March or early April | City Council | Should ask the recorder or some other city official to act as budget clerk and should appoint citizen members of the budget committee to fill expired terms and places vacated by resignation. |
| 2. Early April | Budget Clerk | Should prepare and tabulate required accounting information and distribute departmental budget forms to department heads. |
| 3. Early May | Budget Clerk | Should present tabulation showing budget requests, estimated revenues, and required revenue and expenditure information for previous years, to city council and budget committee. |
| 4. Early May | City Council | Should have met to discuss general fiscal policy. |
| 5. Early May | Budget Committee | Should have familiarized itself with fiscal policy and overall needs of city and have started preparation of the budget. |
| 6. June 15 | Budget Committee | Should have completed preparation of the budget and presented it to the city council. |
| 7. June 20 | City Council | Should post or publish the budget and an announcement of the place and time of the public hearing at which the budget will be discussed with the taxpayers. |
| 8. July 1 | City Council | Should repeat the publication. |
| 9. July 10 | City Council | Should hold the public budget hearing previously announced. |
| 10. July 15 | City Council | Should adopt the budget in final form, make budgeted appropriations, and see that the county clerk and assessor are notified of the tax to be levied. |

[a]Dates given are the *latest* dates by which each step in budget preparation should be completed in order to meet the requirements of the local budget law.

body will be greatly enhanced. Furthermore, it is highly desirable to include in this section of the budget the necessary revenue measures or budget bills needed to finance the cost of operation.

A fifth section of the budget document may be used to include other information such as bonded debt outstanding, statements of tax collections, a tax distribution chart comparing the tax rate for the recreation and park agency to the rates for other governmental units, a proposed long-range capital improvement plan, salary range classifications, and the like.

## The Work Program

In budget preparation the task of estimating the necessary expenditures for the coming fiscal period is most difficult. However, such estimates can be made only by considering the recreation program and work contemplated for the coming year. Only when we view proposed expenditures in the light of work to be accomplished, not on the basis of looking at last year's spending figures—taking those for granted and scrutinizing only the increases and decreases for the coming year—can we be on a firm budget footing. When we view the budget and the preparation of expenditure estimates as a work program, traditional budget routines of dollar guessing do indeed seem crude and inadequate. Hence particular consideration must be given to (1) the elements of a good work program and (2) the mechanics of preparing the expenditure estimates on the basis of work programs.

**Elements of the Work Program.** A work program is an estimate of the work to be accomplished in the coming fiscal year. It shows, for each unit of departmental operation, the proposed amount and character of the work to be performed in the budget year. It is a work estimate rather than a dollar estimate.

Work programs, sometimes referred to as "performance programs," are an essential part of the budget procedure. They set forth the performance plan of each spending unit of government. Hence they take chance out of planning, force each department head to look ahead and plan for the future, and provide the information necessary for real budget execution and control. The composite of the work programs of all units of government comprise the work plans for the government as a whole.

There are a number of factors, or elements, involved in justifying the work program and relating it to dollar estimates. These elements are: (1) standards of service, (2) standards of efficiency, and (3) achievements.

***Standards of Service.*** Every recreation and park department should establish criteria of service. In many instances government units make no effort to establish service standards or policies. While many national welfare and planning bodies provide "model" service ratios or goals to be accomplished, such data provide only a general guide for government authorities and should be used only in the light of local conditions and needs. On the other hand, many recreation and park executives are comparing levels of service and are asking pertinent questions on costs of operation: What level of service should be provided for each activity during the year? Should the city furnish three or five lifeguards for the protection of swimmers at the municipal pool? Should activities catering to few people and having high operational costs be discontinued or reorganized? Should grass be cut once or twice a week? Is the recreation department justified in operating a yacht basin, a civic theater, or a skating rink? Should community centers operate two, three, or six nights per week? These and other questions on service standards affect a department's costs and should be critically reviewed by every unit head of the recreation and park agency. Only by critical examination of the levels of service may the whole program be a valid instrument for service evaluation.

***Standards of Efficiency.*** Once the standard of service has been established, the next step is to determine a standard of efficiency for determination of volume of work to be performed and an appraisal of the method by which the work will have been accomplished. For this, a unit of measure for volume of work is needed, and accomplishments can only be calculated in measurable work units. Unfortunately, many government agencies including recreation departments do not make use of work units. This is due in part to the lack of a cost accounting system that provides unit cost data in detail, and also in part to the lack of a suitable and practical work unit for certain types of activities. For example, public works and other departments have worked out unit cost data in great detail owing to the ease of establishing a suitable unit of measure such as cubic yards of garbage hauled, square yards of paving, cubic feet of gas supplied, number of inspections, number of patient days, or number of man-hours. For many other activities, no unit of measure has been perfected. This is particularly true with respect to the less tangible services where no single unit of measure can be standardized. In these instances the work program usually consists of the nature and scope of the activity, its objective, and the conditions of carrying it out during the fiscal period. Actually, the important thing to remember on unit measures is not to determine or measure the objective of the service but to break this service into its component elements of what is done—the work performed or the service rendered. Thus such measures can be used for comparison purposes. This being the case, all performance measurement units can be classified into three general types: work, service, and production units. *Work units* refer to the operation or work that is done, such as the acres of grass cut, number of trees pruned, number of teen clubs in operation, number of art classes, number of park trees added, and the like. *Service units* measure performance of a service in terms such as number of student class hours, hours of instruction in specific areas, days per activity, and the like. *Production units* are measures of what is produced, such as the number of units built or miles or linear feet of walks and drives paved or benches and fences constructed. Actually, these units apply to performance, what is being done in the expending of funds. As a matter of fact, there is no hard and fast rule as to what constitutes a performance unit, except that it should provide a basic means of measuring the functions or activities undertaken.

Although some students may feel that units may also be classified in terms of purchases or payment for services or obligations, these do not have the validity of performance units. In fact, they closely resemble the items of budget estimates.

Not only must we consider the volume of work performed but must also consider or appraise the work methods. The budget-making period, when achievement goals are contemplated, is an excellent time to review and appraise work methods, departmental organization, and types of equipment used. Can equipment be used more wisely and economically, or can mechanization of an operation cut down budget costs? Can work loads be reshuffled to improve efficiency? Do high unit costs reflect the need for operational changes? By asking these and other questions, many recreation executives have affected substantial improvements and economies.

***Achievements.*** Budget-making time, the time of working out expenditure estimates for proposed work programs, is an excellent occasion for a recreation and park department to establish clear-cut achievement goals. Achievement objectives are related to dollar estimates. If every unit head of government estimates what work he or she plans to do and the results to be achieved, a more adequate means of evaluation and control is possible. Moreover, this places the responsibility squarely on the shoulders of the recreation and park board or legislative body for approving or reducing appropriations for proposed services. In this way such governing bodies know they are cutting out or approving specific service projects rather than dealing only in abstract dollar terms. Also, pertinent questions relating to the "why" of the service in addition to the "how" should be uppermost in the minds of the recreation administrator, because end results, achievements, and objectives are the criteria for having the service in the first place. The taxpayer is interested in achievements, the services actually rendered and needed, not necessarily in how they were rendered.

**Preparation of Expenditure Estimates.** Once the work goals of the recreation and park agency have been tentatively determined for the fiscal year, assembling expenditure estimates is not difficult. The responsibility for the preparation of expenditure estimates is left largely in the hands of the recreation and park executive. In turn, each division, bureau, or other unit head of the department is responsible for submitting the estimates of their expenditure requirements for the fiscal year. In most instances appropriate forms and sheets of instructions are supplied to each department by the finance office or budget-making authority. While the estimate sheets are supplied well in advance to provide adequate time for preparation, specific dates and deadlines for completing the various steps in the budget process are usually fixed in the budget calendar. It is necessary to point out, however, that ample time should be allowed for review and evaluation by the recreation and park executive before the budget schedule is presented for adoption to the recreation and park board or legislative body. Also, expenditure as well as revenue estimates should be compiled as near as possible to the beginning of the new fiscal year, in order to take advantage of using actual cost figures for present operations in appraising needs for the next budget period.

In the preparation of estimates by the spending units of the recreation and park operation, the question arises as to how to keep these units from requesting more money than can possibly be budgeted for the forthcoming fiscal period. Budget padding is untenable in good fiscal practice, as are unrealistic requests for enlarging service that cannot be financed. One means of preventing budget padding is that of placing the burden of proof for budget increases squarely on the shoulder of the operating heads. This device can be used by a city manager with respect to department heads, or it can be used by a director of a large park and recreation department in regard to estimates from division heads.

Through this control device the department executive would let each unit head know the approximate amount of money alloted to him or her for operation for the forthcoming budget period. This would be a tentative allotment that, in the executive's best judgment, would be the money needed to carry on a satisfactory program of service on the basis of present operation. Each subunit head within the department would then be asked to prepare the work program and estimates of costs so as not to exceed the amount of the allotment. In the event the division proposes new work programs or items of expenditure in excess of the allotment, it could do so by sumbitting a supplementary request containing all new expenditure proposed. Any of these additional requests could then be approved, as revenue is found to be available at the time the budget is formulated in final form by the department executive. Moreover, these supplementary items should meet the following conditions.

1. All requests should be listed in the order of their importance and numbered consecutively.
2 No item in the supplementary request should be of greater importance than any item included in the tentative budget allotment.
3. No essential item for division operation should be omitted from the tentative budget and placed as a supplementary request without foolproof support that it had to be treated in this manner.

4. The justification of the need for the work to be performed and of the means of achieving it should be included in every request.
5. The costs of each item should be broken down into their proper account classifications.

This system makes possible a realistic budget based on a work classification as well as on an account classification and a means of not exceeding a fixed amount predetermined by the department executive, except through supplementary requests that would be approved on the merits of each item.

In essence, by using this control procedure, the executive greatly reduces the work of budget preparation; keeps the budget within proper bounds; and greatly facilitates the decision of making or approving new items of service. Certainly, the executive must still make decisions as to addition of items of service as funds become available. But after all, each supplementary request is listed on a priority basis, which greatly facilitates decisionmaking.

The zero-base budget process may also be used to ensure that all items of the agency's programs are evaluated, justified, and ranked in order of importance. The executive may then proceed to fund those programs according to their rank order.

One of the areas of need for better understanding is related to accountability through performance. Hence the necessity for a work classification related to the account classification. For recreation and park executives who plan to use a work classification, the following procedure could be used:

1. The head of each unit of service within the department would prepare estimates of expenditure for the coming budget year, but would also prepare these estimates on the basis of a definite plan of work for each major operation of each activity. For example, the activity of buildings and grounds maintenance in a park department includes operations such as grass cutting, cleaning of catch basins, refuse pickup, leaf sweepings and tree removals, and the like. For each of these operations, a separate and detailed cost estimate could be prepared that presents: (1) the summary estimate showing proposed expenditures for each activity by work classifications based on quantities, or unit costs and (2) the detailed estimate for the activity by account classification. Operations within the recreation division of a park and recreation department could be classified similarly through use of work or service units.
2. Each spending unit would prepare estimates according to the account classification, with each request summarized to show the following information: the amount expended for the two fiscal years immediately preceding, the latest revised budget for the current year, and the amount estimated and requested in the first request for the budget year.

3. The head of each unit of service within the department would submit along with the first request, a supplementary request for new services projected for the coming fiscal period. As was stated previously, these supplementary requests would need to be accompanied by a separate statement showing justification.

Practices by governing authorities vary as to the kinds of estimate forms used in securing budget data from the units of government. The account classification data, which, in turn are supported by work estimates, are excellent. In the final analysis it is the chief executive who establishes the framework for the form of budget estimates and supporting schedules. Thus, each department head conforms to the procedures and forms established. Once the estimates for all the subunits of the department have been received by the department head, he or she should review and evaluate them in terms of the department's immediate objectives and long-range goals. The department head should consult with division heads concerning revisions or new work proposals, but in the final analysis he or she is responsible for all estimates of the department and must fit all requests into the budget plan.

## Account Classification

Reference has been made to account classifications, or classifications of income and expenditures. Such classifications are essential in the budget process. An account classification is merely a systematic grouping of expenditures and income. Rather than a listing of every service or item purchased by a government, a classification of these expenditures makes possible orderly groupings of financial information about the nature of the spending program. The same is true with income classification. In some instances attention is given only to a classification of expenditures, but it should be noted that classification of income is equally important.

All government units use to some extent expenditure and income classifications. While the form and detail vary with the governing jurisdictions, common elements are to be found in each. Expenditure classifications are generally of five types: (1) object, (2) function or program, (3) organization unit, (4) character, and (5) fund classifications.

**Object Classification.** The object-expenditure classification is a part of the object-expenditure budget process previously discussed. As mentioned, the term *object* refers to the things the organization purchases, such as supplies, commodities, property, or services. All government spending, in the final analysis, is made for objects, and hence a high degree of standardization is possible in the classification of objects. The major purpose in an acceptable object classification is to limit the main classes of expenditure into clear lines of demarcation and to develop subclasses that refine the nature of the main class. Such subgroupings should, however, be in suitable detail so that they provide adequate information for control and comparison purposes.

One of the earliest and most widely used expenditure classifications was prepared by A. E. Buck,[3] which continues to be a basis for use today although there have been several modifications. A typical classification form follows.

### Expenditure Classification by Objects

- 1000 Services—Personal
  - 1100 Salaries, regular
  - 1200 Salaries, temporary
  - 1300 Other compensations
- 2000 Services—Contractual
  - 2100 Communication and transportation
    - 2110 Postage
    - 2120 Telephone
    - 2130 Freight and express
    - 2140 Travel expense
    - 2150 Car allowance
  - 2200 Printing, binding, and advertising
    - 2210 Printing and publication
    - 2220 Duplicating services
    - 2230 Photography and blueprinting
    - 2240 Advertising and legal publications
  - 2300 Conferences—Memberships and subscriptions
    - 2310 Conference expenses
    - 2320 Memberships and subscriptions
  - 2400 Professional services
    - 2410 Accounting services
    - 2420 Engineering and architectural services
    - 2430 Other professional services
  - 2500 Utilities
    - 2310 Electric
    - 2320 Gas
    - 2330 Water
  - 2600 Insurance
    - 2610 Fidelity bonds
    - 2620 Worker's compensation insurance
    - 2630 Unemployment compensation insurance
    - 2640 Property insurance
    - 2650 Employee medical insurance
    - 2660 Employee life insurance

[3] A. E. Buck, *Public Budgeting*, New York: Harper & Bros., 1929.

- 2700 Repairs
  - 2710 Equipment service contracts
  - 2720 Repairs to equipment
  - 2730 Repairs to buildings and other structures
  - 2740 Repairs to walks and roads
- 2800 Janitorial, cleaning, waste removal, and other services
- 2900 Other contractual services

3000 Commodities

- 3100 Supplies
  - 3110 Stationery
  - 3120 Office supplies
  - 3130 Books and publications
  - 3140 Maintenance supplies
  - 3150 Cleaning–janitorial supplies
  - 3160 Fuel supplies
  - 3170 Chemicals
  - 3180 Medical supplies
  - 3190 Other supplies
- 3200 Materials
  - 3210 Lumber and building materials
  - 3220 Masonry and road materials
  - 3230 Structural metals
  - 3240 Paints, oils, and glass
  - 3250 Tree and plant materials
  - 3260 Other materials
- 4000 Other expenses
  - 4100 Interest expense
  - 4200 Claims–judgment
  - 4300 Refunds
  - 4400 Transfers
- 5000 Debt retirement
- 6000 Capital outlay
  - 6100 Land purchases
  - 6200 Buildings and fixed equipment
  - 6300 Park construction and improvement
  - 6400 Equipment

The following definitions will assist the reader in the differentiation of the major classes just cited.

1000 Services—Personal
Personal services involve salaries and wages paid to persons employed by the government body.

2000 Services—Contractual
Contractual services involve work performed for the government through agreement or contract by other than employees, as well as the provision of equipment and furnishing of commodities under agreement.

2100 Communication and transportation
These expenses include the cost of postage, telephone, telegraph, freight, express, and expenses for transporting persons and reimbursement of expenses incurred by recreation and park staff for use of their own personal vehicle.

2200 Printing, binding, and advertizing
These expenses include all charges for printing including advertising and printing of legal notices, and expenditures for mimeographing, photography, blueprinting, and binding.

2300 Conferences—Memberships and subscriptions
These expenses include educational and training expenses for personnel, departmental memberships in recreation and park organizations, and subscriptions to newspapers, magazines, and other periodicals.

2400 Professional services
These expenses inlcude all charges incurred for any accounting function such as data processing, cost of engineering and architectural consultants, and other professional services such as legal or financial consultants.

2500 Utilities
These expenses include all charges for heat, light, power, and water furnished by public utility enterprises.

2600 Insurance
These expenses include premiums paid for fidelity bonds, worker's compensation and unemployment compensation insurance, insurance on buildings, equipment and land, and those premiums paid for the employees' medical and life insurance policies.

2700 Repairs
These charges include all repair expenses of a contractual nature for buildings, structures, walks, roads and equipment, and the costs of service contracts for equipment.

2800 Janitorial, cleaning, waste removal and other services
These charges include janitor, towels, waste removal, snow removal, pest control, and other services of a contractual nature.

2900 Other contractual services
These charges include all other services of a contractual nature that are not included in other categories.

3000 Commodities
This group of expenses includes both supplies and materials.

3100 Supplies
A supply is a commodity that is consumed, impaired, or worn out in a reasonably short period of time. Such supplies include stationery, food, fuel, ice, clothing, pool chemicals, fertilizer, medical supplies, cleaning.

3200 Materials
Materials are items of a more permanent nature that may be combined or converted to other uses. Materials include lumber, paints, iron or other building materials, masonry and road materials, fiber products, trees and shrubs, repair parts.

4000 Other expenses
Other expenses are those charges incurred that are not commodities or contractual in nature. They include items such as interest expense paid for the use of money or capital, payments of legal claims and judgments against the recreation and park agency, refunds of overpayments received or when a program with a registration fee is canceled, and to record the legal transfer of money from one fund to another.

5000 Debt retirement
This item includes the amount of annual payments for the reduction of the principal of the governments' debt and may also include the retirement of short-term debts such as tax anticipation warrants.

6000 Capital outlay
These expenses include charges for land, equipment, buildings, structures, and fixed improvements. Equipment here includes movable items that, when used, show little impairment or change and have a definite period of usefulness. Furniture, machinery, motor vehicles, tools, and the like readily fall under this classification.

In practice, the expenditure breakdown would show the amount of money expended in the previous one or two years, the amount budgeted for the current fiscal year, and the proposed amount for the next year. For example, the expenditure detail might look like the following.

**Services Personal**

| | *Expended 1977–1978* | *Expended 1978–1979* | *Budgeted 1979–1980* | *Proposed 1980–1981* |
|---|---|---|---|---|
| 1100 Salaries Regular | | | | |
| 1110 General Manager | $_____ | $_____ | $_____ | $_____ |
| 1111 Director of Recreation | $_____ | $_____ | $_____ | $_____ |
| 1112 Superintendent of Parks | $_____ | $_____ | $_____ | $_____ |
| Totals | $_____ | $_____ | $_____ | $_____ |

**Services Contractual**

| | | | Expended 1977–1978 | Expended 1978–1979 | Budgeted 1979–1980 | Proposed 1980–1981 |
|---|---|---|---|---|---|---|
| 2100 | Communication and Transportation | | | | | |
| | 2110 | Postage | $______ | $______ | $______ | $______ |
| | 2120 | Telephone | $______ | $______ | $______ | $______ |
| | | Totals | $______ | $______ | $______ | $______ |

By using any of these expenditure classifications, we can easily see how the amounts proposed for expenditure can be shown opposite the item under the major classes indicated.

The numbers appearing before each item are used to code for accounting and other purposes. In studying these classifications and the detail of the subclasses, one can readily see that each digit of the number represents the division of the class to which it belongs, until it reaches the final object of purchase. For example, let us look at the item numbered 3154, which could represent soap. The first digit on the left, 3, indicates the main class of the item, "commodities"; the next digit, 1, is "supplies"; the next, 5, "cleaning supplies"; and the last, 4, could be the basic object, soap.

The number system makes possible a coding system that can readily be transformed to punch cards for a machine accounting system. The same coding device is used in determining the expenditures for the various departments, divisions, bureaus, or sections of a government organization. A department such as parks and recreation, for example, may have the code P; the recreation division, R; and the golf courses (bureau of the recreation division), G. Hence the code PRG 2215 would indicate that this item of expenditure was for the golf bureau of the recreation division of the park and recreation department; the digit 2 on the left would be coded for contractual service; the 2, the subclass of printing, binding, and advertising; the 1, the printing item; and the 5, the object, or in this instance, scorecard printing. It can readily be seen how such classifications make possible detailed cost information and how such abbreviated codes facilitate the making of vouchers, invoices, and requisitions.

**Classification by Function or Program.** While most government jurisdictions classify their expenditures by object, something should be said for the importance of listing expenditures by program. People are interested in the cost of the program or function performed, not necessarily the object. In other words, they wish to know how much it costs to operate the swimming pools, the golf courses, the athletic program, or the community centers.

Some recreation and park departments use a functional or program classification of expenditures along with the object classification. Actually, functional groups fit readily into a classification of expenditures by objects. For instance, in the left-hand column of a financial statement, the main classes of expenditure by object—personal services, contractual services, commodities, and current charges—are listed with their subclasses. In the headings extending across the top of the columnar sheet are listed the functions or program operations. By showing the amount of expenditure for the objects in the various functional columns of the statement, one can get a clear picture of the costs of operation by both object and function or program. This combined object-functional or object-program classification has much in its favor and gives the recreation executive invaluable data for comparing the cost of units of service and other cost analysis.

In a report on financial record keeping published by the National Recreation Association, the program or functional classification presented is divided into three parts: administration, facilities, and special services.[4]

Under "administration" are placed those items of expenditure related to the general organization and management of the department that cannot readily be charged to facilities or special services. For example, the recreation and park executive's salary and central office expenses would appear here.

Under "facilities" appear the separate headings that in the eyes of the recreation authorities are most appropriate for cost comparisons or breakdowns. For example, swimming pools, tennis courts, playfields, recreation centers, parks, playgrounds, stadia, museums, camps, golf courses, and the like could each appear as a column for account purposes. It is important to make clear that there is no hard and fast rule as to how detailed the facility breakdown should be. Actually, it is best to consider only major types of facilities that are operated as single or distinct areas.[5] But here again, the objective of having the functional cost data should determine the items for account listings.

Under "special services" appear the program features carried on by activity specialists. Music, drama, athletics, arts and crafts, dance, nature, and the like would fit this category if they are citywide in scope. Seldom are these services limited only to the facilities of the department.

A hypothetical form is shown in Figure 13-1 to help the reader visualize this breakdown of accounts. Only a few of the functional items are shown. The number of functions can be expanded as the department sees fit.

**Organization Unit Classification.** This classification has as its purpose the identification of expenditures on an organizational basis. An organization unit consists of the structural entities that singly or in combination perform the work of a department. The unit may be called a bureau, division, section, or agency. Each has a directing head; each performs a function related to the overall purpose of the department.

[4]National Recreation Association, "Financial Record Keeping," Publication No. P 235.
[5]Ibid.

1100—Recreation Division

| | | Facilities | | | | Special Services | | | |
|---|---|---|---|---|---|---|---|---|---|
| | Adminis-tration | Recreation Centers | Swimming Pools | Golf Courses | Play-grounds | Arts and Crafts | Drama | Athletics | Total |
| A. Services—Personal<br>1. Salaries | | | | | | | | | |
| 2. Overtime | | | | | | | | | |
| 3. Temporary salaries | | | | | | | | | |
| B. Services—Contractual<br>1. | | | | | | | | | |
| 2. | | | | | | | | | |
| 3. | | | | | | | | | |
| C. Commodities<br>1. | | | | | | | | | |
| 2. | | | | | | | | | |
| 3. | | | | | | | | | |
| D. Current charges | | | | | | | | | |
| E. Current obligations | | | | | | | | | |
| Total | | | | | | | | | |

**Figure 13-1**

Classification by organization units is sound only when each basic function is represented by an organization unit. No unit should be established below the level of the function because of the fluctuating nature of the activity or operation comprising the function. Again, as in the case of the functional classification, no uniformity exists in the items for classification. In its favor, however, is the opportunity to allocate spending to the responsible unit of organization.

**Character Classification.** Classification of expenditures by character has as its purpose the organization of expenditures on a time element basis. In other words, items of spending related entirely to the budget year are known as "current expenses," those items that represent past commitments or obligations are designated as "fixed charges," and those the benefits of which fall into the future are known as "capital outlays." As previously mentioned, most government bodies classify their expenditures into only two basic groupings—current operating expenses and capital outlays. Buck[6] advocates that classification by character be on the basis of four main character groups: (1) current expenses, (2) fixed charges, (3) acquisition of property, and (4) redemption of debt. He feels that this grouping can more easily fit into a classification of expenditure by objects and that, furthermore, it acts as an index to the method employed in financing government operation.

**Classification by Fund.** Classification of expenditure by funds has as its purpose the grouping of expenditures against the fund for which authorization is granted for appropriations. Such funds may have restrictions as to expenditure and use, and these restrictions must be kept clearly in mind when the budget is in the process of preparation. Many park and recreation departments have special expendable funds under provisions of law that provide "special mill levies" to support their operation. Such levies may be for the general operation or special items of service or development. In each case the moneys designated for these special funds may only be expended in accordance with the provisions that created them, and recreation and park authorities should be particularly familiar with their limitations and restrictions.

**Summary of Account Classifications.** The object classification of expenditures is the one most generally used by public recreation and park departments.[7] However-

[6]A. E. Buck, op. cit., p. 209.

[7]In a study completed in 1978 by Richard Joseph Schroth it was determined that 58 percent of responding recreation and park departments used a line item/object of expenditure budget with 19 percent using a program or functional budget and all other types of budgets below 10 percent. Richard Joseph Schroth, "Effects of Local Governmental Structure On Budgetary Procedures of Municipal Park and/or Recreation Departments," unpublished doctoral dissertation, Indiana University, Bloomington, Ind., 1978.

ever, all the classifications described are related and may be used either separately or in combination. As indicated, the object-program classification has proved popular with many park and recreation authorities. But the character classification can also be superimposed on this grouping, as can the fund classification or even the organizational unit classification.

From the standpoint of budget making, any classification is satisfactory if it produces significant information. Cost estimates should be presented in a way that gives a clear picture. They should never obscure the facts or confuse the points at issue. If use of the five classifications can give expenditure information valuable to the executive and his or her staff, consideration should be given to using them.

## Classification of Revenue

Just as expenditures are grouped together, so is revenue or income classified. Most cities and other units of local government have rather fixed fiscal classifications superimposed on each respective operating department. In most instances income data are grouped, or classified, by sources. However, here again, it is desirable to combine the source of revenue with the department or other unit of government producing it. For example, one source of income may be fees, rents, or permits. In preparation of cost data, identification of the unit initiating this income source makes valuable fiscal information available. Furthermore, it is well for recreation and park heads to know the funds to which all income is credited.

## Budget Authorization or Adoption

The presentation and adoption of the budget document after it has been prepared constitute one of the most critical stages in the budget process, for it is at this point that decisions will be made by the recreation and park board or legislative body as to how well the projected departmental programs will be supported by public funds. This stage in the budget process is a legislative one. If the budget has been well prepared and reviewed by the executive, its chance for adoption with little change will be excellent.

The budget document should be presented to the recreation and park board or to the legislative body by the executive several weeks before the time specified for adoption in the budget calendar. After receiving the budget, the board or legislative body will usually assign it for review to a committee of the whole, a committee on finance, or some other appropriate committee, for nearly all legislative bodies use the committee system.

Actually, the legislative body is concerned with three factors: (1) review, (2) public hearings, and (3) adoption or approval.

**Review.** The finance committee or the legislative body acting as a committee of the whole may call in the executive to explain and justify items or to provide additional data. In some instances staff members may be asked to explain phases of the work program as proposed. During this process the executive and staff should be prepared to use comparisons, graphs, and other audio-visual techniques to explain the budget. It should be noted that the legislative body or recreation and park board often reflects personal, partisan, or sectional interests. Therefore the executive may at times be asked to modify phases of the budget program to conciliate these influences. To be sure, major alterations reflecting partisan projects that do not fit into departmental objectives should be condemned.

**Public Hearings.** Every opportunity should be provided for public hearings on the proposed budget, but to be effective, these hearings need to be planned for citizen expression. Too often the public is not interested in this phase of the authorization process, for only a few persons attend these hearings. Usually, a representative from a taxpayers' group, a reporter from the local press, and a few interested citizens will be the only citizen representation. Ample publicity should be given these hearings, but if they are to be effective, the public needs to be involved in studying the needs of the recreation program throughout the year. Only in this way will the public manifest interest at this time.

**Adoption or Approval.** At the conclusion of the budget hearings, recommendations are made for formal approval. It is advisable that this approval be given before the beginning of the fiscal year. If this final approval must be given by a government body other than the recreation and park authority, it usually takes the form of approval through passage of an appropriation bill, ordinance, or statute by the legislative body involved.

## BUDGET EXECUTION

Once the budget has been prepared and adopted, it is ready for execution. The recreation and park executive is responsible for administering the budget, but good execution is a team effort. Much of the actual fiscal record keeping and management is delegated to staff members who must be given authority equal to their responsibility for fiscal management. It is also advisable that the agency carry fidelity bonds on all personnel handling money to ensure against loss from misuse of funds. Proper budget execution also requires that a budget or fiscal plan is prepared and followed. The administration and execution of the budget involve establishing (1) a system of budgetary control, (2) an adequate financial accounting system that provides regular reports on departmental expenditures, and (3) an audit of work programs in progress.

## Budgetary Control

All budget execution is involved in budgetary control, but one of the most common devices for controlling departmental expenditures is use of an allotment system. Such a system is merely a means of scheduling the expenditure of funds for work programs on a monthly or quarterly basis. In other words, the annual budget is divided into a series of monthly or quarterly budgets on the basis of the estimated volume of work to be performed during that period. This calls for advance planning of estimates of costs by the operating heads. The allotments for each period would not necessarily be of equal amounts. They would vary, depending on the projected work plans. In this way the executive can receive at the beginning of each month a financial statement that shows the amount of money available in each fund or account on the basis of work performed. Actually, the amount of money expended at any moment and the amount overspent or underspent on the basis of work estimates are available to the executive. This provides the executive with complete control of the departments' spending and hence avoids the pitfall of poor budget execution—overspending.

The use of machine accounting techniques allows many executives to carry the allotment system a step further, by assigning the responsibility for expenditure of funds for work programs to department heads on an annual basis. The executive, by reviewing computer printout reports, can monitor the income and expenditures and take action if problems develop.

## Adequate Accounting System

Any system of budgetary control can only be administered properly if the accounting system gives accurate information on each appropriation item. This means that a system of accrual accounting should be used. Under accrual accounting all encumbrances against each appropriation item are shown. Hence unpaid obligations show immediately against specified accounts, and a true picture of the status of each appropriation item is possible. With an allotment system and accrual accounting, the way is paved for budgetary control and up-to-the-minute reporting of the financial condition of the recreation organization.

## Purchasing

A successful financial system must include good purchasing procedures as the majority of the factual information used in the accounting system is completed during the purchase. This information must be properly recorded and transmitted to the other accounting operations. The best means to accomplish this is to use a purchase order system. Many governmental units not only use purchase orders but will not authorize payment for any item or service unless a purchase order or contract has been issued and approved. It should be noted that a signed purchase order is a legal contract if it is based on an offer and acceptance for a consideration or an obligation and is between competent parties. A sample purchase order, as shown in Figure 13-2, includes the following information.

1. The name of the governmental organization making the purchase.
2. The purchase order number that serves as the identifying reference number.
3. The date of execution, how the item is to be shipped, the FOB site, and who ordered the item. (FOB site or freight on board means that the vendor is to include the price of shipping the item to the site designated. This is especially important when bids are taken as the shipping costs will vary and may make a difference on which bid is low.)
4. The description of the item being purchased, the number of items needed, the unit costs and the total amount. A column for the account number, see description on page 303, is included for recording the purchase in the proper account.
5. The signature of the staff member authorized to make purchases.
6. The place to enter a vendor number may be included. This is used with a machine accounting system. Each vendor is assigned a number and the machine accounting will list the amount of the expenditure under that vendor. The total funds paid that vendor can then be totaled at the end of the month, quarter, year, or other reporting period.

The purchase order authorizes a purchase but it alone cannot be used to authorize payment. The item must be received or the work performed and verified by a staff member. Once verification is complete and an invoice is received, payment authorization may be made. Most recreation and park departments or local governments present a list of bills to the recreation and park board or other legislative board for approval on a monthly or bimonthly basis.

## Bids

Bids or price quotations are used for purchases where there are a large quantity of items or the cost of an individual item is quite high. State statutes or local ordinances sometimes require formal bids for items above a certain amount such as $2500 or $5000. The use of an informal bid procedure or price quotes is advisable for many purchases or contracts under the $2500 level. The quantity

Smithville
Parks and Recreation

910 South Street, Smithville, Oregon, 12345 Phone (305)453–9800

PURCHASE ORDER
No. 14236
Show This Number On All Invoices, B/L, Packages And Correspondence.

TO:

Date
Ship via
F.O.B.
Ordered by

Ship to:

| Qty. | Description | Account | Unit | Amount |
|---|---|---|---|---|
| | | | | |
| | | | | |
| | | | | |
| | | | | |
| | | | | |
| | | | | |
| | | | | |
| | | | | |
| | | | | |
| | | | | |
| | | | | |
| | | | | |
| | | | | |
| | | | | |

All changes must be authorized

Vendor Number

City of Smithville

Authorized Signature

**Figure 13-2**

purchasing of office supplies or the supplies for a summer athletic program, for example, can result in considerable savings when informal bids are requested. For an example of a price quote form and additional hints on informal bids and quotes see Appendix G.

## Salary Payments

The authorization for hiring staff and for salary payments is based on the budget, with the recreation and park executive responsible for the execution of the personnel plan that was derived from the budget. This responsibility is usually delegated to the department heads who hire their staff members and authorize salary payments. Fiscal control is accomplished by proper documentation normally in the form of time cards, a time clock, or other means. It is important, however, that each time card be reviewed and signed by the department head or other authorized staff member before a paycheck is prepared.

A good internal auditing control to ensure that payroll funds are not misused is to require that each employee pick up his or her paycheck in person from the accounting office. This may not be possible in larger departments but in that case the pay checks should be distributed by a staff member other than the one who authorized the time cards.

## Receipt of Funds

The expenditures within the budget document are based on the income, therefore those responsible for departmental expenditures need information on the status of the income received. Thus it is important that all receipts are deposited immediately and reports usually in the form of monthly income summaries should be provided all staff persons responsible for budget execution. Here again, it is important for internal audit control that no one staff member handle all transactions involving income. The staff member receiving the income such as registration fees for a recreation program should not make the deposits and record the amount in the books. Better control results when the money received is transferred to another staff member for depositing. It is important, also, that receipts are given for all phases of the transaction.

## Investing

Recreation and park departments operating within a municipal government will not have much opportunity to invest funds as this is normally done by the city treasurer or finance officer. Special park and recreation districts do have this opportunity, however, and every opportunity should be made to take full advantage of this investment opportunity. The first step is to obtain a qualified investment consultant. Many organizations appoint a treasurer who serves as an advisor to the board or as an officer of the board. Others hire a consultant, usually from the banking community, on a retainer basis. Payments must be standardized to provide maximum opportunity for investment. For example, payroll dates set for the first and fifteenth of the month and a date for payment of all bills on the fifteenth will require large amounts of cash on only two days of the month. For the remainder of the month a small amount of cash can be kept available for emergencies and all other funds can be invested. The interest earned over a year's time can be quite sizable and used to provide additional services.

### Independent Audit

Most states reqiure that all governmental agencies have an annual audit by an independent audit firm. This annual audit can also be very important to the recreation and park board and the executive to ensure that all funds are accounted for and that proper accounting procedures are being used.

### Questions for Discussion

1. Explain the term *budget*. What is the purpose of a budget?
2. Why is a budget calendar important in fiscal planning?
3. Explain fully the three stages of the budgetary process.
4. What is meant by a "work program" in budgeting? Explain the three elements of such a program.
5. Explain the term *account classification* and the types that are commonly found.
6. What is the role of the legislative body in budget administration?
7. Explain what is meant by "budget control."
8. What is meant by "an object classification with a functional or program breakdown"?
9. How are the numbers before each item of the object classification used for accounting purposes?
10. Distinguish between "supply" and "material," between "current charges" and "current obligations."
11. What is meant by "zero-base budgeting." Where can it be used?
12. Why are purchase orders valuable in the budget execution process?
13. Why is it important to have an annual audit?

## SELECTED REFERENCES

Aronson, J. Richard, and Schwartz, Eli, *Management Policies in Local Government*. International City Management Association, Washington, D.C., 1975.

Buck, A. E., *Public Budgeting*. Harper & Bros., New York, 1929.

Butler, George D., *Introduction to Community Recreation* (5th ed.). Mc-Graw-Hill, New York, 1976.

Cheek, Logan M., *Zero-Base Budgeting Comes of Age*. AMACOM, New York, 1977.

Edginton, Christopher R., and Williams, John G., *Productive Management of Leisure Service Organizations: A Behavioral Approach*. Wiley, New York, 1978.

Hjelte, George, and Shrivers, Jay S., *Public Administration of Recreational Services* (2d ed.). Lea and Febiger, Philadelphia, 1978.

Kraus, Richard, and Curtis, Joseph E., *Creative Administration in Recreation and Parks* (2d ed.). Mosby, St. Louis, Mo., 1977.

Lutzin, Sidney, and Storey, Edward H., *Managing Municipal Leisure Services*. International City Management Association, Washington, D.C., 1973.

Lynn, Edward S., and Freeman, Robert J., *Fund Accounting Theory and Practice*. Prentice-Hall, Englewood Cliffs, N.J., 1974.

Moak, Lennox L., and Milhouse, Albert M., *Concepts and Practices in Local Government Finance*. Municipal Finance Officers Association of the United States and Canada, Chicago, 1975.

Pyhrr, Peter A., *Zero-Base Budgeting A Practical Management Tool for Evaluating Expenses*. Wiley, New York, 1973.

Schroth, Richard Joseph, "Effects of Local Governmental Structure on Budgetary Procedures of Municipal Park And/Or Recreation Departments", unpublished doctoral dissertation, Indiana University, Bloomington, Ind., 1978.

Snyder, James C., *Fiscal Management and Planning in Local Government*. Lexington, Health, Lexington, Mass., 1977.

# CHAPTER 14

# Areas and Facilities: Guidelines and Principles of Planning

The types of properties required for recreation and park purposes in a community can be determined only after careful study and planning. Recreation and park planning is an integral part of community planning; it has a direct relationship to other land use including thoroughfares, residential neighborhoods, public schools, and elements that determine the livability of a community.

Recreation and park planning has the purpose of meeting the leisure-time needs of people. While these needs include organized programs and services, people also want equipment, areas, and facilities for their own unorganized use. Any plan for area and facility development therefore must consider the satisfactions, both organized and unorganized, that people seek in their leisure-time hours.

## GENERAL GUIDELINES

Basic to all recreation and park planning is the drafting of a long-term guiding plan for a community or region. Such a plan, known as the "master plan," attempts to show the proper distribution of the parts that make up a community or region. Drafting this plan calls for specific skills, technical information, and experience on the part of the planner.

It is the job of the planner to relate the physical needs of a community to the living patterns of people, which cannot be done on a piecemeal basis. Indeed, careful consideration must be given to all the component parts of a city, as well as to its relation to the region. Planning is a means of giving society what it wants; it takes the component parts of a city and properly distributes them into a design for effective living. The planner has the same relation to a city as an architect has to a building. Each considers the need; each looks at long-range goals. But the planner is primarily concerned with making the city a better place in which to work, live, and play. Churchill succinctly stated this point when he wrote, "A city plan is the expression of the collective purpose of the people who live in it, or it is nothing."[1]

The planning process begins with dividing a city or region into units for measurement purposes. The units most frequently used are the *neighborhood* and the *community*. While the term *district* is sometimes used to designate a part of a city, districts have little social significance, since district boundaries are usually arbitrarily determined for administrative purposes. Therefore in approaching the matter of recreation and park planning, attention will be focused in this chapter on the neighborhood, the community, and principles of planning that act as guides for action.

## WHAT IS A NEIGHBORHOOD?

As our cities have grown in size, areas of definite homogeneous living patterns have tended to emerge. These can be identified as neighborhoods. People living in urban areas and metropolitan regions tend to group themselves with those of similar economic, religious, and racial backgrounds. This merging of people develops a feeling of identification with others, an association that clings to people. Here is therefore the social unit, or "building block," for planning.

To understand the neighborhood concept is one means of comprehending the physical organization of a city. Brunner and Hallenbeck stated that "the neighborhood is the smallest and simplest natural identifiable area within which the people associate together in an informal, face to face manner on the basis of intimate acquaintanceship."[2] A neighborhood is therefore a simple social and physical environment in which the family satisfies a few of its basic wants. These wants center around schools, churches, shopping districts, and centers for recreation and social activities. Families wish their neighborhoods to be attractive, well planned, and efficient.

This emphasis on the convenience of people in a living area provided the basis for the formulation of the concept of the *neighborhood unit* in planning.

[1]Henry S. Churchill, *The City Is the People* (New York: Harcourt, Brace & World, 1945), p. 186.

[2]Edmund Brunner and Wilbur C. Hallenbeck, *American Society: Urban Rural Patterns* (New York: Harper & Row, 1955), pp. 129–130.

First used in the regional plan of New York and its environs and later in the London County plan, this term now has common usage. Clarence Perry defines the *neighborhood unit* in rather specific terms as the populated area that would require the support of an elementary school of between 1000 and 1200 students, which, in turn, would mean a population of between 5000 and 6000 people.[3]

Perry gives six basic principles of neighborhood planning:

> 1. *Size.* A residential unit development should provide housing for that population for which one elementary school is ordinarily required, its actual area depending upon its population density.
> 2. *Boundaries.* The unit should be bounded on all sides by arterial streets sufficiently wide to facilitate its bypassing, instead of penetration, by through traffic.
> 3. *Open Spaces.* A system of small parks and recreation spaces, planned to meet the needs of the particular neighborhood, should be provided.
> 4. *Institution Sites.* Sites for the school and other institutions having service spheres coinciding with the limits of the unit should be suitably grouped about a central point, or common.
> 5. *Local Shops.* One or more shopping districts, adequate for the population to be served, should be laid out in the circumference of the unit, preferably at traffic junctions and adjacent to similar districts of adjoining neighborhoods.
> 6. *Internal Street System.* The unit should be provided with a special street system, each highway being proportioned to its probable traffic load, and the street net as a whole being designed to facilitate circulation within the unit and to discourage its use by through traffic.[4]

While Perry's neighborhhod concept calls for a physical organization or area that is convenient for the family and favorable to the raising of children, he does not expect the social interest or institutional affiliations of people living in the neighborhood to be confined only to this unit.[5]

A pattern of the neighborhood as a part of a larger segment in a city's structure has also been presented by N. L. Engelhardt, Jr.[6] He states,

> The neighborhood unit includes the elementary school, a small shopping district, and a playground; and these facilities are grouped together in such a way that walking distance between them and the home does not exceed one-half mile. A

[3]Clarence Perry, "The Neighborhood Unit," *Neighborhood and Community Planning: Regional Survey of New York and Its Environs*, Vol. VII (New York: Russell Sage, 1939), p. 28.

[4]*City Planning and Urban Development* (Washington D.C.: U.S. Chamber of Commerce, 1952), p. 41.

[5]Clarence A. Perry, *Housing for the Machine Age* (New York: Russell Sage, 1939), p. 261.

[6]N. L. Engelhardt, Jr., "The School-Neighborhood Nucleus," *Architectural Forum*, October 1943, p. 32.

neighborhood unit of seventeen hundred families usually has an elementary school with a standard enrollment of between six hundred and eight hundred pupils.[7]

The neighborhood is basic in community planning. Although its form and size vary, the essential characteristics are similar. Each neighborhood represents a unit of population that has a common need for educational, recreational, and maintenance functions. Also, it is the area that makes for effective expression of civic consciousness. Neighborhoods are a powerful element and a formative tool in our society, which can either frustrate or strengthen the democratic process.

## WHAT IS A COMMUNITY?

A system of parks and recreation comprises properties that make possible a broad program of recreation services for the people of a community. Hence the term *community* needs to be defined. While the *neighborhood unit* is identifiable as a fixed area, the term *community*, as used by planners, is somewhat flexible. The word has many meanings. To some, it is synonymous with "city"; to others, it is a natural grouping of neighborhoods within a city, an area with a clear identity of its own. It is not intended to have a rigid definition. Hence community is used here as a given identifiable area encompassing groups of people living together as a population aggregate. It includes several neighborhoods and has an identity as a recognized area of a city. Nearly all large cities have identifiable communities within their boundaries; in the small town such identification fades and "city" and "community" may become one. In general, the community is served by secondary schools, the neighborhood, by the elementary schools.

The urban community, on the other hand, encompasses all the people in a city living together as a population aggregate. Moreover, it may or may not include people living outside the city limits, depending on whether the interdependence of people makes it a social unit.

## RELATIONSHIP TO RECREATION PROPERTIES

Just as recreation areas of different types serve different functions that are easily accessible to individuals and families in a neighborhood, so should areas and facilities that require much space and unique development be located and developed to serve the needs of a group of neighborhoods or a community. Essentially the same service areas have evolved in parks and recreation as in the public schools. Both serve the neighborhood and the community; both select

[7]Ibid.

sites with reference to accessibility and other factors of location. The playlot, the playground, and the neighborhood park and recreation center serve the neighborhood. The large sports-oriented parks with athletic fields and facilities that require much space may serve either a neighborhood or a community. The large recreation park, the community center, and specialized facilities serve the urban community.

Nearly every city has definite neighborhood patterns. In many instances the boundaries are fixed owing to traffic barriers including railroad rights-of-way, major highways, and roads. Terrain, streams, rivers, and industrial locations are also factors that determine neighborhood boundaries.

Although most well-planned neighborhoods are characterized by groupings of families and individuals who can easily be served, within walking distance, by one or more small shopping districts, elementary schools, parks with playgrounds, and other neighborhood facilities, this ideal pattern cannot always be found. Nonetheless, the goal is an ideal family environment in an area of common standards, and people will tend to merge into such distinct neighborhood patterns.[8]

## BASIC PRINCIPLES FOR PLANNING RECREATION SYSTEMS

Planning for recreation parks and facilities should proceed from agreement on policies or rules that reflect the best thinking of the community regarding the purposes, scope, and general character of the public recreation system. Such policies or rules are commonly called *principles*. Stated clearly as a basis for planning, they largely embody the accepted philosophy of recreation. They assure the citizens that decisions concerning particular facilities will be consistent with the broad concepts that experience has shown should guide the development of the entire public recreation system. By adopting a carefully considered set of principles, the public recreation and park agency or planning agency shows that it values the best in recreation and park planning and wish(es) to proceed with deliberation and wisdom in creating a system of areas and facilities that will best serve the people. It commits itself to a considered course of action and wins public confidence by so doing.

## PURPOSE OF PRINCIPLES

Specifically, principles are needed (1) to determine the general approach to the selection and location of various types of recreation parks and facilities (in-

[8]The remaining text of this chapter is taken, with minor adaptations, from the *Guide for Planning Recreation Parks in California*, by the California Committee on Planning for Recreation, Park Areas, and Facilities (Sacramento: Documents Section, State of California, 1956), pp. 22, 25–32.

cluding neighborhood recreation centers, larger recreation parks, and such special use facilities as sport centers); (2) to establish the relationship of one site to another in the total complex of recreation and park areas; and (3) to establish the relationship of the entire recreation and park system to other physical elements of the city or urban area. For example, a universally accepted principle in recreation planning is that a neighborhood recreation center should be centrally situated in the area it is intended to serve. This principle enables those responsible for the selection of sites to proceed from a common understanding of what is desirable and practicable. Principles are also needed to institute an orderly procedure for planning further development of the public recreation system. In all planning involving land and water areas—whether it be planning for a small town, a metropolitan region, or a vast river valley—similar guiding precepts are necessary to ensure attainment of objectives.

As for objectives sought in recreation, they are almost too well understood to require elaboration. They may be stated simply as the enrichment of living through the constructive use of leisure and the expression of normal human interest in art, dance, drama, music, sports, nature, the world of the mind, and social activities.

## PURPOSE OF OBJECTIVES

Not to be confused with principles are the standards that serve as measures of the quality or adequacy of particular recreation and park areas. A *principle* governs the general location of the neighborhood recreation center, for instance, whereas *standards* concern the concrete details—what size it should be, what facilities it should include, and how large an area it should serve. Principles and standards together constitute the basic tools required for planning a public recreation system.

## GENERAL GUIDE

The following principles, 14 altogether, concern only the broader aspects of physical planning for park areas and recreation facilities; they do not pertain to program planning, financing, or public relations. They are presented as examples of the kinds of guiding policies any agency or jurisdiction concerned with recreation will find useful in shaping a recreation and park system. The committee urges that they be adopted in any particular locality only after careful study and earnest consideration.

Interpretive paragraphs following the statement of each principle explain why it is important and indicate how it should be applied.

## Principles to Apply in the Early Stages of Planning

**Opportunities for All.** *A recreation park system should provide recreation opportunities for all, regardless of race, creed, color, age, or economic status.*

Human beings are alike in needing opportunities to use their leisure pleasurably and constructively. Since our cities have attracted people of diverse origins, faiths, and social traditions, the recreation park system necessarily must include a wide variety of areas and facilities, capable of being adapted to programs designed to appeal to many different kinds of people. It must serve not only those whose families have long been identified with the American scene, but also national groups and racial minorities not yet assimilated into the larger social fabric. It must meet the needs of the middle aged and elderly as well as the needs of children, young people, and adults. No segment of the population, rich or poor, gifted or retarded, should be overlooked in planning the system. Failure to analyze the requirements of some groups can lead to social problems demanding increased police and welfare services.

**Analysis of Facilities, Needs, and Trends.** *Planning for recreation parks and facilities should be based initially on comprehensive and thorough evaluation of existing public facilities, present and future needs, and trends; thereafter periodic review, reevaluation, and revision of long-range plans should follow.*

Areawide planning should constitute the basis for decisions regarding selection and acquisition of particular recreation sites. Careful analysis, neighborhood by neighborhood and community by community, of data on age groups in the population, the incomes of those who are employed, the leisure habits of the people, and the problems arising from the lack of recreation facilities or from inadequate facilities will enable the public recreation agency, in cooperation with the schools and the planning agency, to develop plans for a recreation park system that will be reasonably satisfactory for 20 years or more. Planning can avoid costly mistakes such as those sometimes made in the past, when lack of overall planning resulted in acquisition of sites that were badly located for effective service, too small for the population served, incapable of proper development to meet the specific needs of a neighborhood or community, and too costly to operate.

A site that appears at first glance to be desirable may prove, on consideration of probable changes in the population and in the recreation habits of a neighborhood or community, to be inadequate. Of course, in a society that is opening new scientific and social frontiers as rapidly as ours, not all changes can be foreseen. Yet serious miscalculations can be avoided by periodic reevaluation (at least every five years) of facilities, needs, trends, and long-range plans. As a matter of orderly administrative procedure, the overall concept should be scrutinized at intervals and adjustments and improvements in principles and standards and in administration should be made in accordance with changing circumstances. The job of planning is never finished. It is a continuing creative process.

**Analysis of Private Facilities.** *Facilities and services provided by private agencies and institutions and commercial recreation enterprises to meet leisure needs of the population should be carefully evaluated by the public recreation and park agency and cooperating jurisdictions before plans for new recreation parks and facilities are prepared, so that a proper relationship between private and public facilities may be established and duplication may be avoided.*

The major responsibility for providing areas and facilities for leisure use rests on the public recreation and park agency, since it is supported by all taxpayers and is obligated to serve everyone. The school district shares this responsibility in proportion to its participation in the recreation program. In meeting their obligations to the people, the public agencies would be ill advised to duplicate facilities provided by private agencies. Public agencies must make certain, however, that a private agency with an extensive program is actually serving the surrounding area on a nonexclusive basis.

Coordination between public recreation and park agencies and private agencies that serve everyone is necessary if facilities of both types of agencies are to benefit citizens to the maximum extent possible.

Commercial recreation enterprises, such as theaters, bowling alleys, tennis centers, and roller skating rinks, meet some of the leisure needs of the people. Likewise recreation facilities provided by industrial and retail concerns for the use of their own employees meet the needs of a certain number. Public agencies therefore must ascertain the extent of the unmet need and provide recreation areas and facilities accordingly.

**Public Cooperation.** *Planning for recreation parks and facilities should be undertaken with full cooperation of the citizens, so that the recreation system may reflect their thinking concerning the needs and interests of all groups.*

When planning improvements in existing recreation parks or developing plans for new parks, the recreation and park agency, or the agency and the school district jointly, should invite the views and suggestions of the people served. Besides enabling officials to check their own ideas of what people want with what the people themselves say they want, meetings with citizens will strengthen interest in the recreation program and assure the financial support needed to expand the recreation system.

Holding a series of public meetings is time-consuming and demands great patience, skill, and open-mindedness, but it is unquestionably an effective way to make sure that the plans finally developed will meet with general approval. There is one danger to consider: the thinking of some citizens may be conditioned by what they are familiar with; they may be unaware of trends that will affect recreation in the future. Good leadership can make every public meeting an occasion for presenting a broad and enlightening picture of the problems of leisure in the future. An informed citizenry can be depended on to make sound decisions on policy.

## Principles to Apply in Planning the Overall System

**Unified System.** *Parks and recreation facilities for a city, county, special district or metropolitan district should be planned as related parts of a unified, well-balanced system to serve the entire area of jurisdiction.*

The effectiveness of any particular park depends on its being carefully related to other recreation facilities, since the use of each site affects the use of other sites. Piecemeal planning—the consideration of each site as a separate, unrelated area—almost inevitably results in the selection of sites that are too far apart or too close together or unrelated to school and cultural facilities. In order to obtain the full social utility of each recreation park, all parks should be planned as parts of a unified system, servicing the whole jurisdiction. Comprehensive planning will avoid overlapping service areas, will assure that equal standards of accessibility are applied according to density of population, and will reveal opportunities for relating recreation facilities to one another and to other local service facilities. In metropolitan areas, especially, as many of the larger parks as possible should be physically linked by parkways, interior greenbelts, and scenic trails.

In urban areas that include several jurisdictions, such as one or more municipalities, various areas under county government, several school districts, and perhaps one or more special recreation and park districts, planning for recreation and parks should be on a cooperative basis, particularly to ensure an equable and effective distribution of community, areawide, and special use recreation facilities. Recreation systems planned within separate jurisdictional boundary lines may lack coordination when viewed together as a combination of systems for meeting the needs of a total urban area.

**Integration in General Plan.** *The recreation plan, showing both existing and proposed parks and recreation facilities, should be integrated with all other sections of the general or master plan for the locality.*

The land included in the park system is but one element of the total city, planning area, special district, county, or other jurisdiction. Recreation areas can be located advantageously only by considering their relation to residential areas of all types and population densities, to schools, shopping facilities, industrial areas, and particularly to streets, highways, and transit routes. For example, the California Conservation and Planning Law, which provides for the preparation of master plans by cities and counties, especially acknowledges the desirability of relating all public facilities, witness the following provision:

> During the formulation of master plans, the planning commissions shall inform and, to such an extent as may be necessary, confer and cooperate with such school boards, departments, or agencies as may have jurisdiction over the territory or facilities for which plans are being made, to the end that maximum coordination of plans may be secured and properly located sites for all public purposes may be indicated on the master plan.[9]

Cooperative planning by the recreation and park agency, the school district, and other public agencies and jurisdictions offers opportunities to group public buildings and areas so as to form focal points of neighborhood or community activity. Such arrangements can stimulate increased use of each individual facility.

The process of preparing a comprehensive plan for the recreation system is similar to the process of formulating a general plan for the locality. The best results can be obtained by preparing the two plans simultaneously.

If a locality lacks a general plan and has no planning staff equipped to undertake its preparation, the recreation and park agency and cooperating agencies or jurisdictions would be well advised, before attempting to develop a comprehensive recreation plan, to explore the possibilities of having the city or county initiate a planning program, either by employing a permanent planning staff or by engaging a firm of professional planning consultants.

**Related Areas.** *Planning for parks and recreation facilities should encompass areas beyond a city, county, or other jurisdiction that are related to it.*

Viewed from the air, a city is not just the area within municipal boundary lines. It is the total developed area, including built-up areas outside the city limits. Seen imaginatively as it may be 20 years from now, it is all the presently developed area plus a large surrounding area that is awaiting development.

## Principles to Apply in Planning Individual Parks

**Central Location.** *Each recreation center or park site should be centrally located within the area it is planned to serve and should be provided with safe and convenient access for all residents of the area.*

Neighborhood recreation centers, especially, should be centrally located in the area served and should be well removed from major trafficways, rail

[9]Government Code, Title 7, Section 65475.

lines, and other hazards, since these centers serve primarily young children. If children can reach the center via safe residential streets, their parents will welcome their participation in recreation activities, but if the children must cross dangerous thoroughfares to get to the center, parental permission may be withheld and attendance may suffer.

Although central location is important for community and citywide parks and special use facilities, convenient transportation should be an equally important consideration in site selection. Most people using these larger parks and facilities will travel to them by automobile or by public conveyance rather than on foot. For this reason a deviation from strict application of the principle of central location might be justified, in order to select a site served by major arteries and transit lines.

As far as possible, such natural barriers as lakes, canals, rivers, and steep hills and such man-made features as major trafficways and rail lines should form the boundaries of the service areas of parks.

**Flexibility and Adaptability.** *Within a particular park the location, size, and design of activity areas and facilities should be regarded as flexible, so as to be adaptable to changes in the population served and in the recreation program offered to meet changing needs.*

In the course of 20 years or more the population of an area and its recreation habits will almost certainly change in many ways. The recreation program must change, and this means that the kinds of outdoor and indoor facilities available for use must also change. Since the location of a park is fixed and since size ordinarily cannot be increased, the only flexibility possible is in the arrangement of the various areas and facilities. The larger the site, the more opportunity there will be for redesigning and rearranging the areas and structures as needs require and trends suggest. Structures themselves should be designed originally with the expectation that throughout the years there may be a need to remodel interiors and make changes in the use of rooms.

**Beauty and Efficiency.** *Beauty and functional efficiency should complement each other in parks and recreation facilities and should be equally important goals of planning.*

Public recreation areas should have as much "eye appeal" and should be as well planned as the best present-day shopping centers, garden apartment developments, and suburban manufacturing plants. School properties used for recreation purposes should be as well designed as the school buildings themselves. Today the public demands attractiveness as well as efficiency; and designers are expected to achieve beauty when solving purely practical problems of space arrangement, circulation, and construction.

The public recreation and park agency should employ the talents of skilled architects, landscape architects, and other specialists in park design, for beauty and workability can be achieved only by applying a high degree of design skill to the development of sites and facilities.

Once a project is completed, good maintenance is highly important. Areas planted in lawns, trees, and shrubs require the best of care to develop as the landscape architect conceived them. Buildings must be kept clean, serviced, and in good repair to give the aesthetic pleasure that any well-designed structure can afford everyone who uses it.

## Principles to Apply in Carrying Out Plans

**Advance Acquisition of Sites.** *Land for parks and recreation facilities should be acquired or reserved well in advance of the development of an area, in the same manner as it is reserved for other public purposes.*

Opportunities to select the best sites for parks rarely present themselves once an area has begun to develop. Too frequently land is subdivided, houses are built, and families move in before there is real awareness of the need for public recreation space, when it is perhaps too late to obtain adequate sites, or any sites at all.

Sites for community parks, especially, should be acquired while the land is still undeveloped, because good-size sites are needed, and are difficult to obtain or costly after land has been sold for residential development. The early acquisition of sites for neighborhood recreation centers is no less important. If they have not been reserved or acquired before developers have begun to work out plans for subdivisions, the recreation and park agency, or other agencies and jurisdictions, in cooperation with the planning agency, should quickly point out to the developers the desirability of providing recreation sites and specific proposals should be discussed.[10] Once the street pattern for an area has been outlined, even in preliminary form, the possibility of establishing well-located parks representing proper standards will be greatly diminished.

Advance selection of sites is important for other reasons. Recreation and park agencies and school districts, for instance, can plan together to create related facilities that will provide maximum recreation opportunity for the people. Purchase of raw land or land that is only beginning to be subdivided also saves taxpayers money. Farsighted action, furthermore, can preserve in a natural state certain scenic areas the charms of which would be destroyed by urban development.

[10]Many cities and counties have adopted "Land Dedication Laws," which require developers to give a percentage of the land in a new development to the local government for parks and schools. The Supreme Court of Illinois upheld the constitutionality of the mandatory dedication requirement in the case of *Krughoff versus City of Naperville*, 68 Ill. 2d352, 369N.E.2d892, 12 Ill., December 1977, 185.

A policy of reserving or acquiring land for parks in advance of development presupposes the establishment of a financial program to implement the policy.

**Achieving Space Standards.** *Space standards for parks should be met and the land acquired even if the limited financial resources of a recreation and park agency oblige it to delay complete development.*

There is no substitute for adequate space in a park. Never should a recreation and park agency or agencies cooperating with it select a site that is admittedly inadequate for the long-term needs, because funds are not available at the time for full development of the land. It is better to buy sufficient land and, if necessary, temporarily defer development altogether than to skimp on acreage. Often the site can be developed in stages as funds become available, but at each stage in a site development program carried out during a period of years the quality of development should be such as to contribute to a completed project meeting the highest standards of design and construction. Development by stages, of course, requires an overall plan for the entire area before any work is undertaken.

Some cities shortsightedly make a practice of declining gifts of land for parks when they have no funds on hand for site development or foresee no immediate likelihood of raising funds by a bond issue or some other means. A few reject gifts of land because they regret to see lands removed from the tax rolls. Most citizens realize, however, that recreation areas make positive contributions to general well-being—better health, better social adjustment, better morale—that cannot be assessed in economic terms. Even unimproved land in a built-up city serves a purpose. It relieves congestion and functions as "breathing space." Furthermore, in the past no city has ever acquired too much land; nearly all stand condemned for having acquired too little. Rather than deprive a neighborhood of even the possibility of some day having a park, land should be acquired, by gift or purchase, as the case may be, and held in an undeveloped state until funds are available for its development. Local foundations have been used in some areas to acquire and hold these lands until the park and recreation agency is ready to assume title and to develop the area. The sight of the unimproved land may provide an incentive to the citizens to propose some means of developing it.

**Suitability of Sites.** *Selection or acceptance of sites should be based on their suitability for the intended purpose, as indicated in the over-all plan for the recreation system.*

One of the chief advantages of a comprehensive plan showing the approximate locations of proposed neighborhood recreation centers and community and citywide parks is that it serves as a guide to decision making when gifts of land are offered for recreation purposes or when funds are available for purchase. Nothing is so important as acquiring land that is in the right location—easily accessible to as many potential users as possible and situated so as to avoid duplication of other facilities in the overall system. Likewise important are the size, shape, and topography of the site, as well as its surroundings. Properties that are small, oddly shaped, steep, exposed to strong winds, or near streams that overflow their banks are generally unsuitable for recreational use. However, some of these areas might be considered for other reasons such as conservation, flood control, and the like.

Proper criteria for site selection, adopted as part of the overall recreation plan, should enable the public recreation and park agency and cooperating agencies or jurisdictions to select sites that are of generous dimensions, appropriately varied in topography, situated in quiet, safe surroundings, and protected, through adequate zoning, from the intrusion of commercial and industrial activities that would be detrimental to recreational use.

Considering the length of time that a well-selected site will be serviceable to the people, whatever sum is necessary to acquire one that meets all requirements is entirely justified.

**Perpetuity of Use.** *Parks should be lands dedicated and held inviolate in perpetuity, protected by law against diversion to nonrecreation purposes and against invasion by inappropriate uses.*

Despite population increases and mounting demand for more public recreation space, some cities in recent years have set dangerous precedents by converting long-established public squares to automobile parking areas and by permitting libraries, exhibition halls, utilities substations, municipal office buildings, fire stations, and other structures unrelated to the regular recreation program to be placed in parks and special use recreation areas. In almost every instance the people have been deprived of recreation space that they need.

If, to serve the public interest, an exception must be made to this principle, local legislation should require replacement of land lost by land of equal usefulness, size, and value.[11]

The attitude that publicly owned recreation space constitutes a cost-free land reserve for general governmental purposes is clearly against the public interest. Trafficways, particularly freeways, are often planned through parks solely on the presumption that since the land has no structures on it, it is cheap land, whereas the true value of parkland cannot be measured in "present-day market value." Publicly owned recreation space cannot justifiably be sold or leased for private enterprises, such as hotels, theaters, and office buildings. As

[11] Legislation in Illinois requires that a referendum be approved by the voters before any parkland can be sold or leased for other than park use; however, land may be exchanged for property of substantially the same size or larger and of the same suitability for park purposes. Illinois Revised Statutes, Chapter 105, Article 10, Sections 7, 7a, 7b, 7c.

leisure becomes more important in the daily lives of the people, cities will need every square foot of park space they now have, as well as large additions to the total acreage presently available for recreational use.

Local policy should provide for the protection of existing recreation space from temporary or permanent diversion to other uses, whether public or private. Temporary uses often become permanent.

## Questions for Discussion

1. What is meant by a "master plan"? How is the master plan related to recreation planning?
2. What is meant by a *neighborhood community region?* How are these units related to recreation areas?
3. Perry gives a number of principles of neighborhood planning. What are they? Show their relation to recreation planning.
4. What is meant by the term *principles*? How are they related to recreation planning?
5. Cite the principles you would apply in the early stages of recreation planning. Those you would apply in planning the overall system? Those you would apply in planning individual parks? Those you would apply in carrying out your plans?

## SELECTED REFERENCES

Bannon, Joseph J., *Leisure Resources: Its Comprehensive Planning*. Prentice-Hall, Englewood Cliffs, N.J., 1976.

Black, Russell Van Nest, *Planning for the Small American City*. Public Administration Service, Chicago, 1944.

Butler, George D., *Introduction to Community Recreation* (5th ed.). Mc-Graw- Hill, New York, 1976.

Hoppenfeld, Morton, *Planning Community Facilities for Growing Cities*. Mitchell Van Bourg, Berkeley, Calif., 1956.

Kraus, Richard G., and Curtis, Joseph E., *Creative Administration in Recreation & Parks*, Mosby, St. Louis, Mo., 1973.

McLean, Mary, *Local Planning Administration* (3d ed.). The International City Manager's Association, Chicago, 1959.

Miller, William Dwain, "Planning Parks for Urban Growth: Fort Collins, Colorado," unpublished doctoral dissertation, Colorado State University, Fort Collins, Col., 1973.

# CHAPTER 15

# Areas and Facilities: Steps in Planning

The planning of recreation areas and facilities is a complex task that requires an understanding of many factors. Some of these include the character and distribution of the population, community wants, staff competence, financial resources, building and zoning codes, and natural topographical considerations.

Effective area and facility planning must be preceded by effective program planning. It is totally unrealistic to plan recreation areas and facilities in the absence of clear-cut knowledge of the needs they will serve. In short, parks and recreation facilities must be developed with program goals in mind.

The idea that "program goals determine recreation areas and facilities" implies that such properties should meet present and future recreation needs. Hence recreation and park executives must translate program specifications into functional facilities geared to present and future community use. It is obvious therefore that the recreation and park board or other authority must provide policies on park and recreation development that will aid the course of recreation in the community for decades ahead. Too often, property needs are met on a piecemeal basis. A working formula showing the steps required in planning for areas and facilities is as follows.

## STEP 1—STUDYING COMMUNITY NEEDS AND FACILITIES

It is incumbent on the recreation and park administrator and the board to know the leisure-time needs and wishes of people as well as to help them organize their thinking on the role of the recreation and park agency in the community. These needs and desires should provide the base in formulation of recreation policy and a long-range improvement plan.

A community survey of needs should answer the following questions: What will be the recreation policy for the community? What recreation activities are carried on by all groups at present? What kind of recreation program will best serve the community? What areas and facilities are available at present? What additional areas and facilities will be needed? Where should they be located? To what extent can present areas and facilities be used? What future developments are planned by the youth agencies, school board, state and county agencies, and the like? How much money will be available, and how will it be raised?

The question also arises as to who will conduct the study of facility and area needs. Actually, it may be done in a variety of ways. In some instances it is done by the recreation and park administrator, with or without committees. In other cases it may be performed by a recreation survey specialist or research firm. Also, planning staffs at the local or regional level of government may do it. It is not uncommon to see it done through university and college student projects. Not to be overlooked is perhaps the best means of accomplishing this task, that is, through the establishment of a community advisory planning committee. Many successful recreation and park projects have developed as results of the work of advisory planning committees composed of members representing the many organizations and agencies, as well as leaders in the community. Some of the advantages of having such a committee are:

1. It acts as a coordinating body for looking at long-range facility goals.
2. It prevents overlapping of community facilities or properties and focuses attention on areas of need.
3. It represents strength and encourages support, whereas a less representative action might be overlooked.
4. It provides a leadership resource for getting the job done by lay citizens of ability who are dedicated to the task at hand.
5. It provides a broad outlook toward all areas of community services.
6. It welds into a working relationship diverse groups within the community, and hence encourages tolerance, understanding, and justice.

It is important to plan for sufficient time to do an adequate job when studying community needs and facilities through a community planning committee. Start early enough so that work can be organized and deadlines met. Moreover, such a committee should be given a clear and succinct statement of its responsibility and a deadline for work completion. Assistance and cooperation should be assured and careful consideration given to recommendations.

After all, the committee is performing a valuble service on a voluntary basis. Its role is advisory; hence every effort should be made to see that facilities are available for meetings, information is provided as needed, and necessary staff work is provided by the department. Moreover, the community should be informed of the work of the committee. Let the people know what is going on, inasmuch as public interest is only encouraged through knowledge.

## STEP 2—THE LONG-RANGE PLAN

Once the initial study of community needs and facilities is complete the information obtained should be developed into a long-range or master plan. The recreation and park executive and the recreation and park board or other legislative board share this responsibility. They will have to make decisions on the development policies and local standards that will be used in their community. They must also develop the recreation and park plan so it relates to all long-range development plans of the community including, zoning, transportation, school development plans, and the like. Once completed the plan cannot remain static. It must continually be reviewed and changed as new information becomes available.

The long-range plan is very important in the development of parks and recreation in the community, and as a practical matter it should be noted that a long-range plan is almost always required if the agency wishes to receive state or federal funds to assist in park and recreation facilities development.

## STEP 3—PLANNING THE FINANCIAL PROGRAM

An element in practical facility planning that must be considered is that of determining the amount of money available to do the job at hand. Not many public agencies have the financial resources to provide for capital outlay from current tax revenues. Hence bonding may be necessary. However, in instances where costs are financed from the current budget, it is usual to include only minor items such as the costs of small buildings, remodeling, alterations, or equipment. Some public bodies also make use of a building reserve fund into which money from current revenues is placed to pay for structures or area development over the long-range future. However, the reserve fund method is not very practical. With the pressure for additional revenue for current services, it is extremely difficult to accumulate large sums of money over an extended period of time in the future.

A bonding program is the most common method of financing a long-range capital improvement project. Borrowing by means of a recreation bond issue is discussed more fully in Chapter 12. Within debt and commonsense limitations, the planning, acquisition, development, and management of recreation areas and facilities depend on a sound long-range financing program. Securing of sites and development of facilities thereon cost money. It is imperative that there be farsightedness in projecting a financial base for property acquisition and development for the decades ahead.

## STEP 4—SELECTING AND PURCHASING THE SITE

Recreation services require areas and facilities. Therefore the site for the parkland should be determined and purchase procedures initiated. The size and location of the site, as well as function for which it will be used, should be spelled out in the long-range plan. Land should be selected and purchased well in advance of population growth, if that is possible. Early selection is highly desirable for the following reasons.

1. It keeps cost of acquisitions to a minimum, since land prices tend to rise with population growth.
2. It provides for better choice of suitable parklands. Once land is developed with high-cost structures, property that was suitable may no longer be obtainable. In other words, early selection prevents further loss of desirable recreation land to industry or further subdivisions.
3. It makes possible the securing of property for joint use by two or more agencies, counties, or communities. For example, the schools and the community may join forces in purchasing land for mutual advantage.

The selection of the site is of the utmost importance, for use depends on location. People are sensitive to any physical or psychological barrier that discourages use, such as traffic arteries, railroads, streams, ravines, stockyards, or industrial developments. Furthermore, people wish their recreation areas and facilities to be accessible. In fact, they should be a part of the neighborhood or community, just as the schools are an integral part. While there is no absolute rule by which one site is selected over another, there are certain criteria to bear in mind in the selection process. These are:

1. The site should be accessible and in close proximity to the center of the population to be served.
2. The property should be well drained and meet adequate standards as to size, shape, and needs to be served.
3. The site should be removed as far as possible from excessive traffic hazards, unsanitary conditions, or other undesirable influences.
4. The topography should be conducive to the proposed use of the area with a minimum of expensive grading and development.
5. The location of sites adjacent to schools or community centers should be considered in initial planning.

## Property Acquisition

The vital element in recreation area planning is not only to know the space standards for recreation sites and what constitutes desirable property but also to implement these guidelines through a sound program of land acquisition. Unless careful attention is given to acquiring land in properly situated areas of the city as needs arise, programs cannot function effectively. It is therefore vitally important, as we stated previously, that recreation and park authorities foresee and analyze future property needs before population pressures and increase of land values make their acquisition prohibitive in cost. Without question, this is one of the most difficult problems facing the recreation and park executive. It is particularly difficult to secure land in closely built-up areas such as the densely populated older sections of a city.

If areas and facilities are to be considered basic needs of community recreation, the planning for their acquisition and development should be in terms of the program and services they will provide. On this basis, several considerations face recreation and park authorities as they plan property acquisition. First, the acquisition of property should be seen clearly in terms of long-range planning goals rather than an assortment of scattered areas and facilities distributed without regard to community needs. Second, careful consideration of the ways in which the proposed areas are to contribute to the recreation program should be a prerequisite to determining the size and location of sites and the arrangement of facilities thereon. Third, since the areas selected for acquisition and development will be used for many decades, care should be given to acquire sites that will provide the best service that today's wisdom can conceive, while at the same time keeping the way open for future adaptations and changes as needs dictate. Lands may be acquired for recreation or park use in a number of ways. Those most used are: (1) purchase, (2) lease agreement, (3) transfer, (4) gift or bequest, and (5) easement.

**Purchase.** The most common means of acquiring parkland is through purchase procedure. Direct purchase from the owner by the government body is a simple method of acquiring land. However, it is not always easy to purchase needed land if owners are reluctant to sell or if the price is excessive. Hence the condemnation of land through the right of eminent doman is frequently used to acquire property. The *right of eminent domain* is merely a legal right to acquire land for public purposes. The right provides for just compensation to the owner for any land taken by due process of law. This right is provided through state statute and is based on the premise that land may be acquired by a government body for the common good of all people through judicial process. The purchase price is fixed by the court. In this way, inflated prices are adjudicated by the court and a fair price is determined. Actually, this method cannot be criticized too harshly, inasmuch as the owner is assured a fair price for his property.

Mention is sometimes made of the method of acquiring land through excess purchase. This is merely a device of purchasing more land than is needed for a public improvement and selling the excess when the completed improvement enhances the land value. The proceeds are applied to the cost of the improvement or placed in the government's general fund.

**Lease Agreement.** Land may be acquired for a set number of years through a lease agreement. The extent to which parkland or recreation facilities should be leased will depend, in large measure, on the urgency of acquiring the property in question. Lease agreements are used to advantage between government jurisdictions. In these instances a municipality may lease property from a school district, a county, or even the state or federal government for recreation use. It is not desirable to lease property from private sources when the urgency for land is a permanent one. At best, such lease agreements should be considered temporary measures.

**Transfer.** A common device for acquiring recreation lands or other property is through transfer of use and control from one government jurisdiction or department to another. Frequently, land or other property is no longer suitable for the purpose for which it was intended. Rather than sell this property, a government body may transfer it from one department to another. For example, firehouses, cemeteries, storage areas, reservoirs, river frontage, dumps, warehouses, airports, and even tax-delinquent properties are sometimes available as original use patterns change. This property may be developed for recreation and, not infrequently, it is transferred to the control of the recreation and park authority for redevelopment and use. Careful consideration must be given to many factors before accepting such land. Does it fit into the long-range recreation plan of the city? Will redevelopment be economical? Can the property be easily maintained? These and other questions need to be answered before transfer takes place.

Unusuable public lands including old garbage dumps, swamps, tidelands, marshes, ravines, and the like might be carefully examined for possible reclamation and redevelopment. Good examples exist of fine parks developed from sanitary landfills.[1]

**Gift or Bequest.** Securing park or recreation land by way of gifts or bequests is common practice in American communities. Indeed, most of the acreage of some park systems has been acquired in this manner. Many public-spirited citizens donate land for recreation purposes for sentimental or other reasons. In some instances such property is given as a memorial to a loved one. In other

[1]A "sanitary landfill" is property, once a swamp, dump, or other unsanitary area, that has been filled or covered with dirt and reclaimed for general use or a special use.

cases the incentive to give arises out of selfish motives. Regardless, this is an excellent system for securing properties and should be encouraged.

**Easements.** The use of scenic or conservation easements is an additional means of obtaining the use of land for park and recreation purposes. Land along waterways, railroad right-of-ways and other areas is sometimes protected for public use by obtaining an easement. At times the park and recreation agency may also obtain the right to develop a facility such as a hiking or bicycle trail along the easement.

## STEP 5—SECURING ARCHITECTURAL SERVICES

One of the most often overlooked items in area and facility planning and development is the provision of professional architectural services. While large departments may employ an individual trained in architecture or landscape architecture, this is not always so in the small- to medium-size departments, where planning is concerned more with area development than with building construction. Hence the decision to employ such a person is a crucial one. Whether the buildings, areas, and facilities will meet the needs of the recreation programs and service, whether it will be planned economically and functionally, and whether it will reflect utility, beauty, character, and imagination in design will all depend on this person.

Two types of architectural services are needed: (1) the design and planning of buildings and other structures and (2) the design and planning of grounds. There is usually no question as to the feasibility of selecting an architect for designing or erecting recreation centers, swimming pools, or other major structures. The selection of a landscape architect is, however, too often overlooked. The mere fact that an individual works with trees, shrubs, and plant materials is no indication that he or she has the competence to extend beyond a maintenance function to park or recreation area design. At the same time, there is no reason to expect a person trained in administration and management also to be a specialist in this area. Landscape architecture is today a profession. Hence the need to obtain a landscape architect is apparent, except in those instances when a department has a person in its employ with this competence.

As was implied earlier, the trend today is toward *functional design.* This term means design focused on the function, use, or purpose of an area or structure rather than on appearance or a preconceived plan. Too often, buildings or specialized facilities and other structures are mere copies of similar structures in nearby communities. Significant new developments are occurring in which design is not only functional but also beautiful. Therefore the administrator, in selecting either an architect or a landscape architect, should consider his or her qualifications in functional design as well as giving consideration to the following points.

1. *Is the architect familiar with recreation and park services insofar as they affect planning?* An architect should be familiar with recreation facility trends as well as having general knowledge of recreation practices conducted in a community.
2. *Has the architect had experience in planning and designing recreation structures or areas?* The selection of an architect should be limited to individuals who have had experience in planning recreation facilities and park sites. The structures that have been built should testify to the architect's ability as well as to the reputation he or she has gained. The prospective employer should visit some of the completed projects and discuss his or her work with former clients.
3. *Does the architect work well with clients, recreation advisors, and others?* No architect can plan and design areas and structures alone. He or she needs to be a part of the planning team and take part in all pertinent discussions relating to the job to be done. Therefore it is imperative that the architect be cooperative in his or her relationships and open-minded to suggestions.
4. *Does the architect show originality in design and freedom from stereotyped ideas that are hard to change?* Today the trend in architectural design is toward simplicity in style. Each structure or area presents a new or unique problem. Therefore the architect should be imaginative in thinking and free from preconceived ideas.
5. *Does the architect show a professional spirit by considering what is best for the community?* He or she should be a person of unquestionable character and integrity with a reputation for fair dealing, honesty, and sincerity.
6. *Does the architect have an adequate staff to do the job?* The architectural firm should be sufficiently competent in areas of specialization, and the architect should have executive ability adequate to keep things going harmoniously and on time.

Once the architect is selected, a written contract should be drawn. Such contract forms are available from architectural firms or may be drawn up by an attorney. In any case all negotiations should be handled through legal counsel. The contract should be specific as to the services required of the architect. Items should include the services related to the preparation of preliminary drawings, working drawings, specifications, and detail drawings; the taking of bids; the preparation of contracts; the issuance of certificates of payment; and general management and supervision of the job to be done.

## STEP 6—DEVELOPING THE PLAN

Once the site has been selected and purchased and every consideration given to size, type, function, and needs, it is necessary for the architect to prepare plans and specifications. First, the architect will prepare preliminary plans and sketches. At this point the design is at a fluid stage, for it is here that the architect is attempting to translate the myriad of program needs and ideas into a workable plan involving plants, grass, trees, wood, brick, and other materials. Tentative drawings and conceptions are passed on to the recreation and park executive, to the planning committee, or even to the recreation and park board. There is much give and take at this stage, and once decisions have been made, the preliminary plans should be submitted to the recreation board or other authoritative body for final approval.

The preliminary plans should be studied carefully. If they are not satisfactory, they should be reworked until all groups or individuals involved in the initial stage of planning are satisfied. Only then should the proper authority be asked to approve them and working drawings and specifications authorized and prepared.

It cannot be emphasized enough that changes are easy to make at the preliminary planning and design stage. Once final working plans have been prepared, it is difficult to initiate changes without additional expense and delay.

The preparation of working drawings and specifications is the work of the architect and his or her staff. Such drawings usually include a plot plan, elevations, landscape design, plants and materials, floor plans of buildings or structures, sections, details, plumbing, heating, ventilation, electricity, structural plans, and complete specifications.

On completion, these documents should be carefully examined to see that drawings and specifications are correct in every detail. They are then presented for final approval, after which plans should be made for inviting bids for construction.

### Developing the Site

In development of the site the principles of functional planning should be considered. It is important to develop recreation areas for most effective use within the fundamental planning areas of the neighborhood, community, city, or region. While functional planning involves the evaluation of needs of social groupings of people within identifiable geographical boundaries, experience has shown that all plans in the development of any area or structure must be addressed toward basic considerations such as those presented here.

1. *Program.* Does the site and facility plan grow out of community program needs? Is the space adequate for building and facility development to meet the needs of children, youths, and adults?
2. *Utility.* Are the proposed play areas, facilities, and structures so located with reference to one another that the public can use them with maximum ease and satisfaction? Does the plan provide for

multiple use of the same space? Are adequate toilet facilities, drinking fountains, benches, and other conveniences provided?

3. *Safety*. Are the areas segregated adequately to protect against danger from flying balls, motor traffic, or other hazards? Are the play areas or facilities designed to provide a positive influence toward safe and healthful practices?
4. *Beauty*. Is the area attractive and pleasing in appearance? Does it make judicious use of shrubs, plants, lawns, and good architectural design?
5. *Adaptability*. Can the area or facilities be adapted easily to meet changing recreation demands with a minimum of cost? Will proposed structures or buildings also be flexible to changing needs? Can buildings be extended without destroying the design of the whole?
6. *Use of Natural Features*. Is the area fitted to the site, and does it make use of features such as natural slopes for outdoor theaters or winter sports, groves of trees for picnicking, nature study, or day camping, or land areas for sports, games, and athletics?
7. *Economy*. Are the areas and facilities planned for economy of construction and operation to get the most in recreation use for every dollar spent? Are they also designed for a minimum of maintenance costs?
8. *Interfunctional Arrangement*. Are the areas and facilities so planned that activity relationships are coordinated harmoniously for age groupings and ease of supervision?
9. *Accessibility*. Is the site designed to provide easy access from all sides?

## Summary

The planning of areas and facilities for recreation calls for vision, imagination, and clear purpose with regard to the direction in which the recreation service should move. It is the job of the planner not only to know the objectives toward which all activity is directed but also to know the means of accomplishing these purposes through use of structures and areas. It is not enough to build monuments to aesthetic tastes or civic pride; more is required to open the frontier of recreation for the joy of living and to provide human understanding through participation in creative and cultural activities involving imaginative physical facilities. In short, the areas, buildings, and facilities for recreation should fit the community and accomodate its varied interests and needs.

## Questions for Discussion

1. Explain fully the statement "Program goals determine recreation areas and facilities."

2. In the planning of recreation areas and facilities it is not uncommon to establish a community advisory planning committee. State the advantages of having such a committee. Can you see any disadvantages?
3. What kind of professional architectural service would you use in developing your recreation plan? How would you select the architect?
4. Explain the ways by which land may be acquired for recreation or park use. Which is the most common device for acquiring such land?
6. What is meant by the *right of eminent domain*? Do you feel its use is fair? Why or why not?
7. Cite examples of how land may be acquired through "excess purchases," "sanitary landfills," or "easements."
8. In developing a park area, what basic considerations would you keep in mind to achieve a functional plan?
9. Explain briefly the steps you would follow in planning for community recreation areas and facilities.

## SELECTED REFERENCES

The Athletic Institute and American Association for Health, Physical Education and Recreation, *Planning Facilities for Athletes, Physical Education and Recreation*. The Athletic Institute and American Association for Health, Physical Education and Recreation, Chicago, 1974.

Bannon, Joseph J., *Leisure Resources: Its Comprehensive Planning*. Prentice-Hall, Englewood Cliffs, N.J., 1976.

Buechner, Robert D., *National Park Recreation and Open Space Standards*. National Recreation and Park Association, Washington, D.C., 1971.

Butler, George D., *Introduction to Community Recreation* (5th ed.). Mc-Graw- Hill, New York, 1976.

Kraus, Richard G., and Curtis, Joseph E., *Creative Administration in Recreation & Parks* (2d ed.). Mosby, St. Louis, Mo., 1977.

McLean, Mary, *Local Planning Administration* (3d ed.). The International City Manager's Association, Chicago, 1959.

Whyte, William H., *The Last Landscape*. Anchor/Doubleday, Garden City, N.Y., 1970.

# CHAPTER 16

# Areas and Facilities: Standards and Types

The areas and facilities available for recreation determine to a large extent the forms of recreation activity in which people engage. It is not possible for most individuals or families to acquire parkland or develop recreation areas or facilities. Consequently, it has become necessary for communities to utilize their governmental powers to furnish and develop ample recreation resources.

Before this century little attention was given to the acquisition and development of properly located park and recreation areas of suitable size to serve certain recreation functions, but with the establishment and expansion of recreation programs in many communities, the need for areas and facilities to serve specific uses became apparent. In time, the demand for recreation areas and facilities of different types focused on the need for providing and developing suitable space standards.

In the field of planning the uses of area or space standards are basic. It is impossible to develop a long-range plan of community recreation and park needs and resources without the use of such a guide. Furthermore, only through standards can we evaluate proposals for recreation development. Indeed, a prerequisite of programming planning is the adequacy of areas and facilities.

## PURPOSE OF STANDARDS

The purpose, or objectives, for which standards have been established has been stated by many planning bodies. All such statements have in common the checking of areas and facilities against a working measure or ideal that has official endorsements by a recognized national recreation or planning organization. The dictionary (*Webster's New Collegiate Dictionary*, Seventh Edition, 1976, p. 1133) definition of standard, "something established by authority, custom, or general consent, as a model or example: CRITERION," implies that a standard is a model or exact measurement, but most definitions used by agencies and organizations have in common the goal of reaching a desirable level of operation or attainment.

Most government bodies use standards, which have been proposed in most instances by national organizations concerned with promoting a high attainment of professionalism in recreation planning. The standards proposed by the National Recreation and Park Association are widely used and have been well thought through. The standards of this organization have been used by planning commissions and other planning bodies in many cities of the United States, and have been incorporated in many publications of agencies of the federal government.

Standards must be evaluated in terms of local development and cannot be applied without change, for all municipal and urban areas have different problems and varying local conditions and resources. Therefore standards must be used judiciously as basic norms subject to modification as the need arises.

## STANDARDS VERSUS GOALS

The minimum standards recommended in this chapter are not absolutes. All standards must be evaluated in terms of local development and cannot be applied without change, for all municipal and urban areas have different problems and varying local conditions and resources. For example, it may not always be possible or desirable to follow such standards in certain built-up areas of a city. On the other hand, it is generally these older more crowded parts of a city that are most in need of good recreation facilities, and advantage should be taken in these areas of any opportunity to acquire suitable land that can be turned to recreation purposes.

## GROSS RECREATION ACREAGE

For many years a generally accepted standard for recreation space has been one acre of land for each 100 persons in the community. This standard is still widely accepted and used but many communities and regional planning agencies have increased the standard to two or even more acres of parkland for each 100 residents. The National Recreation and Park Association recommends that three acres of land distributed in neighborhood, district, and large urban and regional parks be set aside for each 100 persons (see Table 16-1). In

**Table 16.1 Recommendations by Classification and Population Ratio**

| *Classification* | *Acres/ 1000 People* | *Size Range* | *Population Served* | *Service Area* |
|---|---|---|---|---|
| Playlots | [a] | 2,500 sq ft to 1 acre | 500–2,500 | Subneighborhood |
| Vest-pocket parks | [a] | 2,500 sq ft to 1 acre | 500–2,500 | Subneighborhood |
| Neighborhood parks | 2.5 | Min. 5 acres up to 20 acres | 2,000–10,000 | 0.25–0.50 mile |
| District parks | 2.5 | 20–100 acres | 10,000–50,000 | 0.50–3 miles |
| Large urban parks | 5.0 | 100 + acres | One for ea. 50,000 | Within ½ hr driving time |
| Regional parks | 20.0 | 250 + acres | Serves entire population in smaller communities; should be distributed throughout larger metro areas | Within 1 hr driving time |
| Special Areas and Facilities | [a] | Includes parkways, beaches, plazas, historical sites, flood plains, downtown malls, and small parks, tree lawns, etc. *No standard is applicable.* | | |

*Source: Robert D. Buechner,* National Park Recreation and Open Space Standards, *National Recreation and Park Association, Washington, D.C. 1971.*

[a]Not applicable.

By Percentage of Area

The National Recreation and Park Association recommends that a minimum of 25 percent of new towns, planned unit developments, and large subdivisions be devoted to park and recreation lands and open space.

addition, The National Recreation and Park Association recommends that 25 percent of the land in new towns, planned unit developments, and large subdivisions be set aside for park and recreation use and open space.

The development of new subdivisions in recent years has reflected some trends that must be reviewed when a community considers open space standards. The traditional pattern of rectangular blocks and parallel streets has lost ground to new planned unit developments where living units are clustered together, leaving much more open space in the development. In fact, many zoning ordinances adopted by cities require a large percentage of the area in a planned unit development be left in open space. Many of these new developments also provide a recreation facility for the homeowners. These facilities may include a club room, game room, tennis courts, swimming pool, and the like. This open space and the recreation facilities will tend to reduce the demands on the public parks and recreation facilities in that community, and therefore, the local recreation and park agency can make adjustments in the requirements of the open space standards.

Standards as adopted by a community should also include a special emphasis on service areas. For example, a city may have more park acres than the standard recommends, but if the acreage is contained in one park on the edge of the city, the citizens are not served as well as by another city with less park acres but with the acreage distributed in several parks throughout the city. This service area requirement is included in the standards recommended by The National Recreation and Park Association, table 16-1.

## STANDARDS BY TYPES OF RECREATION AREAS

As indicated, park and recreation lands should include service area standards, as well as a variety of facilities, classified according to function. These facilities can be classified as mini parks, also referred to as playlots or vest-pocket parks, neighborhood parks, district or community parks, large urban parks, regional parks, and special facilities that include greenbelts, parkways, conservation or nature areas, and historic sites.

### The Mini Park

The mini park, playlot or vest-pocket park is a small area usually 2500 sq. ft. to 1 acre designed to provide recreation opportunities for a small neighborhood area. Generally, the mini park is designed for the play of preschool age children; however, in some cases a mini park may be designed for aesthetic purposes to improve a blighted area. An example of this type of park might be found in a business area where a vacant lot that had become a collecting point for trash and rubble was developed into a small park, greatly enhancing the business area. These parks would not include play apparatus for children but might include sitting areas, floral display areas, walkways, and possibly an ornamental fountain.

The neighborhood playlots would generally include safe and attractive play equipment, an open space area for games, a concrete walk for tricycles and big wheels, a sitting area for parents, and possibly a sand box.

### The Neighborhood Park

The neighborhood park usually ranges from 5 to 20 acres. In fact, 5 acres should be considered as an absolute minimum for a neighborhood park. The service area should be 0.25 to 0.50 mile, thus a person should not have to walk more than 0.50 mile to be served by a park in the community. The neighborhood park is designed primarily to serve the recreation needs of children 6 to 15 years of age. However, the park should include a play area for preschool children, much like the playlot, areas or facilities for senior citizens, and family picnic areas. Table 16-2 shows the space standards for a neighborhood park as recommended by The National Recreation and Park Association.

As indicated in table 16-2, an ideal location for a neighborhood park is adjacent to the neighborhood school. Many recreation and park departments and local school boards have joint agreements calling for the development of joint neighborhood facilities. Some typical agreements may be found in Appendix F. This joint development of park–school sites has also been advanced by land dedication ordinances in many communities. These ordinances require that a percentage of land and/or cash be given for parks and schools from each new housing development. In addition to the park–school possibilities, the recreation and park authorities should also look at the possibilities of joint development with other public agencies such as public housing authorities, libraries, and public health centers.

**Table 16.2 Space Standards for Neighborhood Parks**
Suggested space standards for various units within the park. The *minimum* size is five acres.

| | *Area in Acres* | |
|---|---|---|
| *Facility or Unit* | *Park Adjoining School* | *Separate Park* |
| Play apparatus area—preschool | 0.25 | 0.25 |
| Play apparatus area—older children | 0.25 | 0.25 |
| Paved multipurpose courts | 0.50 | 0.50 |
| Recreation center building | a | 0.25 |
| Sports fields | a | 5.00 |
| Senior citizens' area | 0.50 | 0.50 |
| Quiet areas and outdoor classroom | 1.00 | 1.00 |
| Open or "free play" area | 0.50 | 0.50 |
| Family picnic area | 1.00 | 1.00 |
| Off-street parking | a | 2.30[b] |
| Subtotal | 4.00 | 11.55 |
| Landscaping (buffer & special areas) | 2.50 | 3.00 |
| Undesignated space (10%) | 0.65 | 1.45 |
| Total | 7.15 acres | 16.00 acres |

*Source: Robert D. Buechner,* National Park Recreation and Open Space Standards, *National Recreation and Park Association, Washington, D.C. 1971.*

[a]Provided by elementary school.
[b]Based on 25 cars @ 400 sq ft per car.

## The District Park

The district or community park is generally 20 to 100 acres in size with a service area of 0.50 to 3 miles. This park is designed for youth 15 years of age or older and adults; however, the park should include facilities for preschoolers, elementary age children, senior citizens, and families. In fact, the district park usually includes those facilities listed for the neighborhood and mini parks.

The park may provide the following features: preschool and children's play apparatus areas; sports fields for baseball, softball, soccer, tennis, basketball, volleyball, and the like; a recreation center; swimming pool; bicycle and jogging trails; open turf areas for games and activities; a picnic area, and possibly special areas such as an outdoor theater, horseshoe courts, and the like. A traditional band shell may be included, but many recreation and park agencies today are using movable self-contained stage units instead. These units are less expensive and provide the agency more flexibility, since they can be moved to present performing arts concerts in many parks throughout the community.

The space standards for district parks as recommended by The National Recreation and Park Association are shown in Table 16-3.

## The Large Urban Park

This park is an area, usually at least 100 acres in size, with open meadow, water, or woodland that offers an attractive setting and a suitable environment for people to engage in activities not possible in smaller recreation areas. These parks afford an opportunity for people to get away from an urban environment, to enjoy contact with nature, and to engage in a variety of activities that lend themselves to a large park setting. Opportunities may be provided for family and group picnicking, day camping, nature study, horseback riding, boating, swimming, fishing, winter sports, and the like. If fully developed, the area may provide golf courses, arboretums, botanical and zoological gardens, and resident camps. Its value lies in the effective use of natural features and unusual development.

Some authorities recommend 100 acres as the minimum size for this type of park, although an area of less than 100 acres may be adequate in a small city. Because it is not always possible to acquire sufficient acreage within municipal boundaries, this type of park is often located in outlying areas. Actually, it may serve a number of cities or a section of a metropolitan area, and it is not unusual for people to come from throughout the region to make use of its unusual development or to participate in specialized activities. Every metropolitan community needs a park of this type, and it has been suggested that one be provided for every 50,000 of the population, designed to serve all ages.

**Table 16.3 Space Standards for District Parks**
Suggested space requirements for various units within the park. The *minimum* size is 20 acres.

| | *Area in Acres* | |
|---|---|---|
| *Facility or Unit* | *Park Adjoining School* | *Separate Park* |
| Play apparatus area—preschool | 0.35 | 0.35 |
| Play apparatus—older children | 0.35 | 0.35 |
| Paved multipurpose courts | 1.25 | 1.75 |
| Tennis complex | 1.00 | 1.00 |
| Recreation center building | [a] | 1.00 |
| Sports fields | 1.00 | 10.00 |
| Senior citizens' complex | 1.90 | 1.90 |
| Open or "free play" area | 2.00 | 2.00 |
| Archery range | 0.75 | 0.75 |
| Swimming pool | 1.00 | 1.00 |
| Outdoor theater | 0.50 | 0.50 |
| Ice rink (artificial) | 1.00 | 1.00 |
| Family picnic area | 2.00 | 2.00 |
| Outdoor classroom area | 1.00 | 1.00 |
| Golf practice hole | [a] | 0.75 |
| Off-street parking | 1.50 | 3.00[b] |
| Subtotal | 15.60 | 28.35 |
| Landscaping (buffer & special areas) | 3.00 | 6.00 |
| Undesignated space (10%) | 1.86 | 3.43 |
| Total | 20.46 acres | 37.78 acres |

*Source: Robert D. Buechner,* National Park Recreation and Open Space Standards, *National Recreation and Park Association, Washington, D.C. 1971.*

[a]Provided by Junior or Senior high school.
[b]Based on 330 cars @ 400 sq ft per car.

## The Regional Park

The regional park is very similar to the large urban park in that it is normally designed to include woods, meadows, lakes, and other natural features. The recommended minimum size for the regional park is 250 acres and it should be located within a one-hour driving time of the people it is expected to serve. Like the large urban park, the regional park may include areas for family and group picnicking, nature study, horseback and bicycle riding, boating, swimming, fishing, and winter sports. In addition to day camping, the larger regional parks may include facilities for overnight camping. The park may also include special facilities such as golf courses, arboretums, botanical and zoological gardens, shooting ranges, cross-country ski trails, and sports complexes.

Many highly developed suburban parks having adequate acreage are of this type. They consist of areas within driving or boating distance where people go for a day's outing of fun and relaxation. Also included should be the large highly developed metropolitan park. These parks are both local and regional in scope. In fact, some of them draw more people from outside the city limits than from within them. Moreover, some may attract thousands of visitors daily. It can be readily seen that these parks are unique in development. Examples are Golden Gate Park in San Francisco, Griffith Park in Los Angeles, Hermann Park in Houston, Fairmount Park in Philadelphia, Central Park in New York City, and Grant Park in Chicago.

## The Parkway

A parkway "is essentially an elongated park with a road running through it, the use of which is restricted to pleasure traffic."[1] This type of property, when found in a regional park system, is intended to serve as a pleasure drive from the city to the surrounding area, and in some instances it provides a link with the various units of the park system. Parkways are usually limited-access highways that are planned as an integral part of a metropolitan highway system. They usually follow the course of natural boundaries such as streams, rivers, lakes, or coastal areas. To enhance the scenic effect, an effort is made to preserve the natural beauty of the landscape on each side of the highway. Two hundred feet is considered the minimum desirable width for a parkway, but it should be sufficiently wide along certain sections to provide for scenic stops, a picnic area, or other special recreation features as needed.

## Greenbelts

Greenbelts are open spaces that are developed along rivers and streams, irrigation canals, abandoned railroad and transmission line right-of-ways, and floodplain areas. Park development may include jogging, bicycle and horseback riding paths, fishing and picnic areas, and general aesthetic improvement.

## Conservation and Nature Areas

These areas are usually large tracts of land kept primarily in their natural state. They are used to preserve natural areas such as forests, deserts, marshlands,

[1]George D. Butler, *Introduction to Community Recreation*. New York: McGraw-Hill, 1949.

mountain areas, and other lands of recreational and scenic interest. Properly developed and controlled, they may provide for winter sports, hiking, camping, horseback riding, picnicking, nature study, fishing, swimming, boating, and other water sports. In many instances these areas are provided by the county, state, or federal government rather than by the municipality. Such large reservation areas usually consist of 500 to 1000 acres or more.

## Historic Sites

The U.S. bicentennial created great interest in historic sites at the national, state, and local levels. Organized movements in many communities have been successful in saving historic buildings such as railroad depots, early banks, homes, and even several blocks of a downtown area. Many local recreation and park departments have become the recipients of these sites. In many instances the recreation and park department has been the leader in the historic movement by applying for and receiving a matching fund grant to purchase and/or rehabilitate the facility. These sites have been redeveloped and opened to the community where they have become a center for community pride and involvement. As such, they can be valuable assets to the recreation and park department.

## Specialized Areas and Facilities

Specialized facilities may be developed separately or as part of a larger park development, depending in part on the particular facility. There are a great variety of these specialized recreation facilities, and park planners should give careful consideration to the wide range of possibilities in deciding which facilities are going to be provided. Table 16-4 gives the space requirements of a number of outdoor facilities. Such areas or facilities are often incorporated as parts of district, large urban, and regional parks. Nonetheless, the special facilities described in the table are important to every community.

**Recreation Center.** The recreation center may be found at the neighborhood level and in the district and large urban parks. At the neighborhood level the recreation center is usually constructed to be between 15,000 and 20,000 sq. ft. It is designed to serve approximately 8000 people and generally will include multipurpose rooms, an arts and crafts area, game room, kitchen, lounge and lobby, rest rooms and office. If a gymnasium is not available in a neighborhood school, the recreation center may also include a gymnasium and locker room facilities.

The recreation center in a larger park that serves a community or city area will be considerably larger, from 20,000 to 40,000 sq. ft., and will include several multipurpose rooms, gymnasium, shower and locker rooms, game room,

**Table 16.4 Outdoor Recreation Facilities—Space Requirements**

| *Facility* | *Area* | *Dimensions Recommended* | *Standard per Population* |
|---|---|---|---|
| Arboretum | — | — | 1 for each 10,000 |
| Archery range | 300′ | 50′×450′ | 1 for each 1,500 |
| Baseball diamond | 90′ diamond | 300′×300′ | 1 for each 6,000 |
| Bicycle trail | — | — | 1 for each 2,500 |
| Bridle trail | — | — | 1 for each 2,500 |
| Bowling green | 14′×110′ | 120′×120′ | 1 for each 1,500 |
| Boating facility | — | — | 1 for each 2,500 |
| Band shell | — | — | 1 for each 10,000 |
| Botanical garden | — | — | 1 for each 10,000 |
| Basketball court | 50′×94′ | 60′×150′ | — |
| Boccie area | 18′×62′ | 30′×80′ | — |
| Croquet area | 30′×60′ | 30′×60′ | — |
| Casting pool | — | — | 1 for each 2,500 |
| Camp | — | — | 1 for each 10,000 |
| Football field | 160′×360′ | 180′×420′ | — |
| Handball court | 20′×34′ | 30′×45′ | 1 for each 1,500 |
| Horseshoe area | 40′ spacing | 12′×50′ | — |
| Recreation pier | — | — | 1 for each 2,500 |
| Iceskating area | — | — | 1 for each 2,500 |
| Shuffleboard court | 6′×52′ | 10′×64′ | 1 for each 1,500 |
| Golf course (9 holes) | 50 acres | — | 1 hole for each 3,000 |
| Golf course (18 holes) | 100 acres | — | — |
| Soccer field | 210′×330′ | 240′×360′ | 1 for each 1,500 |
| Softball diamond | 60′ diamond | 250′×250′ | 1 for each 3,000 |
| Tennis court | 27′×78′singles<br>36′×78′doubles | 50′×120′<br>60′×120′ | 1 for each 2,000 |
| Shooting range | — | — | 1 for each 1,500 |
| Volleyball court | 30′×60′ | 50′×80′ | — |

arts and craft area, an auditorium or areas for performing arts, class or club rooms, kitchen, large meeting room, restrooms, office, lounge or lobby, and some specialized areas such as a ceramics workshop or weight room.

An important consideration in all recreation facilities is to provide adequate storage space.

**Performing Arts Center.** Many recreation and park agencies, especially in the larger cities, operate a performing arts center or facility. The center might be a building with a large auditorium and stage and smaller theatre areas, or it could be a major outdoor facility such as the Philadelphia Dell Performing Arts area or the Hollywood Bowl. Many smaller communities feature a band shell in their central park area, or they may provide programs in outdoor amphitheaters. The use of mobile stage units and other mobile recreation units to present recreation programs in many locations throughout the community is also growing very rapidly.

**Swimming Pools.** The generally accepted standard for swimming facilities is that each should be able to serve 3 percent of the population at one time and that 15 sq. ft. of water surface be provided in the community for each 3 percent of the population. The deck space should be considerably larger, preferably double the size of the water area. It is highly desirable to include adequate parking and land for related activities. Bathing beaches, in particular, should provide for easy access, parking, and sports and game areas usually associated with a beach.

**Tennis Centers.** Indoor tennis and racquetball facilities are being constructed by many recreation and park agencies, especially in areas where weather restricts play on outdoor courts during the winter months. The facilities include the court area with eight or more courts, entrance lounge, locker room facilities including sauna and whirlpool installations, and concession areas. Many facilities also include weight and exercise areas, a nursery to take care of small children, and a pro shop.

**Ice Skating Centers.** Ice skating centers like the tennis centers have been increasing very rapidly. Some recreation and park agencies operate these centers on a year-round basis conducting hockey, figure skating, and learn-to-skate programs even during the summer months. These centers include one or more ice areas, one of which should be regulation hockey size. Seating areas should be provided, especially for the hockey areas, and the facility should also include locker rooms, a pro shop, an area for renting and sharpening skates, a concession area, restrooms, and if possible, separate team dressing rooms for the hockey program.

**Golf Courses.** Golf courses need 50 acres or more of land for nine holes and at least 100 acres for an 18 hole course. It is desirable for the land to have topographical features such as uneven terrain, to be partly wooded, and to be within a reasonable distance of the population to be served. Property located near or beyond the city limits is easiest to acquire for this activity. Furthermore, the development of a suitable clubhouse that can be used for social and other indoor activities by community groups is desirable. Imaginative planning could foresee construction of a suburban unit including golf, tennis, swimming, and diving facilities, shuffleboard courts, and the many other outdoor facilities commonly associated with private country clubs located strategically in urban or metropolitan areas.

Not to be overlooked is the par-3 golf course, sometimes referred to as the "pitch-and-putt" course. This facility has proven to be extremely popular. Reducing a nine-hole golf course from 50 to 20 acres, more or less, makes possible participation for those who wish to spend less time on the course.

**Athletic Field or Stadium.** The athletic field or stadium is usually located on a high school or college campus. However, such units are found in a number of public park and recreation departments. The minimum space needed for an athletic field is 5 acres, but 10 to 20 acres are preferable. To seat spectators and provide parking, dressing rooms and rest room facilities would require 20 or more acres. The athletic field usually provides for a football field, and/or soccer field, space for track and field events, and, occasionally, a baseball diamond.

**Baseball and Softball Diamonds.** The interest in a community toward baseball and softball usually indicates how extensively facilities are developed for these sports. However, the generally accepted standard calls for one baseball diamond for every 6000 inhabitants and one softball diamond for every 3000. Here again, such facilities are usually included in the neighborhood and district or community parks. The diamonds in the neighborhood parks generally serve elementary age children while those in the district and larger parks generally serve the organized sports leagues and as such are usually lighted.

**Tennis, Volleyball, and Basketball Courts.** Tennis, volleyball, and basketball courts are generally provided in all parks serving the community with the possible exception of tennis courts in some smaller neighborhood parks. A tennis court (27' x 78' single) and (36' x 78' double) requires 7200 sq. ft. of space. It is recommended that they be built in units of four or more courts, since this reduces the initial construction cost and provides for better program operation. A city should have one tennis court for every 2000 inhabitants. It is also recommended that a city provide one basketball court for every 500 residents. There is no accepted standard for volleyball courts but this sport is growing in popularity and like tennis and basketball, they should not be overlooked in recreation and park planning. It is particularly desirable that some courts be lighted for night use.

**Public Camps.** Until recent years, camping as an individual, group, or family activity was of little concern to park and recreation authorities at the municipal level. Today, however, many municipal recreation agencies have acquired and developed both family and resident camp properties. Sites for camps are sometimes included as a part of the large recreation park or the reservation. Separate campsites located in areas far from the city are sometimes acquired. The purpose of the camp should determine the type of property to be acquired, although outdoor living should be the major concern of all camp activity. Day camps should be located close to the people to be served. Organized resident camps for children and youths may be located some distance from the urban area, and so may family camps. The generally accepted standard for camp space is that 1 acre of land should be available for each camper in the organized resident camp. It is desirable, of course, that such land be in, or contiguous to, large expanses of public lands. The site should meet the needs of the campers, but it is preferable that the property be secluded, conducive to outdoor living, near a body of water for boating and swimming, and advantageous for health and safety.

**Other Areas.** Other areas that should be considered in planning park and recreation facilities based on the community needs and interests are:

Archery area
Bicycle paths
Boccie courts
Community gardens
Horseshoe courts
Jogging paths
Model airplane areas
Soccer fields
Artificial ice rinks
Bridle paths
Cross-country ski course
Marinas
Shuffleboard courts
Trap & skeet shooting areas
Roller skating areas

## SPECIAL REQUIREMENTS FOR HANDICAPPED

It is important that recreation and park planners consider the special needs of handicapped persons when designing recreation facilities. In fact, state and federal laws make it mandatory in many cases for the facility to be designed to serve handicapped persons. Even without the mandatory requirement, a little extra planning can open up new worlds for the handicapped person. The recreation and park executive should obtain a listing of the state and federal requirements and other guides that will help in this planning.[2]

## STANDARDS AND PLANNING RELATIONSHIPS

Attention has been given to types of recreation areas and facilities. While it is desirable to look at the parts to understand the whole, care must be taken to point up the need for an integrated park and recreation system. The planning units of the neighborhood, community, city, and region are related and must be considered as an integrated unit in metropolitan planning. The large recreation park is not divorced from a regional concept, nor is the mini, neighborhood, or district park. Their only distinction has to do with their function. The same groups are served by all. Indeed, the major criteria are based on frequency of use, travel distance, and wider recreation opportunities. On the other hand,

[2]An excellent list of standards and requirements for handicapped persons may be found in the publication, *Planning Facilities for Athletics, Physical Education and Recreation*. Chicago: The Athletic Institute and American Association for Health, Physical Education, and Recreation, 1947, p. 77.

the public schools and other public properties including armories, fairgrounds, public auditoriums, libraries, and the like must also be considered in overall recreation and park planning. While each has a primary function other than public recreation, the concept of integrating these physical resources with public recreation and park properties is not new. Great strides have been made in the direction of designing school plants for school and community recreation use. Much attention has been given to the "community park–school" concept. The rationale underlying this planning idea is based on the principle that integrated planning for recreation use of public areas and facilities makes for economy of operation. In other words, duplication of facilities and areas can be eliminated, thus ensuring maximum returns on the tax dollar.

## Planning Units

Urban areas are commonly divided into units for planning purposes. The "neighborhood" and the "community" are generally chosen as basic planning areas because of their identifiable nature involving definite physical boundaries and social groups. In addition to these two units, the "city" and the "region" have identifiable characteristics and are also considered as planning units. Using these four divisions of urban areas, one can classify recreation areas and facilities into these planning units. For example, the mini park, and the neighborhood park serve the area known as the "neighborhood"; the district or community park serves the "community"; the large recreation park serves the "city"; and the regional park, conservation or natural areas, and the like serves the "region." In addition to those areas there are the special use facilities listed earlier. This simple grouping of four basic types of areas and parks has merit in urban planning.

## Questions for Discussion

1. Define the term *standard*.
2. Cite the space and service area standards for a neighborhood park, district park, and large urban park.
3. What recreation areas and facilities are found at the neighborhood level? the community level? the regional level?
4. What distinguishes a regional park from a city or large urban park?
5. Describe the types of mini parks.
6. What facilities should a neighborhood recreation center include?
7. Describe some types of special recreation facilities.
8. Why is it important to consider special requirements for handicapped persons when planning recreation and park facilities?

## SELECTED REFERENCES

The Athletic Institute and American Association for Health, Physical Education and Recreation, *Planning Facilities for Athletics, Physical Education and Recreation*. The Athletic Institute and American Association for Health, Physical Education and Recreation, Chicago, 1974.

Bannon, Joseph J., *Leisure Resources: Its Comprehensive Planning*. Prentice-Hall, Englewood Cliffs, N.J., 1976.

Buechner, Robert D., *National Park Recreation and Open Space Standards*. National Recreation and Park Association, Washington, D.C., 1971.

Butler, George D., *Recreation Areas: Their Design and Equipment* (2d ed.). McGraw-Hill, 1959.

Butler, George D., *Introduction to Community Recreation* (4th ed.). McGraw-Hill, New York, 1967.

Gold, Seymour M., *Urban Recreation Planning*. Lea and Febiger, Philadelphia, 1973.

# CHAPTER 17

# Office Management

Every recreation and park system makes use of an office, and every office must perform a service function or it has no reason for existing. The office is a preparing, record-keeping, filing, and communicating center. It handles many details and performs a multitude of tasks that help keep the flow of work moving and increase the overall efficiency of the organization. Every recreation and park administrator needs to know the elements of efficient office organization and administration. The administrator cannot delegate this entire function to subordinates regardless of the size of the organization. The administrator must be concerned with reading and research, writing and correspondence, communication, receiving of office visitors, meeting with committees and staff, study of reports, storage of information, ordering of supplies and equipment, and preparation of memoranda and reports. Therefore he or she must develop efficient office procedures in performing these tasks.

## OFFICE SPACE AND LOCATION

It is highly desirable that the recreation and park office be centrally located for good accessibility to the people to be served. In many instances this location is predetermined; it is one of the city offices in the municipal building or, if a county or school operation, is located in the county courthouse or school administration building. There are many advantages to having the general administrative offices grouped together in the same building. This is also true for the larger departments that have separate offices for program, maintenance, planning, finance, and the like.

Some recreation and park systems have their offices located in parks, recreation centers, stadia, or commercial office buildings. Still others have renovated an abandoned home, school building, fire station, or warehouse to serve this function. Regardless of location, if the office serves the purpose of the organization, is accessible to the public, and has adequate space, it can perform its function to advantage.

## FUNCTIONS OF THE OFFICE

The functions that an administrator needs to understand for efficient office management, mentioned in the previous section, will be discussed briefly under (1) office routine, (2) reading and research, and (3) reports.

### Office Routine

Much of the work in an office is of a routine nature. These routines are common to all offices and call for efficiency and smoothness of operation. In fact, the efficiency and dispatch by which these routines and details are handled reflect the quality of departmental service. Such routines are explained under office hours, correspondence, filing, sources of information, telephone use, appointments, and office visitors.

**Office Hours.** In all recreation and park systems definite office hours should be maintained. This is especially important for the smaller departments that have limited or no secretarial help. On the other hand, the larger departments should also maintain an office schedule whereby the public knows when the staff is available to provide assistance or information. Some departments make no effort to have the office open at a regular time and visitors may come to a closed office that gives no indication of when it will be open. A posting on the door of office hours and a telephone number to call in case of emergency is very helpful. This is particularly needed if the office is closed for the noon hour, closed before 5:00 P.M. or closed on Saturday.

**Correspondence.** Every administrator receives letters and other communications, many of which require a reply. Dispatch in answering correspondence is one mark of a good administrator. Not to answer promptly is a sign of discourtesy and inefficiency, in the eyes of the sender. Consequently, it is good practice to answer correspondence within 24 hours, or 48 hours at the most.

Additional suggestions on the handling of correspondence are as follows.

1. The letter should be given to the person who is best able to answer it. In other words, not every letter needs to be answered personally by the recreation administrator, unless it calls for his or her attention.

2. All letters should be answered in a clear, concise, complete, and courteous manner. Needless to say, they should be written in good English.
3. Letters of a routine nature should be delegated generally to the secretary or clerical force for reply. However, what constitutes a routine reply should be clear to them. A standard form is sometimes used in these instances.
4. All correspondence should be copied and copies properly filed.
5. All outgoing letters should be read for possible errors before a signature is affixed. Many administrators feel that it is not necessary to affix their degree titles after their signatures or on letterheads, except where it is common practice at institutions of higher learning.
6. Letters should be dictated into the dictaphone whenever possible, since it saves the secretary time. If not used, a definite time for dictation should be established as this allows the secretary to plan a work schedule.
7. Incoming mail should be opened, sorted, and placed on the proper desk or in the mailbox center by the secretary or clerical staff. "Incoming" and "outgoing" desk trays are used in some offices to facilitate the handling of correspondence. Outgoing mail should be placed in one location where it can be picked up and processed for mailing.
8. All outgoing mail should be stamped by a postage meter, if one is available. A postage meter is faster, reduces the chance of stamp theft, and permits a special message to be printed on the envelope if desired.

**Filing.** Filing is the system of storing information so that it will be readily available when needed. It is essential for every office. In fact, securing file information at a moment's notice is a work of efficiency. It is important that the responsibility for filing be clearly defined. Generally speaking, the two types of material stored for future reference are correspondence and sources of general information. Both may be filed according to a variety of filing methods. However, it is common practice in recreation and park organizations to use the following filing systems: (1) alphabetical, (2) numerical, or (3) subject.

1. *Alphabetical.* The alphabetical system is a simple and popular method of filing correspondence and other materials under a writer's name, organization, or subject. Materials are placed in the file according to an alphabetical arrangement.
2. *Numerical.* The numerical system designates a number to each item to be filed according to a predetermined index or code. Thus materials are placed in the file according to the numbers attached to the

folder or divider. Actually, the numerical file system makes for easy reference, and there is little cause for errors. However, it is not widely used, except in large offices that have extensive filing and trained file clerks.

3. *Subject.* The subject filing system is the method of filing material according to an alphabetical subject arrangement. It is necessary here to classify a list of subject headings and to develop an index of these headings to prevent overlapping and duplication. Material on budgets, committees, contracts, publicity, personnel, and the like can easily be filed by subject.

Every office should have inactive, or storage, files in addition to active files. Periodically, it is desirable to transfer materials from the active to the storage files. Too often, files are cluttered with useless out-of-date material that is seldom, if ever, used. Such materials can be either destroyed or placed in storage. Close scrutiny should be given to file materials that are transferred, to prevent valuable documents from being misplaced.

Attention should also be called to the borrowing of information from files. No file material should ever be borrowed without a record being kept of what was taken, when it was taken, and the name of the taker.

**Sources of Information.** Every recreation and park system needs to make available to its staff certain kinds of reference material and information. Such material is usually located and available at the main office. Books, magazines, pamphlets, reports, maps, directories, policies, laws, regulations, and operational manuals should be found here. Some departments have a special library for use of the staff and include these materials as well as other pertinent reference works that are frequently needed. Staff responsibility for maintaining the library should be assigned and a system of checking out such materials provided.

**Telephone Use.** Much departmental business is transacted over the telephone. Hence it is an essential instrument in the efficient operation of an office. But its use requires certain rules.

1. Attending to telephones is necessary. In the small office that does not have a secretary on duty, the telephone can easily be answered by someone in another office or even by an answering service after a certain number of rings. The telephone company can easily adjust the instrument to make this possible.
2. Answering telephones properly is an important part of personnel training. The name of the department or organization should be given to the caller by the person answering or when the call is switched to another office. Also, it is proper for the person answering to give his or her name and possibly title.

3. Informing the caller that the person being called is away from his or her desk is important. For example, the secretary can say, "Mr. Smith is away from his desk at the moment. May he call you when he returns? Or can someone else help you?" Or, if the person being called is away for an extended period, the secretary can say, "Ms. Jones will not be in until this afternoon. Do you wish to leave a message or have her call you? Or can someone else help you?" When calls are taken by a secretary or switchboard operator, a form to record the message should be provided to be placed on the desk of the person being called. A simple 3″×5″ form that provides space for inserting the time, date, name of caller, phone number, name of person taking call, and message will usually suffice.
4. Asking the caller if the call can be returned if the person being called is not available is proper procedure. It is extremely discourteous for a person to accept routine telephone calls when he or she is in conference with another individual. Unless the call is urgent, the caller has no priority to interrupt the business being carried on at the moment. It is important that the call be returned as soon as possible following the conference.
5. Identifying the caller to the person being called is not always advisable. For example, "May I tell Mrs. Smith who is calling?" This can result in bad public relations, particularly if the secretary informs the caller that Mrs. Smith is not in now. The caller has an image of Mrs. Smith telling the secretary that she does not want to talk to the caller. It is desirable to put the phone call right through without asking who is calling.
6. Placing his or her own telephone calls, the park and recreation executive frees the secretary for other work and speeds up the dialing process, particularly with the phone equipment available today.
7. Answering the telephone courteously or discourteously creates a positive or negative image of a department. The public relations value of receiving every telephone call in a friendly and cheerful manner is great, for, after all, a person calling an office can only judge that organization by the way his request or communication is received. Every telephone company has a public relations service that will assist organizations, without charge, in improving their use of the telephone.

**Appointments.** An important task for every professional staff member is that of keeping appointments or meeting important obligations. Hence to keep appointments, most administrators and staffs make use of an appointment calendar. This may be in the form of a desk calendar, an appointment book, or a daily appointment schedule. Actually, any method that is infallible in reminding the individual of the obligation at hand is satisfactory. It is good practice to have a secretary also keep a record of appointments and to remind the staff members of important engagements or obligations. Keeping regular office hours helps maintain a routine of meeting appointments; as does an annual calendar that calls attention to events, obligations, and routine functions that recur each year. Needless to say, appointments or meetings can be arranged by a secretary without using valuable time of the executive or other leadership staff.

**Office Visitors.** Every recreation and park office receives visitors. Some seek information; others call on matters relating to specific or general program services; but regardless of their widely diverse reasons for coming, each feels his or her mission is important and wants fair, courteous, and prompt consideration of the problem. It is the job of the office personnel to make every visitor feel welcome. Courteous procedures in helping them with their problems pay dividends in higher respect for the organization. Indeed, kindness and consideration are signs of good manners; no office should exist without them. In instances where visitors have an appointment, word should be promptly passed on that the caller is present. If necessary, the secretary should feel free to interrupt a conference or interview to let the executive or supervisor know that someone is waiting for an appointment. This can be done in a courteous manner without offending those present.

## Reading and Research

It would be a mistake to give the impression that the office is only concerned with details and routines. Certainly, much of the work is of this kind, but a prodigious amount of preparation, computation, and research is involved in the efficient operation of every office. Decisions or judgments must be made and swiftly implemented in the interest of the public. This requires that professional staff members do considerable reading and research. New procedures, techniques, and operations are constantly appearing that require careful study. Not to keep up-to-date on the latest developments in the field is an indication of a poor professional attitude. Furthermore, effort should be focused toward professional growth through participation in research projects, individual study, and conferences or institutes. All of these chores call for preparation; this involves time, and to budget time during the workday is a must. Moreover, it is not proper that all reading and research be done after office hours, although many find that this is the only uninterrupted time available to do these tasks. If the staff worker asks the question "Is what I am doing in the best interest of the organization?" and if the answer is "yes," few people in the community would expect all of this work to be done away from the office. Therefore office time can very well be set aside to do those research tasks that advance the interests of the recreation system.

### Reports

The rendering of reports is an important task of every recreation and park administrator. Preparations of agendas, progress reports, budgets and financial statements, departmental procedures, publicity releases, rules and regulations, statistical comparisons, and the like are important functions that cannot be completely delegated to others. This work must be done, and the recreation administrator must play a major role in performing it. Access to file information, reference works and other books, reports, and records is necessary for performance of these tasks, as is the assistance of clerical help and staff. Such reports take considerable time, and it is an important function of the office to perform this work as efficiently and expeditiously as possible.

## OFFICE PERSONNEL

Good office management requires that certain tasks related to filing, typing, taking of dictation, duplication of materials, keeping of records, and similar work be performed by skilled clerical, secretarial, or office personnel. Many small park and recreation systems have the services of only one full-time secretary or perhaps, only a part-time one, while larger departments may have sizable office forces. A full-time employee is preferable to a part-time one, and every recreation and park system should endeavor to have at least one full-time clerk or secretary, if the load of work permits. It is a "penny wise" policy to hire a professional administrator and then have that person use his or her energy on office details.

The importance of the secretary cannot be overemphasized. The secretary generally presents the voice of the system, acts as the expediter, buffer, host or hostess, diplomat, and confidante of the employer. Actually, not enough can be said of the value of filling the position with a person having the highest personal and professional qualifications. The secretary should be thoroughly competent in typing, shorthand, bookkeeping, filing, duplicating, and office procedures; should be cooperative, cheerful, friendly, tactful, and loyal; and should have a flair for public relations and know the effect of his or her actions on the "image" of the organization.

## OFFICE PROBLEMS

Problems of minor or major significance appear in any office. But if prompt action is taken to reach the source of the problem and eliminate it with dispatch and good judgment, efficiency will not be impaired. When an office becomes the source of problems, inefficiency, and poor service, the trouble can usually be attributed to one or more of the following factors.

### Failure to Understand the Role of the Office

Unfortunately, some office employees fail to understand that their function is one of service and that their efforts should be focused toward furthering the recreation service by caring for the multitude of office details that are a part of every organization. The purpose of a recreation and park system is not to keep an office functioning; it is the office that serves the system. Poor service and discourteous and critical attitudes toward the professional staff are signs of this lack of understanding. It is the job of the administrator to educate the office staff as to its role in the structure of the organization.

### Poor Supervision

As with other sections, good supervision must be provided for the office staff. Guidance, assistance, and evaluation of work is needed, and the recreation and park administrator should organize the office in a manner that provides for such supervision. Without it, problems arise and petty jealousies develop.

### Poor Organization

Regardless of the size of the office and the number of persons employed, it is sound practice to follow some basic organizational rules:

First, every person in the office should have a definite job to perform, and this should be stated in a job description. Second, every office worker should know to whom he or she is responsible in performing the office tasks. Responsibility to more than one supervisor merely creates confusion and conflicting orders, unless the office worker is assigned to one person who is the clearing head for all the work from others. The assignment of responsibility for allocating and checking work, assigning duties, and evaluating and supervising the office staff will vary according to the size and type of the organization. In a large park and recreation department it is advisable to employ a full-time office manager. In smaller departments this responsibility may be assigned to one member of the office staff and in some cases to the secretary for the administrator. However, this latter assignment may be a mistake, since he or she cannot do justice to his or her primary responsibility and be an office manager too. Third, office duties of similar nature should be grouped together. In other words, if similar tasks can be given to the same staff member, he or she can become more proficient or specialized in performing them and greater efficiency develops as the employee develops pride in performing these tasks. However, too much specialization can lead to monotony; therefore some variety is recommended. Care should also be taken to provide staff with some knowledge of other tasks to ensure backup support for those times a staff person may be out of the office for vacation, sickness, or other reasons. Fourth, the administrator should establish a clear policy as to what decisions are to be made by employees at each level in the office structure. Too often, such decisions are made only by the recreation administrator or office manager, when actually such decisions could be made by an office staff member, if there was a written policy as a guide. Decisions as to the use of recreation buildings, courts, picnic areas, equipment, and the like, as well as interpretation of rules and regulations, are examples of decisions that need not be passed to the top of the organization.

Fifth, each office employee should know precisely where his or her authority and responsibility end. Power complexes on the part of some office employees should be checked immediately. It is unfortunate that some clerks, office managers, and even secretaries lose sight of their functions after years of service, or before, and assume authority that was never delegated to them.

### Office Funds

In some organizations there is a tendency for special office funds to proliferate very rapidly. A coffee fund, flower fund, stamp fund, party fund, and others are set up to serve many needs. The growth of these funds should be discouraged as they can create many problems. For example, if an office has a party fund, when are the parties and what or for whom are they? Some staff members may not wish to participate in the parties and therefore should not be required to donate to the fund. If it is necessary to have a special fund, special care should be exercised in the operation of the fund. Responsibility for the fund should be assigned to one person and limits should be set for the amount of money carried in the fund.

### Failure to Modernize

Many offices grow without thought being given to new techniques of office management. A desk is added for each new employee; additional file cabinets and other office equipment are purchased as needed; but as the office expands, so do the needs for advanced office techniques and operational practices. It is important to evaluate office procedures periodically, for the purpose of simplifying operations. Calling in a consultant on office management is a good practice, just as it is in other professional endeavors.

## OFFICE EQUIPMENT

The kinds of office equipment provided an employee are related to the quality and quantity of the work that can be accomplished. To spend large amounts on labor costs and then follow a "penny-pinching" attitude by not providing adequate office equipment only leads to poor employee morale and less employee output. Every recreation and park system therefore should give careful attention to the matter of purchasing furniture, equipment, and supplies needed to make the job easier for the employees.

Office equipment is constantly being improved, and the recreation and park administrator should be alert to these changes. After all, if new improvements can increase output and efficiency, the initial cost is soon absorbed, and savings result.

Perhaps the flat-top desk is the most widely used piece of office equipment. Actually, it is the "work bench" of the employee having office responsibilities. Desks come in a wide assortment of sizes and shapes. Steel desks (34″ × 60″) have proved popular and are found in many offices, although wood is still used in some executive offices. Office equipment, including furniture, should be selected on the basis of its use.

Other items of equipment and furniture that are found in most recreation offices include the following.

Dictating and transcribing equipment
Tables
Typewriter
Typewriter desk and chair
Duplicating equipment
Card files
Safe or vault
Calculator
Copier
Addressing machine
Collator
Radio communication equipment
Pencil sharpener
Desk lamp
Office safe
Paper punch
Chairs (visitor and desk)
Bookcases
Telephone
File cabinets
Calendars
Clock
Desk pad
Stapler
Postal scale
Postage meter
Waste basket
Paper cutter
Scissors
Memo pads
Coat rack
Letter trays
Miscellaneous items: paper, rulers, clips, pencils, pens, paste, tape dispensers, ink, rubber bands, rubber stamps, envelope moistener

## PHYSICAL ENVIRONMENT

The physical environment of the office should not be overlooked in office management. Consideration of such factors as provision of adequate ventilation and lighting, elimination of noise, and good office layout can do much to increase the efficiency of the office and establish good employee morale.

## Ventilation and Lighting

Office lighting can be inadequate without the office staff being aware of it. Eyestrain and fatigue are the consequences of improper illumination, and it would be wise for the recreation administrator or office manager to check the adequacy of the office's lighting. Ventilation and temperature control are also important. While an air-conditioning unit will minimize the problem, it should be noted that they are high energy users and thermostats should not be set too low. Use of thermostats, draft deflectors, fans and good cross ventilation helps to alleviate this situation to some extent.

## Noise

Noise is distracting, and every effort should be made to eliminate it. This can be done by eliminating the source or by soundproofing the office. The use of soundproof or acoustical building materials in ceilings and walls is helpful in eliminating noise, as is the careful placement of furniture, files, and other office equipment. The installation of carpet can also be helpful in reducing noise. It is wise to investigate all sources of noise and to do whatever is necessary to eliminate them.

## Office Layout

The "layout" of the office refers to the way the building space is utilized for office purposes. The ways in which the furniture and office equipment are arranged and space is allocated to employees are important to work efficiency. Some of the basic factors that should be considered in office layout will now be discussed.

**Use of Space.** Whenever possible, use one large area for clerical and office personnel, rather than a number of individual offices. Not only does it make for better use of space, but it also makes for better supervision and arrangement of office equipment.

**Uniformity.** Provide furniture and office equipment of a similar type. This gives a better appearance and eliminates clutter and employee distraction. Uniform desks, desk lamps, trays, chairs, and even aisle space are examples of such uniformity. The use of soft pastel colors will provide a pleasing effect in the office.

**Desk Layout.** Desks should be arranged so that employees do not face each other as this eliminates distractions and visiting that might occur. If possible, desks should be located in such a manner that the flow of work can move from one desk to another with a minimum of effort and foot traffic. The desk of the office manager or supervisor should be located to provide the best observation of the office operation and, if possible, to provide some privacy for conferences.

## OFFICE RECORDS

Every recreation and park system makes use of records and reports, each of which should have a useful purpose; if it does not, it should be discarded. Records are used to (1) provide data for evaluation, (2) show the scope of program service, (3) give statistical comparisons of service, and (4) provide the basic information needed to attain the long-range objectives of the recreation and park system. Records are usually kept in the central office, except for those that are needed for current use in the offices of the recreation buildings and park operations facilities. Although records are of many types and have many uses, they can be classified into six categories: administrative; program; personnel; property, equipment, and supplies; finance; and miscellaneous.

### Administrative

Full information is needed to perform many administrative and policy-making functions. Records of board, committees, councils, legislation, legal opinions, departmental policy, agreements and contracts, rules, regulations, procedures, organizational plans, insurance, correspondence, and publicity are examples of the kinds of materials of an administrative nature that should be readily available. As a protective measure, important documents such as charters, minutes, ordinances, and the like should be microfilmed with the film filed in a separate location.

### Program

It is necessary to keep records of the services provided by the recreation and park system. This involves reports on programs, program checklists and evaluations, and reports on the number of participants and their classification by age, sex, and type of activity or service. Some of the specific records that are needed can be classified under the following headings.

1. Program checklists (equipment needed, costs, etc.).
2. Program evaluations.
3. Registration and/or attendance reports.
4. Special events.
5. Service to special groups or organizations.
6. Roster of clubs, associations, teams, and classes (as needed).

7. Award rosters.
8. Permits granted.
9. List of officials and judges.
10. Present and past activity schedules.

### Personnel

Every recreation and park system needs personnel records of its staff, its board or commission members, and its advisory councils and committees. Personnel data on volunteer leaders and others, who assist the department in any way, should also not be overlooked. Comprehensive personnel records and reports on all employees of the department should be kept. These comprehensive records should be kept in a locked safe and made available to administrative supervisory staff or the individual employee.

### Property, Equipment, and Supplies

Records and reports on all matters pertaining to property, equipment, and supplies are also needed. Included here would be

1. List of recreation properties (including all data pertaining to each).
2. Maps.
3. Surveys and long-range plans.
4. Blueprints and drawings.
5. Specifications.
6. Inventory of equipment and supplies.
7. Land acquisition schedules.
8. Planting plans.

### Finance

Income and expenditure records are found in every recreation and park system. Such financial records include requisitions, bids, appropriations, budget data, funds, assessed values, property costs, purchase orders, vouchers, payrolls, and unit costs. All income and cost data should be carefully classified and available for ready reference.

### Miscellaneous

Records that do not fall into these categories may be classified as miscellaneous. Records and reports should be on file in the office on conferences, institutes, research, reading lists, directories, names of government agencies and staff members, interdepartmental plans, and the like.

## Questions for Discussion

1. How should a park and recreation administrator budget his or her office time? How rigid or flexible should he or she be in this matter?
2. How does an efficiently managed office affect the operation of the recreation and park system? An inefficient office?
3. Explain the factors you would consider in locating an office. What type of office layout would you consider? Why?
4. Explain what is meant by "office routine." List some routine office procedures, and indicate why they are important.
5. Draw a diagram of an office layout for a recreation and park department of a size determined by yourself.
6. Explain three methods of filing. What are the advantages and disadvantages of each.
7. What are some common frictions and problems that develop in office management? What are the symptoms of these difficulties, and how would you alleviate them?
8. List some of the essential records and reports that are needed in every recreation and park system.
9. How does conduct in the office relate to good public relations?
10. What office equipment should be in a medium-size recreation and park office?
11. What are the advantages and disadvantages of having a central office, as opposed to decentralized offices, in the larger city?
12. What are the duties of a good secretary? of good clerical workers? How would you provide for clerical or stenographic help in a very small department? How large should the department be before you employ a full-time secretary? Why?
13. List, from your experience, the faults commonly found in offices. What would you do to prevent these situations from arising?

## SELECTED REFERENCES

Anderson, R. G., *Organization and Methods*. Macdonald & Evans, London, 1973.

Butler, George D., *Introduction to Community Recreation* (5th ed.). McGraw-Hill, New York, 1976.

Hjelte, George, and Shivers, Jay S., *Public Administration of Recreation Services* (2d ed.). Lea and Febiger, Philadelphia, 1978.

Kahn, Gilbert; Yerian, Theodore; Stewart, Jr., Jeffrey R., *Progressive Filing*. (7th ed.). McGraw-Hill, New York, 1961.

Meyer, H. D., and Brightbill, C. K., *Recreation Administration*. Prentice-Hall, Englewood Cliffs, N.J., 1956.

Needy, J. R., *Filing Systems, Management Aids*. National Recreation and Park Association, Oglebay Park, Wheeling, W. Va., 1966.

Neuner, John J. W., *Office Management Principles and Practices* (4th ed.). South-Western, Cincinnati, 1959.

Raush, Edward N., *Principles of Office Administration*. Merrill, Columbus, Ohio, 1964.

Terry, George R., *Office Management and Control*. Irwin, Homewood, Ill., 1970.

Wylie, Harry L., *Office Organization and Management* (3d ed.). Prentice-Hall, N.J., 1953.

# APPENDIX 

# Enabling Acts, Charters, and Ordinances

## General Statutes of the State of Kansas

Chapter 12, Article 19
Public Recreation and Playgrounds

***12-1901. Operation by city or school district; employees.*** Any city or school district may operate a system of public recreation and playgrounds, acquire equipment and maintain land, buildings or other recreational facilities, employ a superintendent of recreation and assistants, vote and expend funds for the operation of such a system.

***12-1902. Recreation commission; responsibility for operation.*** Any city or school district may operate such system independently, or may cooperate in its conduct in any manner mutually agreed upon, or may delegate the operation of the system to a recreation commission created by either or both of them, but the programs and services within such recreation system shall not be conducted by both the city and school district each acting independently of the other. In any city or school district where a recreation commission has been established, said recreation commission shall have responsibility to operate the system and all programs and services thereof.

***12-1903. Property: gifts.*** Any city, school district or commission given charge of the recreation system by this act is authorized to conduct the activities of the system on any property under its custody and management, or, with proper consent, on any other property and upon private property with the consent of the owners, and may receive gifts from any source whatsoever.

***12-1904. Petition; election; tax levies.*** Whenever a petition signed by at least five percent of the qualified and registered voters of the city or school district shall be filed with the clerk thereof, requesting the governing body of the city or school district to provide, establish, maintain and conduct a supervised recreation system and to levy an annual tax therefor not to exceed one mill, it shall be the duty of the governing body of the city or school district to cause such question to be submitted to the qualified voters thereof to be voted upon at the next regular or special election of the city or school district to be held more than thirty days after the filing of such petition.

***12-1905. Same; joint meeting of governing bodies; notice.*** The petition mentioned in section four (12-1904) of this act may be directed to the governing bodies of said city and school district jointly, and in such case it may be filed with the clerk of either the city or school district. Upon receipt of the petition, the clerk shall set a day not less than five nor more than ten days thereafter for the joint meeting of the two governing bodies for the consideration of the petition and if the petition be found sufficient the proposition shall be submitted to the electors within the corporate limits of the city or school district, whichever is the larger, the same to be submitted as provided by law for such city or school district. Notice of the receipt of the petition and the date and place of the joint meeting shall be given immediately by the clerk to the executive officer of the city and school district by registered mail.

***12-1907. Recreation commission; appointment; terms; officers; vacancies; powers; audits.*** All recreation commissions shall consist of five (5) members to be appointed as follows: Upon the adoption of the provisions of this act by the city or school district acting independently in the manner provided in K.S.A. 12-1904 and 12-1905, or any amendments thereto, the governing body of such city or school district shall appoint four (4) members, the first appointee to serve for four (4) years, the second for three (3) years, the third for two (2) years, and the fourth for one (1) year, and the fifth member who shall also serve for four (4) years shall be appointed by the four (4) appointee members of such commission. Upon the adoption of the provisions of this act by the city and school district acting jointly in the manner provided in K.S.A. 12-1904 and 12-1905, or any amendments thereto, the governing bodies shall each appoint two (2) of its electors to serve as members of the recreation commission, and the persons so selected shall select one additional person, and all of said persons shall constitute the recreation commission.

Of the members of said commission first selected by the school district, one (1) shall serve for a term of one (1) year, and one (1) for a term of four (4) years; one (1) of those first selected by the governing body of the city shall serve for a term of two (2) years, and one (1) for a term of three (3) years. The additional member shall serve for a term of four (4) years. Thereafter, the members of said commission shall be selected in the same manner as the member he or she is succeeding and the term of office of each shall be four (4) years. Whenever a vacancy shall occur in the membership of said commission an elector shall be selected to fill such vacancy in the same manner as and for the unexpired term of the member he or she is succeeding. Said commission shall elect a presiding officer and secretary. Said commissioners are hereby empowered to administer in all respects the business and affairs of the recreation system. The amount received from the tax herein provided shall be set over to said commission and used by said commission for the purposes herein set out; and shall be held by the treasurer of the city or school district who shall be ex officio treasurer of said commission. All financial records of such commission shall be audited as provided in K.S.A. 75-1122, and a copy of such annual audit report shall be filed with the governing body of any city or school district which is involved in the operation of such recreation system.

***12-1908. Certification of budget; tax levy; election to revoke; budget increase; procedure; tax levies; protest petition; election; reorganization authorized.*** (a) Except as otherwise provided in subsection (b) of this section, when the provisions of this act shall have been adopted by an election the commission shall annually, and not later than twenty (20) days prior to the date for the publishing of the budget of such city or school district, certify its budget to such city or school district, which shall levy a tax sufficient to raise the amount required by such budget, but in no event more than one (1) mill or the amount set out in the petition provided for in K.S.A. 12-1904. When said petition shall have been submitted to a city and school district jointly said budget shall be certified to the city or school district whichever shall be the larger, and the tax levied by such city or school district, but such levy shall not be deemed or considered a levy of such city or school district in determining the aggregate levy of such city or school district under any of the statutes of this state. After three (3) years' operation the authority to levy the tax provided for in this section may be revoked by a majority of the electors voting at an election called in the same manner as the election authorizing the same. Upon such revocation all property and money belonging to such commission shall become the property of the city or school district levying the tax under this section.

(b) After any city or school district has begun to operate such a supervised recreation system, it appearing to the satisfaction of the recreation commission of a particular school district or city or of a city and school district jointly, that the budget should be increased so as to adequately meet the needs of the city or school district, such recreation commission may submit a proposed program with the budget for carrying out the same to the levying authority which may then levy a tax sufficient to raise the amount required by the expanded budget, but not to exceed one (1) mill, which levy shall be in addition to the one (1) mill authorized by K.S.A. 12-1904. Any city of the first class or any school district located in any county having a population of more than twenty-seven thousand (27,000) and not more than thirty-two thousand (32,000) or any school district operating a recreation commission within any city of the first class located in a county having a population of more than sixty-four thousand (64,000) and less than one hundred thousand (100,000) may levy for a recreation commission located therein a tax in an amount not to exceed one (1) mill in addition to those levies authorized by K.S.A. 12-1904 shall not be deemed or considered a levy of such city or school district in determining the aggregate levy of such city or school district under any of the statutes of this state but shall be in addition to all other levies authorized by law and shall not be subject to limitations prescribed by law.

(c) Any recreation commission established by a city, school district or both, acting jointly, which has been operating for at least three (3) years on the maximum levies authorized by K.S.A. 12-1904 and subsection (b) of this section, may submit a proposed program, with the budget for carrying out the same, to the levying authority, which may then levy a tax sufficient to raise the amount required by such budget, but not to exceed one (1) mill. Such levy shall be in addition to the levies authorized by K.S.A. 12-1904 and subsection (b) of this section and shall not be deemed or considered a levy of such city or school district under any of the statutes of this state, but shall be in addition to all other levies authorized by law and shall not be subject to any limitations prescribed by law.

(d) In any city or school district in which a recreation commission has been established, before the levying authority shall make any additional levy authorized by subsection (c) of this section, it shall adopt a resolution reorganizing the recreation commission as follows: (1) Where the recreation commission was established by a city acting independently, five (5) members shall be appointed as provided in K.S.A. 12-1907, two (2) members shall be duly elected members of the city governing body and the city governing body shall appoint one (1) member of the board of education of each school district whose boundaries encompass any portion of the city; (2) where the recreation commission was established by a school district acting independently, five (5) members shall be appointed as provided in K.S.A. 12-1907, two (2) members shall be duly elected members of the board of education of such school district and the board of education shall appoint one (1) member of the governing body of each city whose boundaries encompass any portion of the school district; and (3) where the recreation commission was established by a city and school district jointly, five (5) members shall be appointed as provided in K.S.A. 12-1907, two (2) members shall be duly elected members of the board of education of the school district. Ex officio members of the recreation commission shall serve without compensation.

(e) Before the tax levying authority shall make any additional levy authorized by this section, the city or the school district, or both, shall adopt a resolution authorizing the making of the levy. Such resolution shall state the purpose for which the levy is made and shall be published once in the official city newspaper, whereupon, the tax levy may be made without an election unless a petition in opposition thereto signed by a number of the qualified electors of the city or the school district, equal in number to five percent (5%) of the qualified electors of the city or the school district who voted at the last preceding regular city or school district election, shall be filed with the city clerk within sixty (60) days after the publication of the resolution. If a valid petition is signed, no levy shall be made in excess of that being made prior to the adoption of the resolution unless and until such proposition has been submitted to and approved by a majority of the electors voting thereon at the next regular city or school district election or at a special election called for the purpose. When an election is held and a majority shall vote in favor of levying the tax, such tax may thereafter be levied.

***12-1909. School district defined.*** As used in this act, "school district" means a school district of any kind.

***12-1910. Act inapplicable to certain first-class cities and school districts therein.*** The provisions of this act shall not apply to any city which has established a board of park commissioners under the provisions of K.S.A. 13-1346 and any amendments thereto nor to any city school district within the boundaries of which is located any such city.

## Sample Ordinance—General Law Cities[1]
## Recreation or Recreation and Park Services

An ordinance of the City of ________, County of ________, State of California, establishing a recreation (and park) commission, and defining the organization, powers and duties: establishing a recreation (and parks) department and defining the responsibilities and duties: and creating the position of superintendent of recreation (and parks): and establishing a recreation and park fund.

The City Council of the City of ________, does ordain as follows:

## Section I

### Recreation (and Park) Commission

There is hereby created a Recreation (and Park) Commission consisting of five members, both men and women. The Recreation (and Park) Commission members shall be appointed as follows:

**a.** ________ members to be recommended by the Mayor for appointment by the City Council;

**b.** ________ members to be recommended by the High School District Board of Trustees for appointment by the City Council;

**c.** ________ members to be recommended by the ________ Elementary School District Board of Trustees for appointment by the City Council;

**d.** ________ ex-officio members, without vote, may be appointed as follows: Superintendent of the ________ High School District; Superintendent of the ________ Elementary School District; a representative of the City Planning Commission; and the City Manager (or Administrator).

Members of the Commission shall serve for a period of three years. The first Commission to be appointed shall at its first meeting so classify its members by lot that ________ shall serve for one year; ________ shall serve for two years; and ________ shall serve for three years. At the expiration of each of the terms so provided for, a successor shall be appointed by the City Council for a term of three years.

To aid the City Council in selecting members for appointment, it shall be the duty of the Recreation (and Park) Commission to suggest qualified and interested persons eligible for appointment for each vacancy to be filled.

[1]These sample ordinances are taken from the State of California Recreation Commission Publication 56-4, *Public Recreation and Parks in California*, June 1957, pp. 32–40.

Vacancies in said Commission occurring otherwise than by expiration of term shall be filled in the manner herein set forth for the unexpired term of the commissioner leaving the Commission.

## Section II

### Organization of the Commission

Within fifteen days after their appointment the members of the Commission shall meet in regular session and elect from their members a chairman, a vice-chairman, and a secretary. Their duties shall respectively be such as are usually carried by such officers. Officers shall hold office for one year, or until their successors are elected.

The Commission shall adopt rules and regulations to govern procedure and shall by vote set a time for regular meetings which will be held at least once each month and shall determine the manner which will be held at least once each month and shall determine the manner in which special meetings may be held and the notice given. A majority of the regular members constitute a quorum. Absence from __________ consecutive regular meetings, without formal consent of the Commission shall be deemed to constitute the retirement of such member and the position declared vacant.

Minutes of the Commission shall be filed with the City Clerk and City Manager, the Superintendent of the Elementary School District, the Superintendent of the High School District.

## Section III

### Duties and Responsibilities of the Recreation (and Park[s]) Commission

The duties of the Recreation (and Park[s]) Commission shall be to:

**1.** Act in an advisory capacity to the City Council, the Boards of Trustees of the School Districts and the Superintendent of Recreation (and Parks) in all matters pertaining to (parks and) public recreation and to cooperate with other governmental agencies and civic groups in the advancement of sound (park and) recreation planning and programming.

**2.** Formulate policies on recreation services for approval by the City Council.

**3.** Advise the City Council and the School Boards of Trustees as to the minimum qualifications for the position of Superintendent of Recreation (and Parks); formulate a job description for the position; and recommend one or more candidates for appointment to the position by the City Council or City Manager.

**4.** Advise with the Superintendent of Recreation (and Parks) on problems of administration, development of recreation areas, facilities, programs and improved recreation services.

**5.** Recommend the adoption of standards on organization, personnel, areas and facilities, program and financial support.

**6.** Make periodic inventories of recreation services that exist or may be needed and interpret the needs of the public to the City Council, the Boards of School Trustees and to the Superintendent of Recreation (and Parks).

**7.** Aid in coordinating the recreation services with the programs of other governmental agencies and voluntary organizations.

**8.** Make periodic appraisals of the effectiveness of the Superintendent and staff in administering the program.

**9.** Interpret the policies and functions of the Recreation (and Parks) Department to the public.

**10.** Advise the Superintendent of Recreation (and Parks) in the preparation of the annual budget and a long-range recreation capital improvement program.

## Section IV

### Recreation (and Parks) Department

**1.** The Recreation (and Parks) Department is hereby established for the City of __________.

**2.** The functions of this Department are to provide opportunities for wholesome, year-round public recreation service for each age group and to develop and maintain in an attractive and safe manner the recreation (and park) areas and facilities of the City of __________, and to insure that such facilities are suitable for a wide variety of recreation purposes.

## Section V

### Superintendent of Recreation (and Parks)

**1.** The City Council (or the City Manager) shall appoint, upon the recommendation of the Recreation (and Parks) Commission a Superintendent of Recreation (and Parks) to administer the recreation (and parks) program.

**2.** The Superintendent of Recreation (and Parks) shall attend meetings of the Commission; may serve as secretary to the Commission and shall make such reports to the Commission or to the City Council or Boards of Trustees of the School Districts as shall be required.

**3.** The Superintendent of Recreation (and Parks) shall:

**a.** Employ required personnel, such as assistants, supervisors, leaders and clerical and maintenance employees in accordance with civil service proceedings and supervise them in the performance of their various duties.

**b.** Administer, operate and maintain existing recreation (and park) areas and facilities and plan for the acquisition, development and operation of the proposed facilities in accordance with policies approved by the Recreation (and Park[s]) Commission and the City Council.

**c.** Inform the general public of the services and facilities being provided by the Recreation (and Park[s]) Department.

**d.** Solicit suggestions from the general public to improve or increase the effectiveness of the service.

**e.** Cooperate with governmental and voluntary organizations and agencies in the furtherance of recreation opportunities.

**f.** Prepare manuals, bulletins and reviews on recreation problems.

**g.** Provide, upon request, assistance of a technical nature to community agencies and organizations having problems relating to recreation (and park) areas and facilities and programs.
**h.** Counsel with officials of public and private organizations and interested groups concerning community recreation and leisure activities and assist them in the promotion of recreation services.
**i.** Conduct studies of local conditions and needs for recreation services and assist with recruitment and training of recreation personnel.
**j.** Prepare a budget for the approval of the Recreation (and Park[s]) Commission and present the budget to the City Council.

### Section VI

#### Recreation (and Park[s]) Fund

There is hereby established a Recreation (and Park[s]) Fund. The Recreation (and Park[s]) Commission shall submit annually a budget to the City Council for its approval in whole or in part. Said budget should provide not only for recreation services (park maintenance) and operating costs for the ensuing year, but should also contain estimates and recommendations for such long-term capital outlay as may be necessary to provide for an orderly and coordinated development program.

### Sample Ordinance—General Law Counties Recreation or Recreation and Park Services

An ordinance of the County of __________ establishing a recreation (and park[s]) commission, a department of recreation (and park[s]), the position of director of recreation (and park[s]) and prescribing their powers and duties; and authorizing the expenditure of funds, under authority of the Statutes of the State of California and including the Education Code, Division 12, Chapter 4, Sections 24401-24411.

The Board of Supervisors of the County of __________ do ordain as follows:

### Section I

#### Establishment of a Commission

There is hereby created and established a commission consisting of five (5) members, to be known as the Recreation (and Parks) Commission.

### Section II

#### Appointment of Members

The members of said Commission shall be representative of the five supervisorial districts with each Supervisor nominating one member for appointment by the Board of Supervisors; not more than four (4) members of the Commission shall be of the same sex. Ex-officio members shall be appointed to serve without vote (may include two (2) or more of the following: Chairman of the Board of Supervisors, Chief Administrative Officer, County Superintendent of Schools, County Planning Director or Chairman of County Planning Commission, County Engineer).

## Section III

### Compensation

All of the members of said Commission shall serve without compensation. Members may be reimbursed for their actual expenses, including travel, lodging and meal expenses incurred while on official business of the Commission, which has had prior approval by the Board of Supervisors.

## Section IV

### Terms

The terms of the members first appointed shall be determined by lot at the first meeting of the Commission with one member serving two years; two members serving three years; and two members serving four years. Terms of office thereafter shall be for four years. All vacancies shall be filled for the unexpired terms of the member whose office is vacant in the same manner as such member received the original appointment.

## Section V

### Organization of Commission

Immediately after appointment and qualification, or until their successors qualify, the Commission shall organize by electing a chairman, vice-chairman and a secretary who shall serve for one (1) year. Upon the appointment of a Director of Recreation (and Parks) by the Board of Supervisors, the Director may be appointed secretary to the Recreation (and Park[s]) Commission by the Commission.

## Section VI

### Meetings

Regular meetings shall be held at least once a month at a regular date and time to be fixed by the members, and special meetings may be held upon the call of the Chairman. Any three (3) members of the Commission may call a meeting after giving seven (7) or more days' notice of such meeting to all members.

## Section VII

### Quorum

Three (3) members shall constitute a quorum for the transaction of business, but a lesser number may adjourn from time to time.

## Section VIII

### Lapse of Membership

After a member of the Commission fails to attend four (4) consecutive meetings, unless excused by the vote of said Commission, his membership shall automatically terminate, and his successor shall be appointed in the same manner as his predecessor. A member of the Commission may be removed from office by a four-fifths (4/5) vote of the Board of Supervisors.

## Section IX

### Powers and Duties

The Commission shall have the following powers and duties:

**1.** Act in an advisory capacity to the Board of Supervisors in promoting, aiding, encouraging and conducting public recreation, including the development of recreation and park facilities and programs.

**2.** At the request of the Board of Supervisors, recommend to the Board the names of candidates who are qualified in the administration of public recreation services for appointment by the Board to the position of Director of Recreation (and Parks).

**3.** Act in an advisory capacity to the Board of Supervisors and to the Director of Recreation (and Parks) in the planning, maintenance, development and operation of all recreation areas and facilities owned, controlled, or leased by the County of __________.

**4.** Formulate, and recommend to the Board, general policies related to the purposes of the Commission; and adopt by-laws, rules, and regulations, subject to the approval of the Board, as the Commission may require to facilitate the operation of a recreation (and parks) system.

**5.** At the request of the Board of Supervisors, cause a budget to be prepared and submitted to the Board annually, on or before the first day of May, providing for the costs of maintenance and operation of the recreation (and park) facilities and programs for the ensuing year. The budget shall contain estimates and recommendations for such long term capital outlay projects as may be necessary to provide for an orderly development of recreation (and park) areas and facilities.

**6.** Study and make recommendations on the acquisition and development of recreation areas and facilities, such as playgrounds, parks, beaches, pools, campsites, concessions and other centers of recreation.

**7.** Interpret the function and operation of recreation and park services to public officials and to the general public to the end that the services receive adequate financial support from public and private sources.

## Section X

### Recreation (and Park[s]) Department

A Recreation (and Park[s]) Department is hereby established to be administered by the Director of Recreation (and Park[s]) with the advice of the Recreation (and Park[s]) Commission and subject to the approval of the Board of Supervisors.

## Section XI

### Expenditure of Funds

The Board of Supervisors may appropriate and expend funds from the county general fund and/or levy and collect a tax in the same manner as taxes are levied for other public purposes (Education Code, Section 24403) to operate a recreation program; to acquire, develop, maintain and operate public parks and recreation facilities; and for the operation and maintenance of public parks, playfields, land and water areas, structures for centers of recreation, and for the payment of salaries of persons employed for said work.

A Recreation (and Parks) Fund may be established as a depository for all monies received for recreation and park purposes from the procedures from all gifts, legacies or bequests and all monies derived by the county from fees for recreational services administered by the Board of Supervisors and/or under the management of the Director of Recreation (and Parks).

## Section XII

### Director of Recreation (and Parks)

**1.** The position of Director of Recreation (and Parks) is hereby established.

**2.** The Board of Supervisors shall appoint a Director of Recreation (and Parks) to administer recreation and park facilities and programs and may request the Commission to recommend for this position one or more candidates who are qualified in the administration of public recreation services.

**3.** The Director of Recreation (and Parks) shall attend meetings of the Commission; may serve as secretary to the Commission, and shall make such reports to the Commission and to the Board of Supervisors, as shall be required.

**4.** The Director of Recreation (and Parks), upon request, may advise with officials of local jurisdictions and community organizations in the County of __________ concerning the expenditure of public funds for recreation and parks; acquisition, design and development of recreation facilities and areas; and shall maintain effective and cooperative relations with officials of incorporated cities, special recreation districts, State agencies and with local, state and national voluntary recreation organizations.

**5.** The Director of Recreation (and Parks) shall:

**a.** With the approval of the Board and in accordance with civil service procedures, employ required personnel such as assistants, recreation supervisors and leaders, maintenance and clerical personnel; and supervise them in the performance of their various duties.

**b.** Administer, operate and maintain existing recreation and park areas and facilities and plan for the acquisition, development and operation of proposed facilities in accordance with policies formulated by the Recreation (and Parks) Commission and approved by the Board of Supervisors.

**c.** Prepare an annual budget, with the advice of the Recreation (and Parks) Commission, for presentation to the Board of Supervisors.
**d.** Inform the general public of the services and facilities being provided by the Recreation (and Parks) Department; address professional civic and lay groups on recreation subjects; solicit suggestions from the general public increasing the effectiveness of the recreation program; cooperate with governmental and voluntary organizations and agencies in the furtherance of recreation problems; and provide, upon request, assistance of a technical nature to community agencies and organizations on problems related to recreation and park facilities and programs.
**e.** Assist community organizations in the promotion of recreation services; conduct studies of local conditions and needs for recreation services; and assist with the recruitment and training of professional recreation personnel and volunteer leaders.

## Oregon Enabling Act
### Establishing Park and Recreation Districts

***266.010. "County court" defined.*** As used in this chapter, "county court" includes board of county commissioners.

***266.020. Canvass, certification and return of vote.*** The vote cast at any and all elections under this chapter, shall, unless otherwise provided, be canvassed, certified and returned within the time and in the manner provided by the laws relating to elections in irrigation districts in this state.

***266.110. Petition for organization.*** (1) Whenever not less than 25 per cent of the resident freeholders of the proposed district desire to form a park and recreation district within a county, they may present to the county court a petition in writing signed by them, stating the name of the proposed district, the number of members to be on the district board, whether three or five, setting forth the boundaries thereof, and praying that the lands included within the boundaries be organized as a park or recreation district under this chapter.

(2) Each of the petitioners must be a resident and freeholder within the proposed district.

(3) When any part of the proposed district is within the incorporated limits of any city or town, the petition shall be accompanied by a certified copy of the resolution of the governing body of the city or town, approving formation of the district.

(4) The petition shall be accompanied by a good and sufficient undertaking in form and amount to be approved by the county court, conditioned that the petitioners will pay all expenses of the organization of the district, including publication of notices as required, expenses of preparation and delivery of ballots, fees of election officers, and any and all expenses which may be incurred on the part of the county in the formation, election and organization of the park or recreation district.

(5) The petition must be verified by the affidavit of one of the petitioners.

(6) The petition must be published once a week for at least two weeks preceding the hearing thereof, in some newspaper of general circulation published in the county, together with the notice stating the time when the petition will be presented to the county court, and that all persons interested therein may appear and be heard. (Amended by 1957 c.57 § 1)

*Note:* Proceedings pursuant to this section to include land within an incorporated city in a park and recreation district did not violate the rule against superimposing one municipality on another. The proceedings were not invalidated by lack of express charter authority to create or acquire a park within the city. *State v. James*, 189 Or. 268, 219 P. 2d 756 (1950).

***266.120. Hearing on petition.*** At the time stated in the notice, the county court may hear the petition and adjourn from time to time. The county court shall not modify the boundaries of the proposed district as set forth in the petition so as to exclude from the proposed district any land which would be benefited by formation of the district, nor shall any lands which will not, in the judgment of the court, be benefited by the district be included within the district.

***266.310. Officers of district; qualifications.*** (1) The officers of the district shall be a park and recreation board of three or five members, to be elected by the duly qualified electors of the district at large, and a secretary, to be appointed by the board.

(2) Every qualified elector resident within the proposed district for the period requisite to enable him to vote at a general election is qualified to be a member of the board or officer of the district. (Amended by 1957 c.57 § 2)

***266.320. Nomination and election of board members at organization election; terms of office.*** (1) At the election for the organization of the district there shall be elected the first members of the park and recreation board. The number to be elected shall be three or five, according to the number set forth in the petition for organization.

(2) Candidates for members of the board may be nominated by the petition of not less than 10 resident freeholders within the limits of the proposed district. The nominating petition shall be filed with the county court at least 15 days prior to the election.

(3) The successful candidates and their respective terms shall be determined in the manner provided in subsections (6) and (7) of ORS 450.045.

(4) Each of the directors shall hold office until election and qualification of his successor. (Amended by 1957 c.57 § 3)

***266.325. Increasing number of board members.*** A district having a three-member board may increase the number of members to five in the manner set forth in ORS 450.062, except that the annual election and commencement of terms shall be as provided in ORS 266.330. (1957 c.57 § 7)

***266.330. Annual election of board members.*** (1) An election shall be held in the park and recreation district on the second Tuesday in November of each year, at which a successor shall be elected for each of the members of the park and recreation board whose terms regularly expire on the following first Tuesday in January. If one board member is to be elected, the candidate receiving the highest vote shall be elected. If two board members are to be elected, the candidates receiving the first and second highest vote shall be elected.

(2) Each officer elected shall hold office from the first Tuesday in January next after such election, for three years, and until his successor is elected and qualified. (Amended by 1957 c.57 § 4)

***266.340. Oath of office of board members; surety bond.*** A board member so elected shall take the oath of office within 10 days after receiving his certificate of election. He shall provide a good and sufficient surety bond, to be furnished by a surety company authorized to do business in this state, in the sum of $1,000, the premium on which shall be paid by the district. The bond shall be approved by the county court, recorded in the office of county clerk and thereafter filed with the secretary of the park and recreation board. All official bonds provided for in this section shall be in the form prescribed by law for official bonds of county officers.

***266.350. Compensation of board members.*** Every member of the park and recreation board shall receive for each attendance of meetings of the board $2, and shall receive no other compensation from the district.

***266.360. Filling vacancies on board.*** A vacancy in the membership of the park and recreation board shall be filled for the unexpired term by appointment by a majority of the members of the board. If a majority of the membership of the board is vacant, the vacancies shall be filled promptly by the county court of the county in which the district, or the major portion of the area thereof, lies. (Amended by 1957 c.57 § 5)

***266.370. Board as governing power; president and secretary; signing documents; meetings.*** (1) The park and recreation board shall be the governing power of the district and shall exercise all powers thereof.

(2) At its first meeting or as soon thereafter as may be practicable, the board shall choose one of its members as president and shall appoint a secretary who need not be a member of the board. In case of the absence, or inability to act, of the president or secretary, the board shall, by order entered upon the minutes, choose a president pro tempore, or secretary pro tempore, or both, as the case may be.

(3) All contracts, deeds, warrants, releases, receipts and documents of every kind shall be signed in the name of the district by its president and shall be countersigned by its secretary.

(4) The board shall hold such meetings either in the day or evening, as may be convenient, requisite or necessary.

***266.410. General district powers; penalty for violating board regulations.*** Every park and recreation district formed under this chapter shall have power:

(1) To have and use a common seal.

(2) To sue and be sued by its name.

(3) To construct, reconstruct, alter, enlarge, operate and maintain such lakes, parks, recreation grounds and buildings as, in the judgment of the park and recreation board, are necessary or proper, and for this purpose to acquire by purchase, gift, devise, condemnation proceedings or otherwise such real and personal property and rights of way, either within or without the limits of the district as, in the judgment of the park and recreation board, are necessary or proper, and to pay for and hold the same.

(4) To make and accept any and all contracts, deeds, releases and documents of any kind which, in the judgment of the board, are necessary or proper to the exercise of any power of the district, and to direct the payment of all lawful claims or demands.

(5) To assess, levy and collect taxes to pay the cost of acquiring sites for and constructing, reconstructing, altering, operating and maintaining any lakes, parks, recreation grounds and buildings that may be acquired or, any lawful claims against said district, and the running expenses of the district.

(6) To employ all necessary agents and assistants, and to pay the same.

(7) To make and enforce all necessary and proper regulations for the removal of garbage and other deleterious substances, and all other sanitary regulations not in conflict with the Constitution or the laws of Oregon.

(8) To make and enforce rules and regulations governing the conduct of the users of the facilities of lakes, parks, recreational grounds and buildings within the district. Violation of any such regulations or ordinances is a misdemeanor punishable upon conviction by a fine not to exceed $100 or imprisonment not to exceed five days, or both.

(9) To prohibit any person violating any rule or regulation from thereafter using the facilities of the district for such period as the board may determine.

(10) To call, hold and conduct all elections, necessary or proper after the formation of the district, including but not limited to special elections for annexation of territory, and in holding all elections within the district:

(a) To give notice thereof by posting at least three notices in public places within the district not less than 15 days prior to the date of the election, setting forth in the notice the resolution of the board calling the election, and stating the time, place and purposes of the election.

(b) To divide, by resolution, the district into one or more voting precincts for the purpose of each election.

(c) To appoint three judges and a clerk of election for each voting precinct thus created.

(d) To provide proper ballot boxes and facilities for voting, keeping the polls open from 2 P.M. to 7 P.M.

(e) To canvass the votes cast at each election within 10 days thereafter at a meeting held for that purpose.

(f) To declare the results of such election by resolution.

(11) To enlarge the boundaries of the district by annexation of territory, including territory located in whole or in part within the limits of any city, but any such annexation must be after proceedings had as required by ORS 222.110 to 222.180 so far as applicable, and no territory located within the limits of a city may be annexed unless prior to the holding of the election affecting such territory

the common council or governing body of such city has by resolution approved the inclusion of such territory in the territory to be annexed, and a certified copy of the resolution has been filed with the clerk of the district board.

(12) To compel all residents and property owners within the district to connect their houses and habitations with the street sewers, drains or other sewage disposal system.

(13) Generally to do and perform any and all acts necessary and proper to the complete exercise and effect of any of its powers or the purposes for which it was formed.

***266.420. Levy and collection of taxes.*** On the first Monday of June of each year the board shall meet at its usual place of business within the district and, by resolution, determine and fix the amount of money to be levied and raised by taxation, for the purposes of the district, designating the number of dollars and cents so to be raised. The total amount in dollars and cents shall not exceed 10 mills on the dollar on all taxable property in the district. The resolution shall separately state the amount of money to be raised for acquisitions of sites, for construction, reconstruction and alteration and for operation and maintenance. The provisions of ORS 310.050 to 310.070 and 310.090 shall apply in the levying and collection of taxes on the real and personal property in the district, so far as applicable. All provisions of Oregon laws as to the collection of taxes and delinquent taxes and the enforcement of the payment thereof, so far as applicable, shall apply to the collection of taxes for park and recreation purposes.

***266.430. Sinking funds.*** The park and recreation board, by resolution duly adopted, may establish sinking funds for the purpose of defraying the costs of acquiring land for park and recreation sites, and for acquiring or constructing buildings or facilities thereon or therein. Any such fund may be created through the inclusion annually within the tax budget of the district of items representing the yearly installments to be credited thereto. The amount of these items shall be collected and credited to the proper fund in the same manner in which taxes levied or revenues derived for other purposes for the district are collected and credited. The balances to the credit of the funds need not be taken into consideration or deducted from budget estimates by the levying authority in preparing the annual budget of the district. None of the moneys in such funds shall be diverted or transferred to other funds, but if unexpended balances remain after disbursement of the funds for the purpose for which they were created, such balances, upon approval by resolution of the park and recreation board, shall be transferred to the operation and maintenance fund of the district.

***266.440. Handling by county treasurer of tax moneys collected for park and recreation purposes.*** The tax collector shall pay over to the county treasurer all moneys collected by him for park and recreation purposes, as fast as collected. The county treasurer shall keep the moneys in the county treasury as follows:

(1) He shall place and keep in a fund called the operation and maintenance fund of park and recreation district (naming it) the moneys levied by the board for that fund.

(2) He shall place and keep in a fund called the construction fund of park and recreation district (naming it) the moneys levied by the board for construction, reconstruction and alteration.

(3) The treasurer shall pay out moneys from said funds only upon the written order of the board, signed by the president and countersigned by the secretary, which order shall specify the name of the person to whom the money is to be paid and the fund from which it is to be paid, and shall state generally the purpose for which the payment is made. The order shall be entered in the minutes of the park and recreation board.

(4) The treasurer shall keep the order as his voucher, and shall keep a specific account of his receipts and disbursements of money for park and recreation purposes.

(5) The treasurer and sureties upon his official bond shall be liable for the due performance of the duties imposed upon him by ORS 266.420 to 266.440 and 266.470.

***266.450. Entry, publication or posting and taking effect of general regulations and orders of board.*** (1) Any general regulation of the board shall be entered in the minutes, and shall be published once in some newspaper published within the district, if there is one, and if there is no such newspaper, then the regulation shall be posted for one week in three public places within the district. A subsequent order of the board that such publication or posting has been duly made shall be conclusive evidence that publication or posting has been made properly.

***266.460. District attorney to aid board; special counsel.*** The board may instruct the district attorney of the county to commence and prosecute any and all actions and proceedings necessary or proper to enforce any of its regulations or orders, and may call upon the district attorney for advice as to any park and recreation subject. The district attorney shall obey such instructions and shall give advice when called on therefor by the board. The board may at any time employ special counsel for any purpose.

***266.470. Disposition of fines.*** All fines for violation of any regulation or order of the board shall, after expenses of the prosecution are paid therefrom, be paid to the secretary of the board, who forthwith shall deposit the same with the county treasurer, who shall place the same in the operation and maintenance fund of the district.

***266.480. Power to conduct bonded indebtedness for certain purposes.*** Every park and recreation district organized under this chapter shall have the power to contract a bonded indebtedness for the purpose of providing funds with which to acquire land, rights of way, interests in land, buildings and equipment; to improve land and develop parks and recreation grounds; to construct, reconstruct, improve, repair and furnish buildings, gymnasiums, swimming pools and recreational facilities of every kind; to acquire equipment of all types, including vehicular equipment necessary for and in the use, development and improvement of the lands and facilities of the district; to pay the costs, expenses and attorney fees incurred in the issue and sale of the bonds; to fund or refund outstanding indebtedness, or for any one or combination of any such purposes, and to provide for the payment of the same as set forth in ORS 266.490 to 266.580.

***266.490–266.580. Park and recreation district bond procedure.*** The governing board may vote or ten legal voters may petition the board to submit the question of issuing bonds to the voters of the district. The sections include the form of the petition and notice for the bond election. Procedure for newspaper publication of the election notice or posting of notice is prescribed. If the bonds are approved by the voters, they shall be issued by the governing body. The sections provide for advertisement, sale, registration and delivery of the bonds and disposition of the proceeds. The governing board shall levy taxes to pay interest on the bonds which shall be retained by the county treasurer in a separate fund. Provision is made for the redemption of bonds and the civil liability of the county treasurer for funds entrusted to him. Payment procedure of the bond principal and interest and a provision prohibiting payment of a commission for collection thereof is also included.

***266.590. Validation of certain bond issues.*** All proceedings taken prior to March 18, 1949, in the authorization and issuance of bonds by any park and recreation district pursuant to ORS 266.480 to 266.580 hereby are validated, ratified, confirmed and approved, notwithstanding any defects and irregularities in the proceedings or any part thereof, and notwithstanding that the amount of the bonded indebtedness to be incurred was not stated upon the ballot used in the election authorizing the issuance of the bonds.

*Note*. Thus curative statute is not invalid as special legislation. *State v. James*, 189 Or. 268, 219 P. 2d 756 (1950).

## CHARTER OF THE CITY OF DETROIT

### HOME RULE CITIES

The power to adopt this Charter was conferred by Act 279, Public Acts 1909 (now Chapter 49, of the Compiled Laws of the State of Michigan, 1929, and Chapter 49 of Michigan Statutes Annotated)

**Adopted by Vote of the People of the City of Detroit**
November 6, 1973

Filed with the Secretary of State
November 30, 1973

EFFECTIVE JULY 1, 1974

• • •

## Article 7. The Executive Branch: Programs, Services and Activities[1]

• • •

### Chapter 13. Recreation.[2]

### Sec. 7-1301. Department

The recreation department shall operate recreational facilities, offer and carry on organized programs of recreational activities in the city, and, to the extent possible, coordinate all recreational programs and facilities being offered in the city.

### Sec. 7-1302. Advisory commission.

An advisory commission for recreation, comprised of 1 representative from each of the not fewer than 8 districts, shall be created under section 7-103.

Municipal Code
City of Detroit
Affecting
Detroit Recreation Department
December 1979

## Article 1. In General

### Sec. 42-1-1. Definitions

For the purposes of this article the following words and phrases shall have the meanings respectively ascribed to them by this section:

***Commission.*** The commissioners of the recreation department.

***Department.*** The recreation department.

***Parks, public places, and boulevards.*** Unless specifically limited, such terms shall be deemed to include all parks, parkways, park lots, grass plots, golf courses, playgrounds, recreation centers, athletic fields, open places, squares, lands under water and other areas which are now owned by the city or under city control or may hereafter be acquired by purchase, gift, devise, bequest, loan or lease. (C. O. 1954, ch. 260, section 1.)

### Sec. 42-1-2. Boulevards

The following roadways, from lot line to lot line, are hereby declared to be boulevards for the purpose of this article and within the meaning of chapter 9, title 4, of the Charter:

Arden Park. From Woodward to Oakland.
Boston Boulevard. East and West from Oakland to Linwood.

[1]**Home Rule Act.** For state law as to city's authority to establish any department that it may deem necessary for the general welfare of the city, and for the separate incorporation thereof, see M.S.A., § 5.2083(1).
[2]**City Code.** For ordinance provisions as to parks and recreation generally, see ch. 42 of City Code.

Chicago Boulevard. From Jefferson to Mack.
Chicago Boulevard. From Woodward to Dexter.
Dexter Boulevard. From West Grand Boulevard to Joy Road.
Douglas MacArthur Bridge.
Grand Boulevard, East and West. From the mainland terminus of the Belle Isle Bridge to the terminus of the West Grand Boulevard and West Jefferson.
Lafayette Boulevard. From West Grand Boulevard to Green.
LaSalle Boulevard. From West Grand Boulevard to Boston Boulevard.
Oakman Boulevard. From Kendall Avenue to Ewald Avenue and from Chicago Boulevard to the boundary line of the city of Dearborn.
Oakman Court. From Oakman Boulevard to Linwood.
Pontchartrain Boulevard. From McNichols Road to Palmer Park.
Second Boulevard. From Clairmount to the southerly line of the City of Highland Park.
Washington Boulevard. From Convention Hall Building to Park Avenue.

No roadway or highway hereafter acquired by the city, otherwise than by condemnation, shall be deemed to be boulevards without legislative action by ordinance by the common council, even though such roadway or highway be designated a boulevard in the plat, deed or other instrument by which the city may acquire the same. (C. O. 1954, ch. 260, sections 4, 4(a), 5; Ord. No. 379-F, section 1.)

### Sec. 42-1-3. Jurisdiction of commission over roadways in parks.

The commission shall have jurisdiction over the roadways and drives within the boundaries of all city parks, playfields, playgrounds and public places. All parks, playfields, playgrounds or public places hereafter acquired by the city, within or without its boundaries, shall be deemed to be parks within the meaning of this article. (C. O. 1954, ch. 260, section 3.)

### Sec. 42-1-4. Duties of commission generally; establishment of rules and regulations relative to parks; police powers of employees departments; investigation of claims by police department.

It shall be the duty of the commission to improve and beautify in appropriate manner the parks, parkways, boulevards and public places, placed in the jurisdiction of the department by virtue of this article and Charter of the city. The commission shall establish reasonable rules and regulations for the protection of rights and property vested in the city and under control of the department, for the uses, care, maintenance and management of all parks and their dockage, bridges on Belle Isle, public grounds and boulevards and concerning waters surrounding Belle Isle, subject to control of the United States Department of Defense. It shall be the duty of the commission to enforce, through the employees of the department, the provisions of this article. The powers and duties of police officers are hereby conferred upon such employees of the department as are selected by the commission, upon taking appropriate oath as peace officers; provided, that it shall be the duty of the commissioner of police to assign a sufficient number of police officers to properly police the parks, public recreation areas and boulevards of the city and to cooperate with the employees of the department in the enforcement of this article. The police department shall investigate the complaints made by the commissioners of the department of recreation relative to violation of this article and shall take such action therein as the facts may warrant. (C. O. 1954, ch. 260, section 6.)

## Ordinance No. G-1165

## AN ORDINANCE AMENDING CHAPTER 2 OF THE CODE OF THE CITY OF PHOENIX, 1969, BY ADDING THERETO A NEW ARTICLE XVI CREATING A PARKS AND RECREATION BOARD; SETTING FORTH ITS COMPOSITION, TERMS OF OFFICE, POWERS AND DUTIES; AND DECLARING AN EMERGENCY.

WHEREAS, at a special Charter amendment election held in the city of Phoenix on November 9, 1971, Section 1 of Chapter XXIII of the Charter of the City of Phoenix relating to the creation, composition, qualifications, compensation, etc. of the Parks, Playgrounds and Recreational Board was amended by repealing said Section 1, and adopting a new Section 1, to read:

Sec. 1. Creation; composition

The city council shall create by ordinance a Parks and Recreation Board consisting of five or more members to be appointed by the council for terms as specified by said ordinance.

WHEREAS, at said election Section 2 of Chapter XXIII relating to the powers and duties of said Parks, Playgrounds and Recreational Board was also amended by repealing said Section 2 and adopting a new Section 2, setting forth the powers and duties of the Parks and Recreation Board; and

WHEREAS, upon the canvass of the returns of said election, it was determined that the amendment of said Sections 1 and 2 of Chapter XXIII by repealing and adopting new Sections 1 and 2 carried;

NOW, THEREFORE, BE IT ORDAINED BY THE COUNCIL OF THE CITY OF PHOENIX as follows:

***Section 1.*** That the Code of the City of Phoenix, 1969, is hereby amended by adding to Chapter 2 thereof a new Article XVI, entitled "Parks and Recreation Board", to read as follows:

### Article XVI. Parks and Recreation Board

Sec. 2-206. Parks and Recreation Board—Creation; Composition; Term of office.

There shall hereby be created, constituted and established a Parks and Recreation Board of the City. The Parks and Recreation Board shall consist of eight members including the director of the Parks and Recreation Department who shall be a non-voting, ex-officio member, and seven other members to be appointed by the city council to serve for a period of five years. The initial appointments of the board members shall be for the following terms:

Two (2) members shall be appointed to serve for the term of five (5) years;

Two (2) members shall be appointed to serve for the term of four (4) years;

One (1) member shall be appointed to serve for the term of three (3) years;

One (1) member shall be appointed to serve for the term of two (2) years; and

One (1) member shall be appointed to serve for the term of one (1) year.

All subsequent appointments shall be for the full term of five (5) years. In the event of the death, resignation, removal or disqualification of any member of said Parks and Recreation Board, the city council shall appoint on said board a member who shall serve for the unexpired term thus vacated.

Sec. 2-207. Parks and Recreation Board—Quorum; adoption of rules; compensation.

(a) Four members of said Parks and Recreation Board shall constitute a quorum for the exercise of the powers and duties hereinafter conferred upon it, and the concurrence of the majority of said Parks and Recreation Board members shall be necessary to render a decision.

(b) The members of said Parks and Recreation Board shall, at their first meeting, elect one (1) of their members chairman and make such rules for the administration and proper functioning of said board as they deem expedient, but said rules shall not be inconsistent with the laws of the State of Arizona, the Charter of the City of Phoenix or ordinances of the City of Phoenix.

(c) The members of the Parks and Recreation Board shall serve without compensation.

Sec. 2-208. Parks and Recreation Board—Powers and duties.

"The Parks and Recreation Board shall have the following powers and duties:

**1.** As trustee or trustee and successor in interest to its predecessor in interest for the City of Phoenix, to receive, accept or acquire by grant, gift, bequest or devise, property of every kind, real, personal or mixed, wheresoever situate, in fee, trust or otherwise, subject to any provisions made by the donor of such grant, gift, devise or bequest, or to the terms of any trust instrument by virtue of which the property may have been acquired.

**2.** Establish operating policies for recreational facilities and services within and without the city as the developing public recreation needs may require.

**3.** Enter into contracts to grant concessions, licenses and permits for the use of the recreational facilities of the city and to contract with others for the use of recreational facilities needed by the city.

**4.** Establish schedules of charges for miscellaneous recreational facilities and to advise the city council on fees to be set by council on golf courses, tennis centers and swimming pools.

**5.** Advise the council concerning recreational needs and recommend acquisition, location, and nature of facilities to meet said needs.

**6.** Receive and consider the proposed annual budget for parks and recreation purposes during the process of its preparation and make recommendations with respect thereto.

7. Perform such other duties as may be prescribed by ordinance not inconsistent with the provisions of this charter.

*Section 2.* WHEREAS, the immediate operation of the provisions of this ordinance is necessary for the preservation of the public peace, health and safety, an EMERGENCY is hereby declared to exist, and this ordinance shall be in full force and effect from and after its passage by the Council as required by the City Charter and is hereby exempted from the referendum clause of said Charter.

PASSED by the Council of the City of Phoenix this 28 day of March, 1972.

______________________________
MAYOR

ATTEST:
______________________________City Clerk

APPROVED AS TO FORM:
______________________________City Attorney

REVIEWED BY:
______________________________City Manager

# APPENDIX B

# Sample Organization Charts

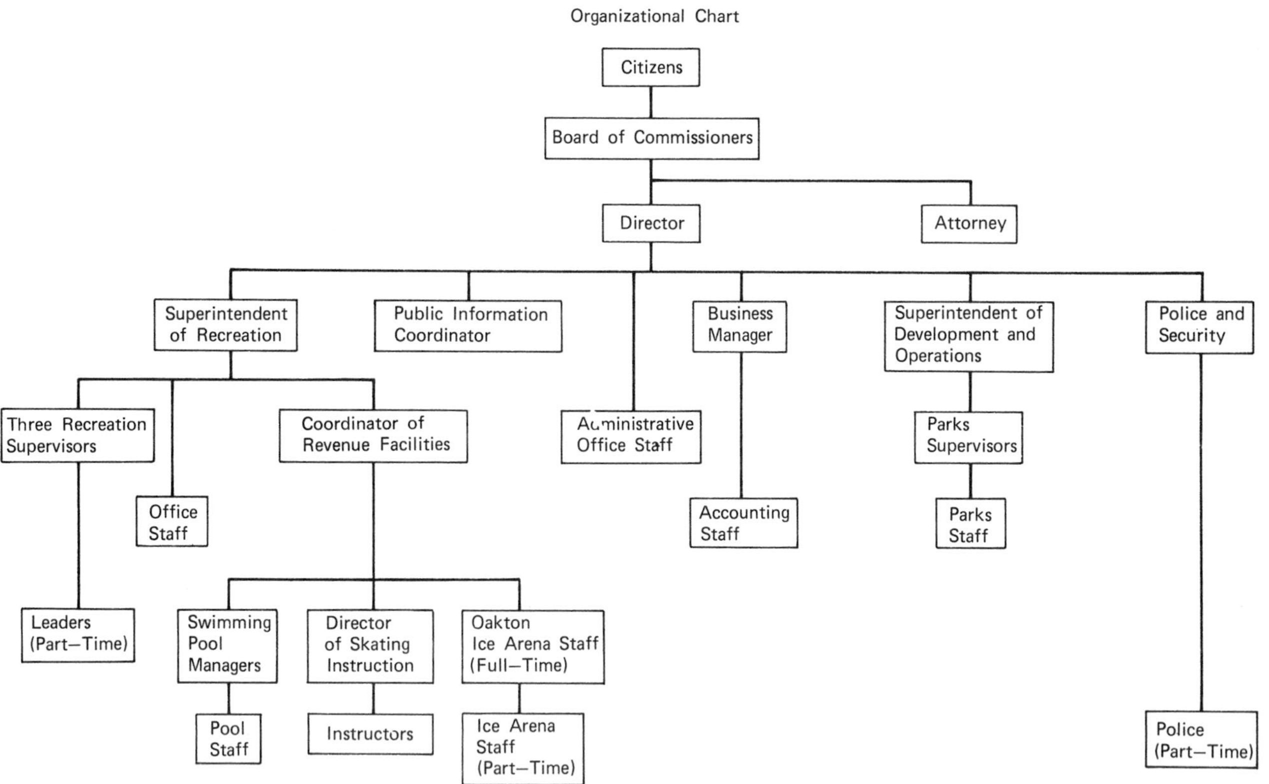


Park Ridge Recreation and Park District. *(Source. Table of Organization/Park Ridge Recreation and Park District/no copyright/permission by William C. Neumann, Superintendent, Development and Operations, Park Ridge, Illinois, Recreation and Park District.)*

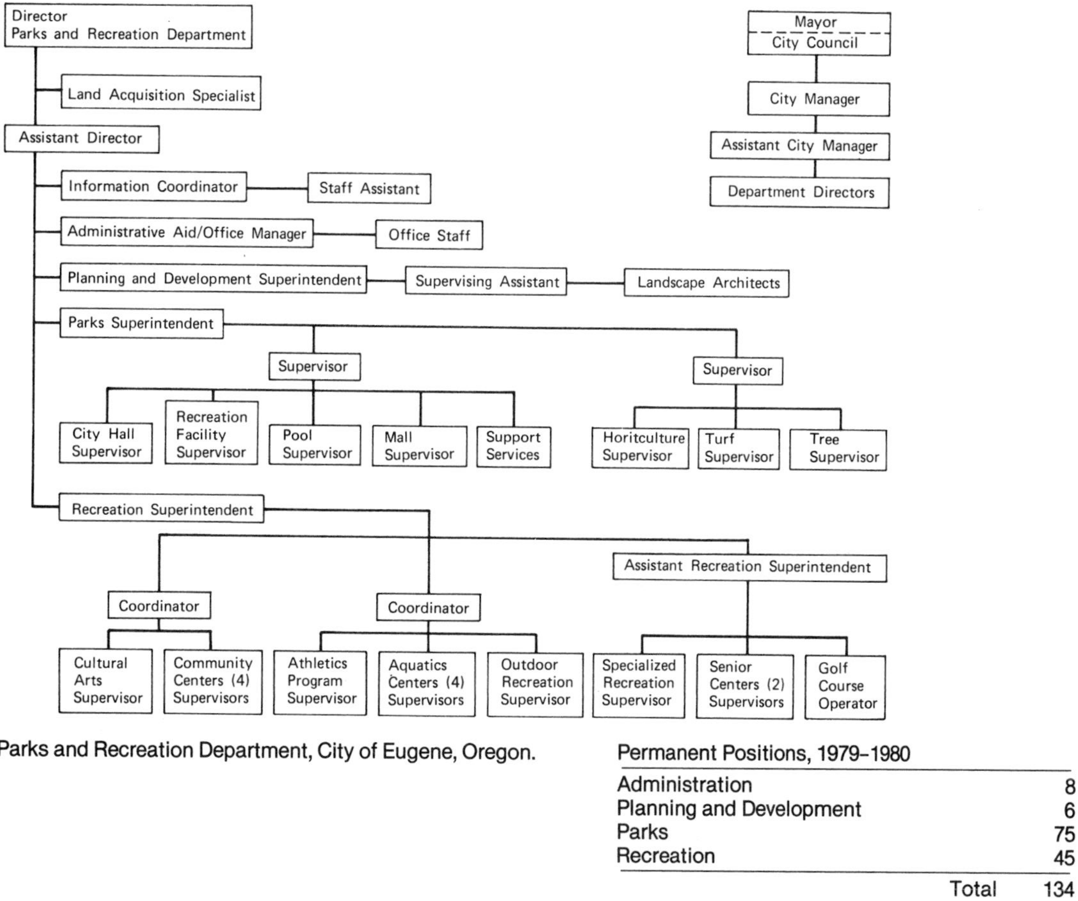


Parks and Recreation Department, City of Eugene, Oregon.

| Permanent Positions, 1979–1980 | |
|---|---|
| Administration | 8 |
| Planning and Development | 6 |
| Parks | 75 |
| Recreation | 45 |
| Total | 134 |

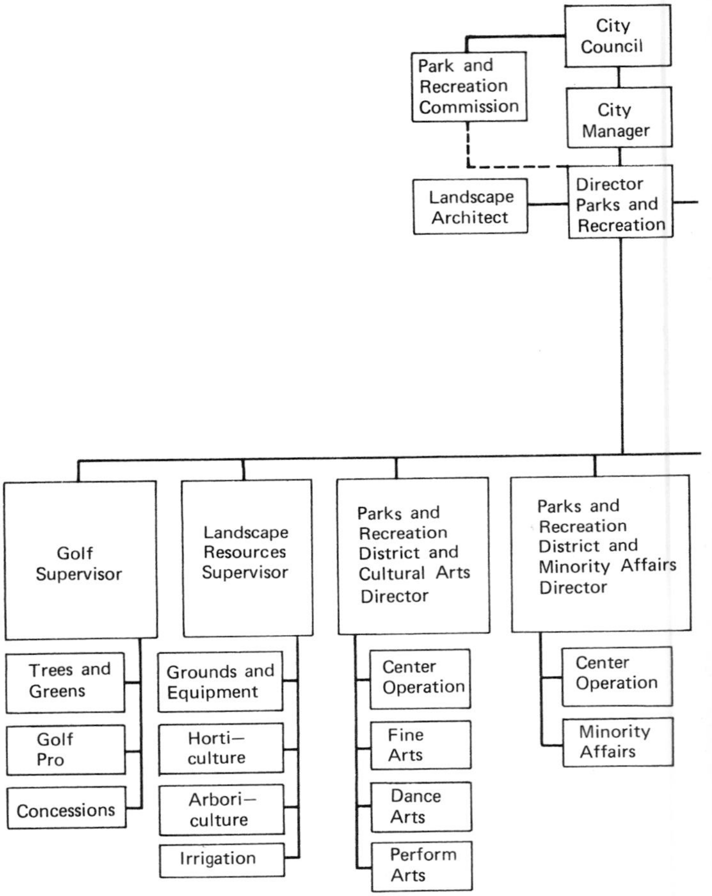

City Council

Park and Recreation Commission

City Manager

Landscape Architect

Director Parks and Recreation

Golf Supervisor
- Trees and Greens
- Golf Pro
- Concessions

Landscape Resources Supervisor
- Grounds and Equipment
- Horti–culture
- Arbori–culture
- Irrigation

Parks and Recreation District and Cultural Arts Director
- Center Operation
- Fine Arts
- Dance Arts
- Perform Arts

Parks and Recreation District and Minority Affairs Director
- Center Operation
- Minority Affairs

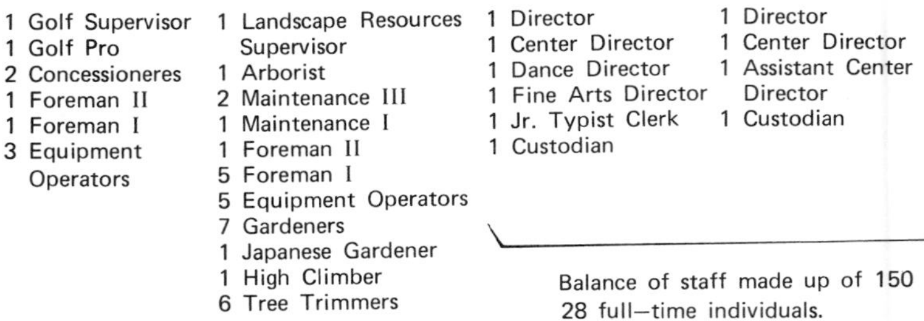


City of San Mateo Parks and Recreation Department.

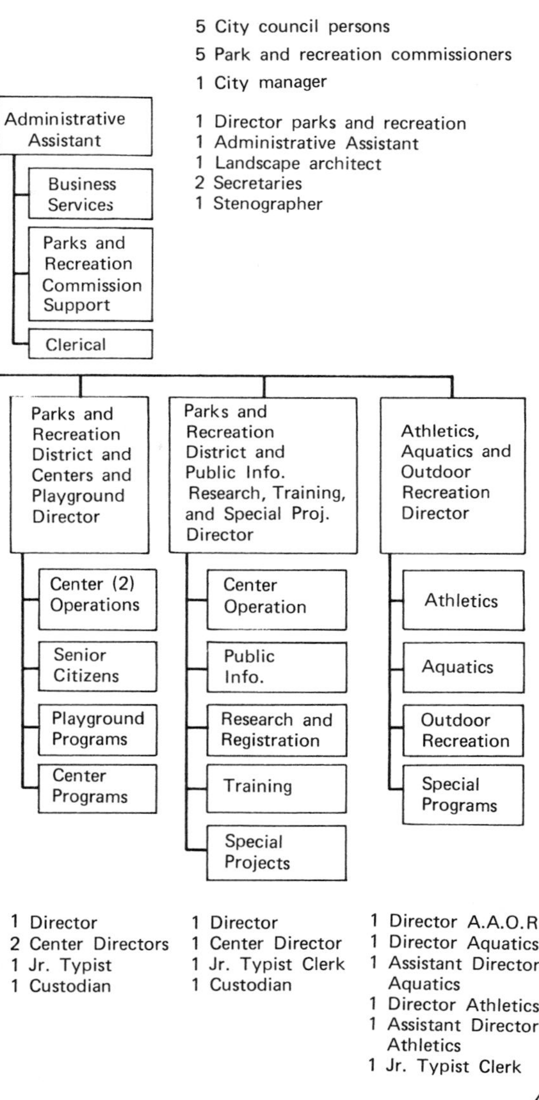


part—time personnel, which is the equivalent of

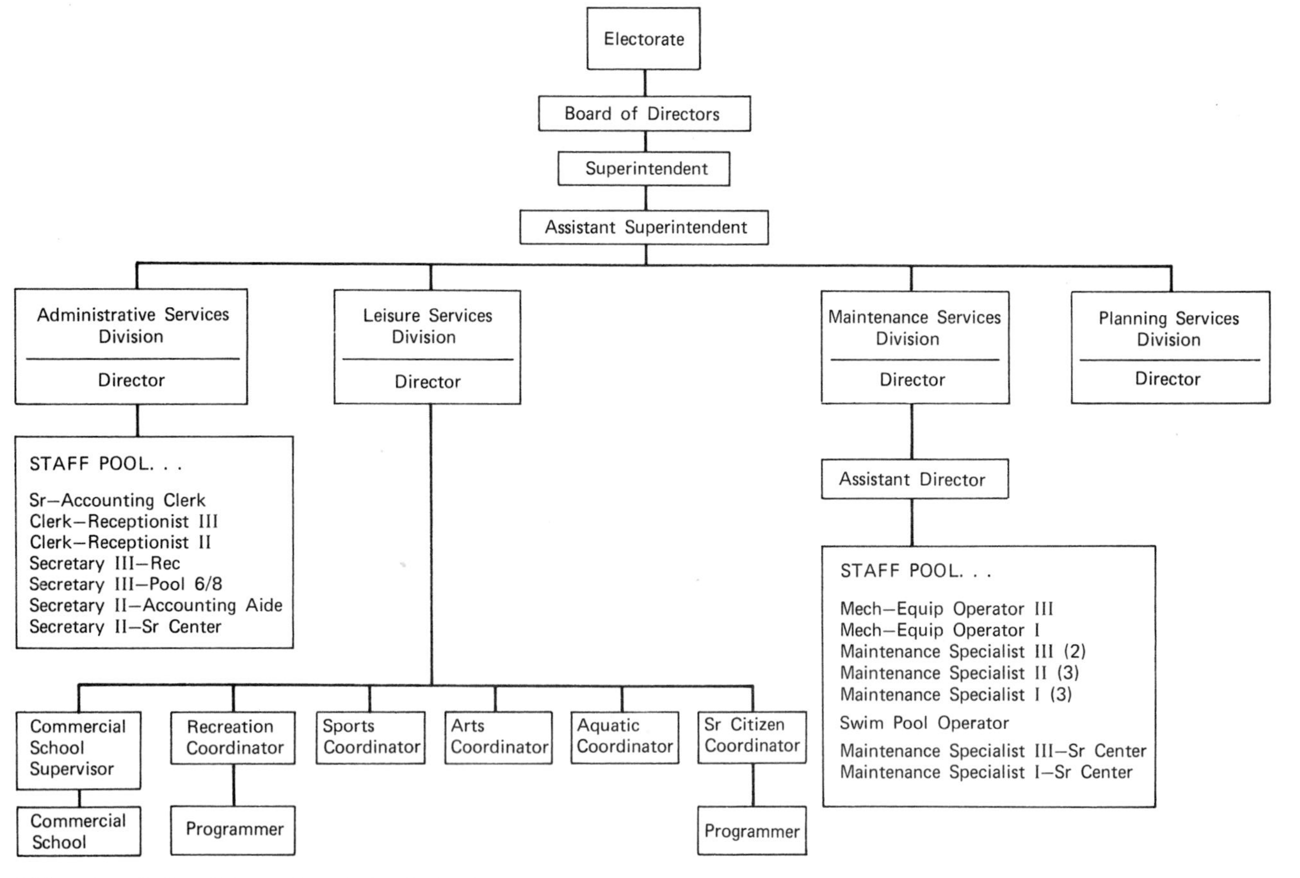


Willamalane Park and Recreation District, Springfield Oregon. Organization Chart for 1979–1980 budget review.

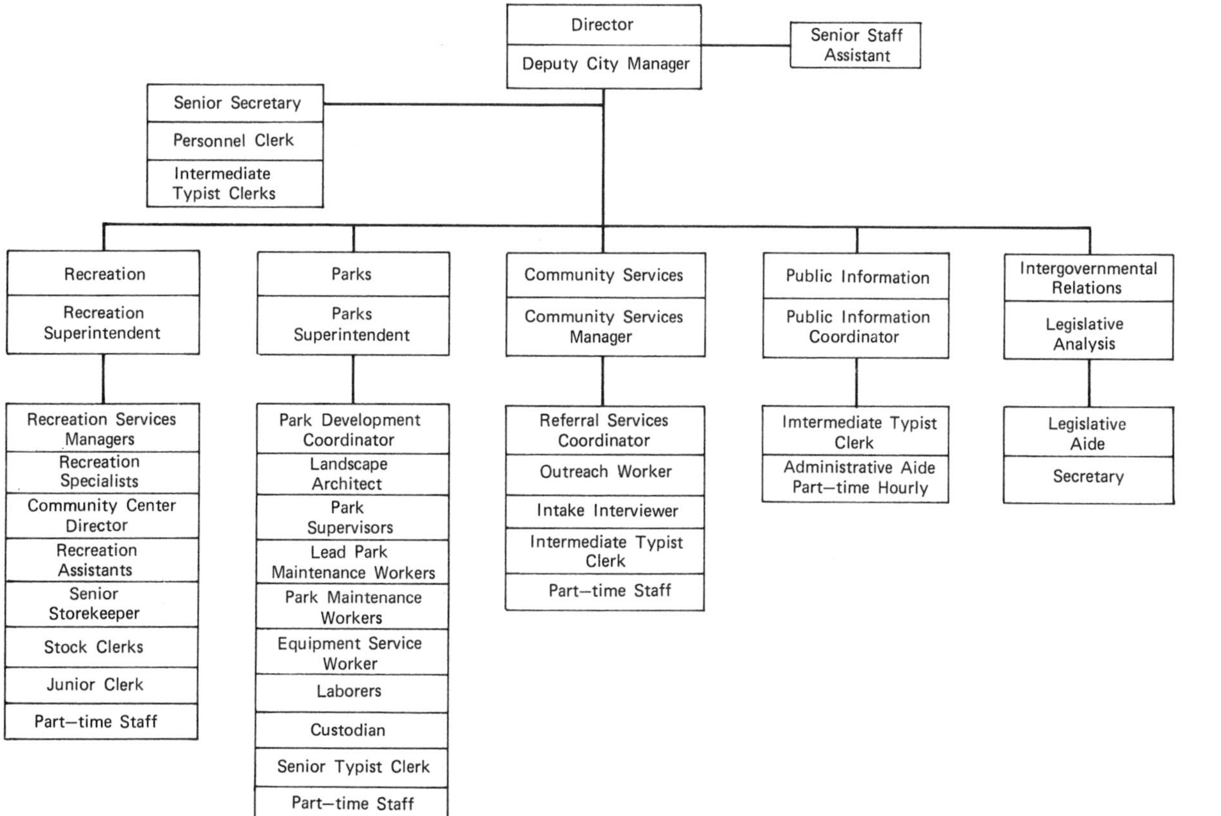


Anaheim Parks, Recreation and Community Services Department, Anaheim, California. *(Source. Table of Organization/Anaheim Parks, Recreation and Community Services Department/no copyright/permission by James D. Ruth, Deputy City Manager, City of Anaheim, California.)*

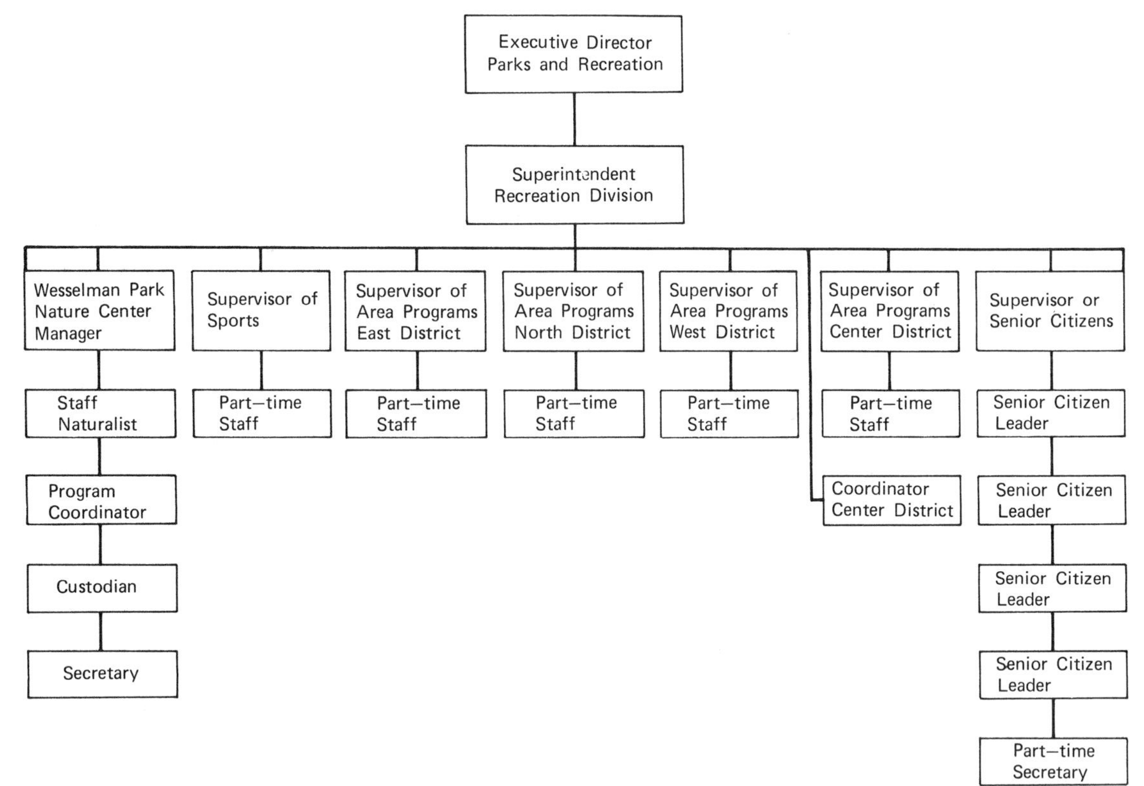


Evansville Department of Parks and Recreation, Evansville, Indiana. *(Source. Table of Organization/Evansville Department of Parks and Recreation/no copyright/Permission by Vern J. Hartenburg, Executive Director, Department of Parks and Recreation, City of Evansville, Indiana.)*

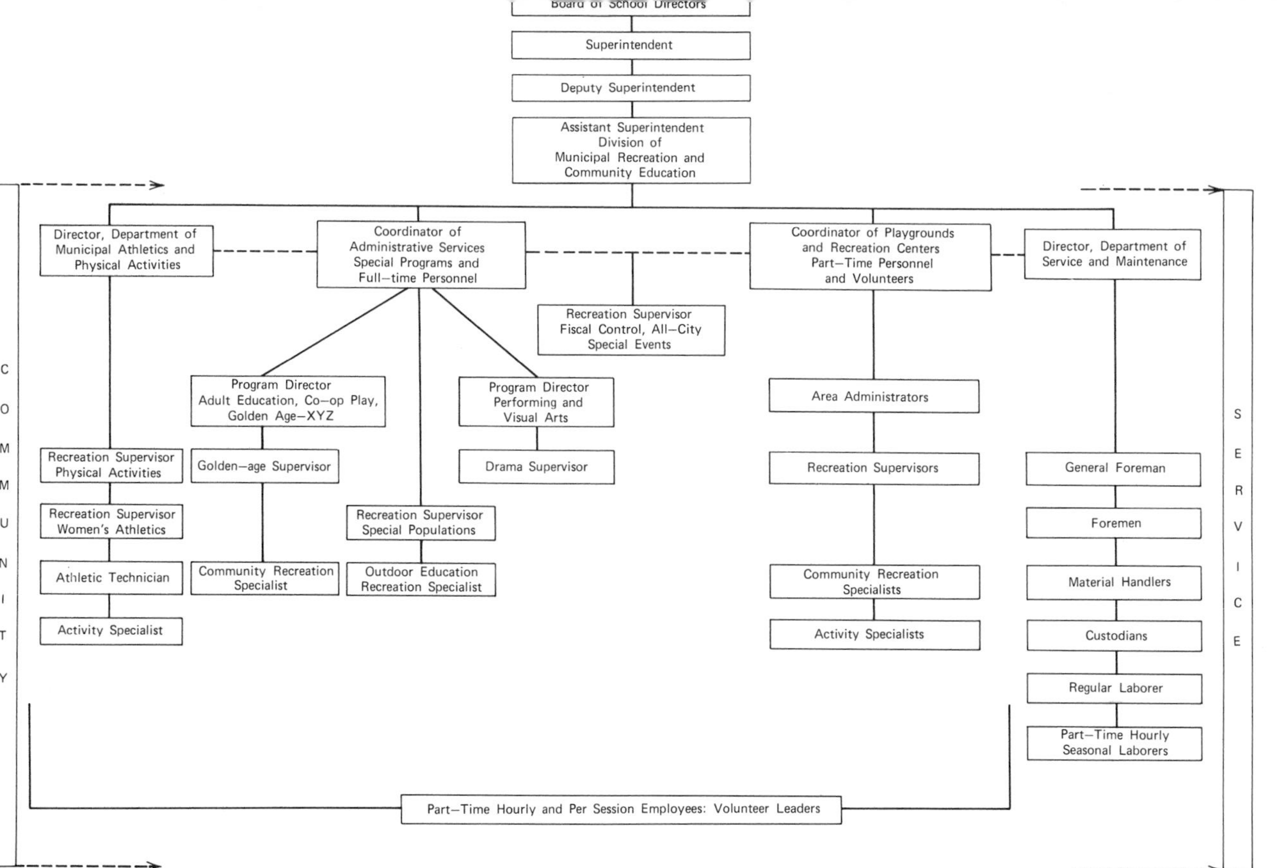


Milwaukee public schools. Division of Municipal recreation and community education. *(Source. Table of Organization/Milwaukee Division of Municipal Recreation and Community Education/no copyright/Permission by Michael Magulski, Assistant Superintendent of Schools, Division of Municipal Recreation and Community Education, Milwaukee, Wisconsin.)*

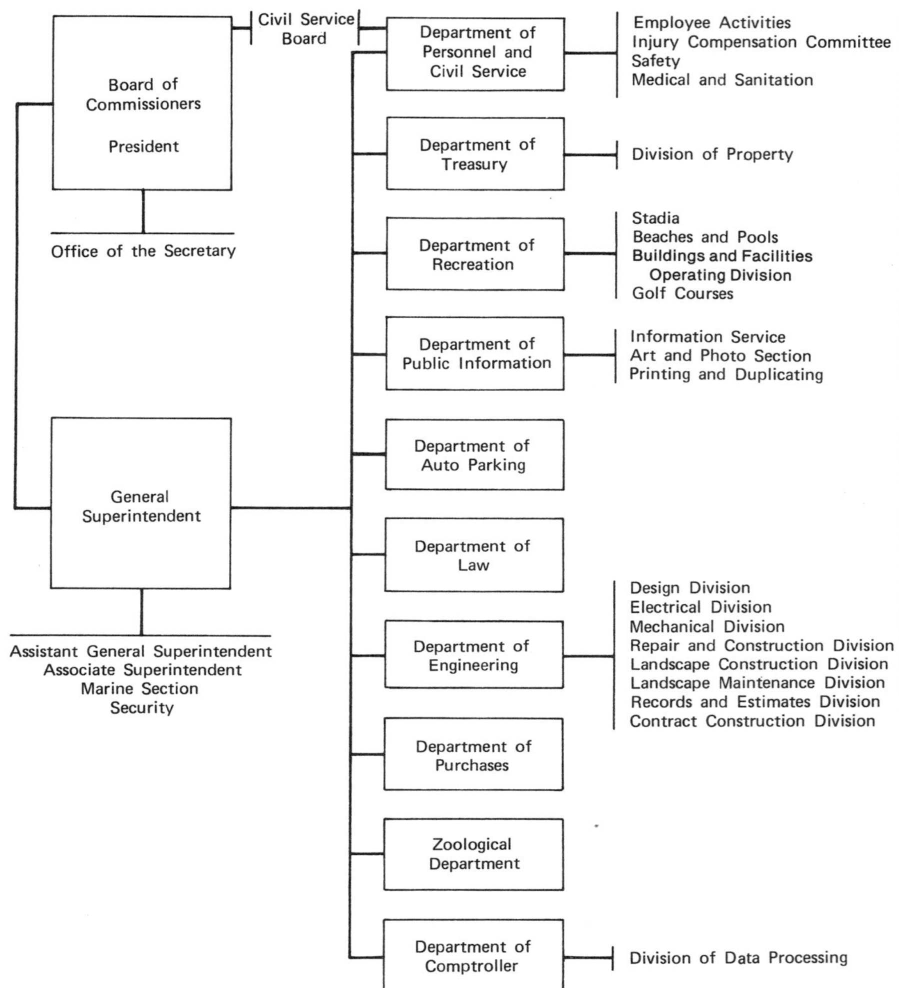


Chicago park district. *(Source. Table of Organization/Chicago Park District/no copyright/Permission by Edmund L. Kelly, General Superintendent, Chicago Park District.)*

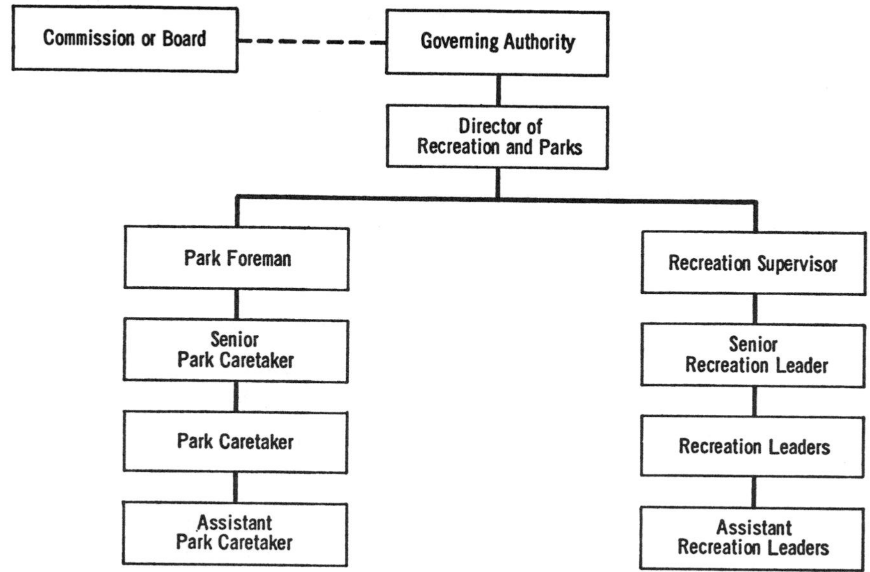


Public recreation and park agency structure (small community). *(Source. Standards for Recreation and Park Personnel (Sacramento: Division of Recreation, Department of Parks and Recreation, State of California, 1962), p. 18.)*

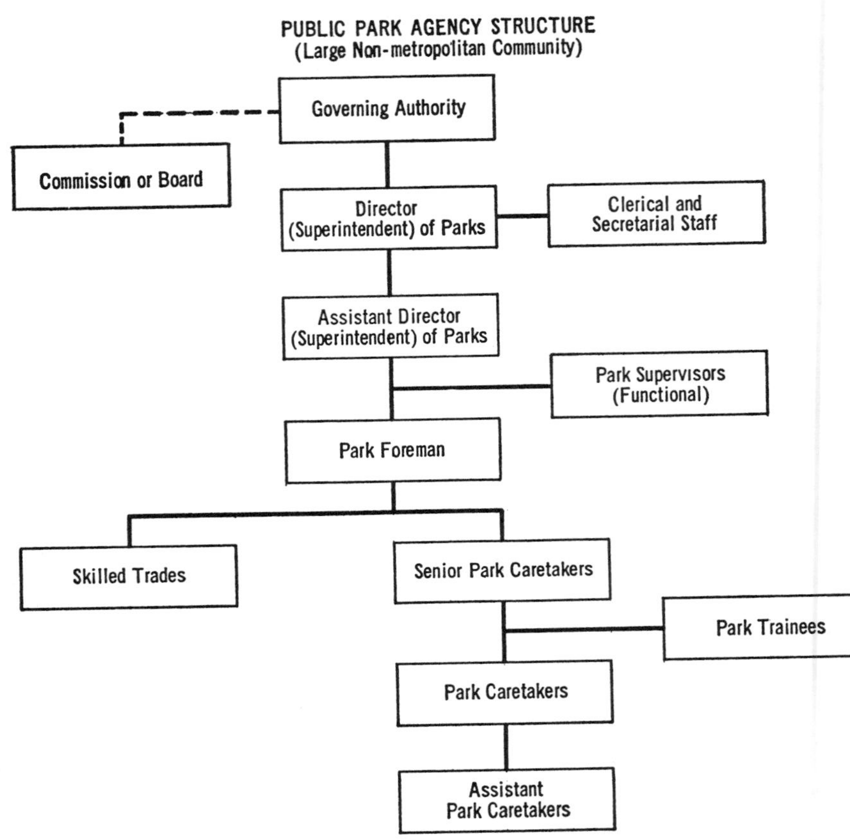


Public park agency structure (large nonmetropolitan community). *(Source. Standards for Recreation and Park Personnel (Sacramento: Division of Recreation, Department of Parks and Recreation, State of California, 1962), p. 25.)*

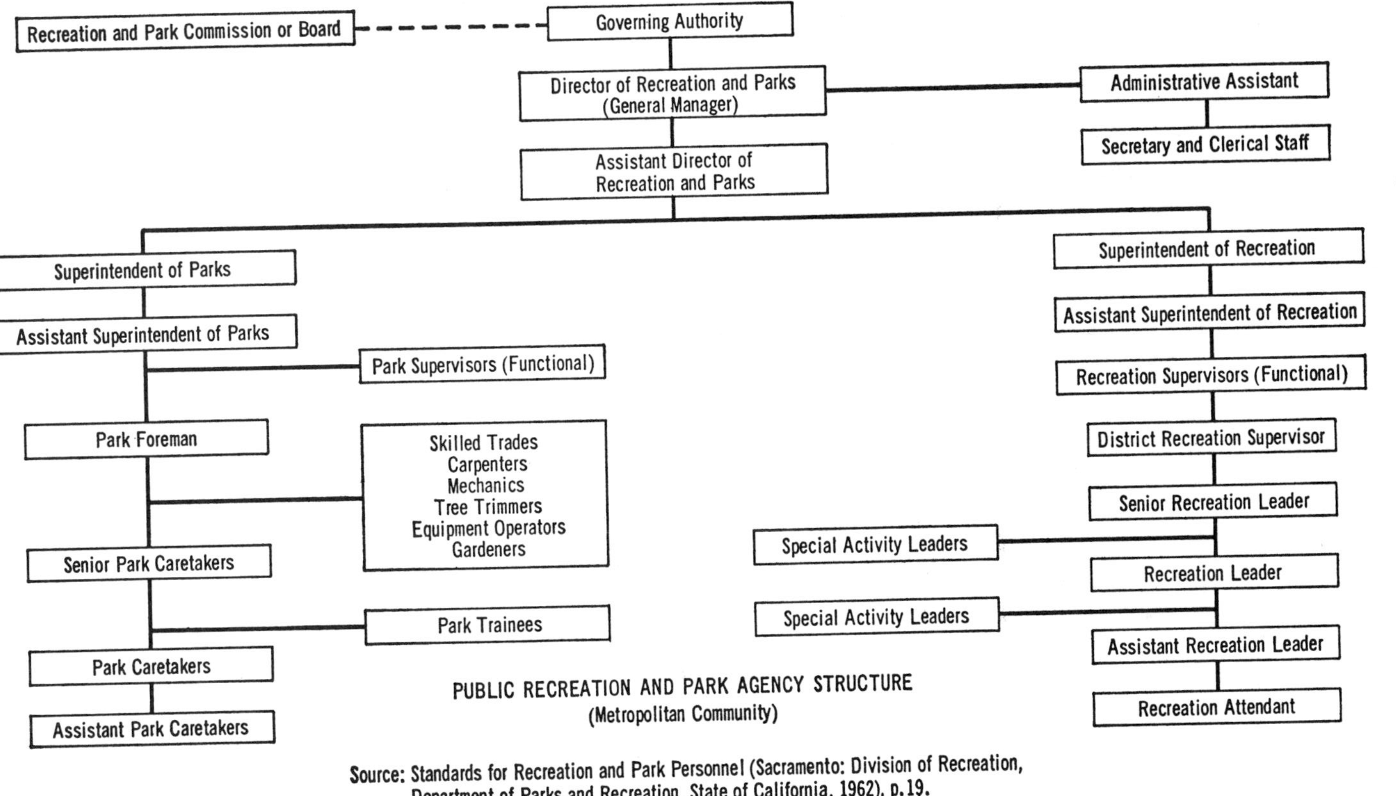


Public recreation and park agency structure (metropolitan community). *(Source. Standards for Recreation and Park Personnel (Sacramento: Division of Recreation, Department of Parks and Recreation, State of California, 1962), p. 19.)*

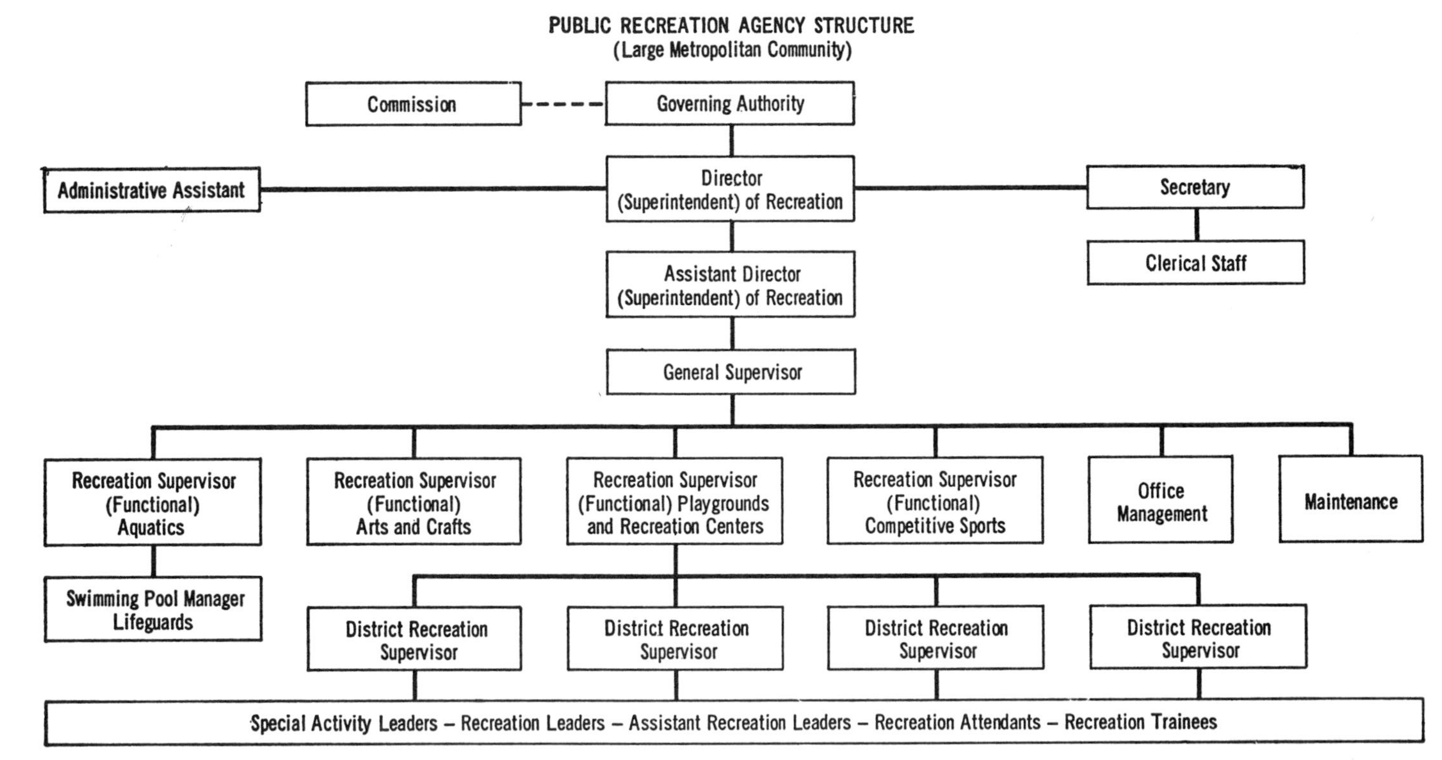


Public recreation agency structure (large metropolitan community). *(Source. Standards for Recreation and Park Personnel (Sacramento: Division of Recreation, Department of Parks and Recreation, State of California, 1962), p. 23.)*

**PUBLIC PARK AGENCY STRUCTURE**
**(Large Metropolitan Community)**

Commission or Board

Governing Authority

Director (Superintendent) of Parks

Secretary

Administrative Assistant or Public Information Officer

Clerical Staff

Assistant Director (Superintendent) of Parks

Park Supervisors (Functional)

Park Foreman

Skilled Trades

Senior Park Caretakers

Park Trainees

Park Caretakers

Assistant Park Caretakers

Source: Standards for Recreation and Park Personnel (Sacramento: Division of Recreation, Department of Parks and Recreation, State of California, 1962). p. 26.

Public park agency structure (large metropolitan community). *(Source. Standards for Recreation and Park Personnel (Sacramento: Division of Recreation, Department of Parks and Recreation, State of California, 1962), p. 26.)*

# APPENDIX C

# Sample Bylaws

REVISED BY-LAWS
of the
TOPEKA PARKS AND RECREATION DEPARTMENTS' ADVISORY BOARD[1]

(As amended to December 13, 1960)
(As amended to January 13, 1965)
(As amended to July 8, 1976)

CITY OF TOPEKA, KANSAS

THE ADVISORY BOARD BE GOVERNED BY
GENERAL STATUTES OF KANSAS 1949
ARTICLE 19: STATE OF KANSAS RECREATION ENABLING ACT

### Article I. Meetings

#### Section 1. Annual Meetings

The annual meeting of the Topeka Parks and Recreation Departments Advisory Board[2] shall be held at a date and time to be designated by the Board.

#### Section 2. Regular Meetings

The regular meeting shall be held on the third Wednesday of each month at the hour of 4:00 p.m. at the Parks and Recreation Departments Office or a place designated in the call.

[1]Revised By-Laws of the Topeka Parks and Recreation Departments' Advisory Board/no copyright/Permission by William McKinney, Superintendent of Recreation.

[2]Hereinafter referred to as the Board.

### Section 3. Notice of Meeting

Notice of all regular meetings shall be mailed to each member of the Board by the Secretary at least five days prior to each meeting.

### Section 4. Special Meetings

Special meetings may be called at any time by the Secretary, at the request of the Chairperson of the Board, or by three members of the Board.

### Section 5. Place of Meeting

The place of the meeting shall be the Parks and Recreation Office unless otherwise stated in the call.

### Section 6. Quorum

A majority of the Board shall at all times constitute a quorum.

### Section 7. Rules of Order

General parliamentary rules, as given in Robert's Rules of Order as modified by rules and regulations of the Parks and Recreation Advisory Board shall be observed in conducting meetings of the Board.

### Section 8. Order of Business

The following shall be the Order of Business of the Board, but the rules of order may be suspended and any matters considered or postponed by action of the Board.

1. Call to order
2. Roll call
3. Consideration of minutes of last regular meeting and of any special meetings held subsequently and their approval or amendment.
4. Election of officers
5. Reports of officers
6. Report of Standing Committees
7. Report of Special Committees
8. Report of Commissioner of Parks and Recreation
9. Unfinished business
10. Petitions and communications
11. New business

## Article II. Officers

### Section 1. Election of Officers

At the annual meeting, the Board shall elect by ballot the following officers to serve for one year from date of election and until their respective successors have been duly elected and shall have qualified:

A Chairperson
A Vice-chairperson
A Secretary

### Section 2. Special Elections

In the event of a failure for any reason to elect any of the said officers, or in case a vacancy shall occur in any of the said offices for any reason, then an election may be held at any regular or special meeting, a notice of such election having been given in the notice of the call of the meeting.

### Section 3. Election of Special Officers and Agents

The Board shall recommend to the Board of City Commissioners the appointment of the Superintendent of Recreation and the Superintendent of Parks. The Superintendent of Parks and the Superintendent of Recreation shall not be members of the Board, nor of the City Commission. The Superintendent of Parks and the Superintendent of Recreation shall be allowed to attend all regular meetings of the Board and to participate in discussion but shall not be entitled to vote. The Board shall report its recommendations for employment of the Superintendent of Parks and the Superintendent of Recreation to the elected Commissioner of Parks and Recreation prior to the recommendation to the Board of City Commissioners.

The Treasurer of the taxing body shall be the ex-officio Treasurer of the Board.

### Section 4. Duties of Chairperson

The Chairperson of the Board shall preside at the meetings of the Board, and shall perform the other duties ordinarily performed by that officer.

### Section 5. Duties of the Vice-Chairperson

The Vice-Chairperson of the Board, in the absence of the Chairperson, shall perform all duties of the Chairperson of the Board. In the absence of both the Chairperson and the Vice-Chairperson, the Board shall elect a Chairperson pro tempore who shall perform the duties of the Chairperson of the Board.

### Section 6. Duties of the Secretary

The Secretary shall perform the usual duties pertaining to this office.

The Secretary shall keep or cause to be kept a full and true record of all meetings of the Board and of all meetings of standing committees of the Board and of such special meetings as shall be requested of this person.

The Secretary shall be the custodian of all documents committed to his/her care.

The Secretary shall issue or cause to be issued notice of regular meetings of the Board and calls for special meetings as hereinbefore provided.

## Article III. Committees of the Board

### Section 1. Standing Committees

There shall be four (4) standing committees of the Board, namely,

1. Committee on Finance
2. Committee on Buildings and Grounds

3. Committee of Program
4. Committee on Personnel Affairs

## Section 2. Appointment of Committees

The standing committees shall be appointed by the Chairperson of the Board at the annual meeting, or as soon thereafter as possible, and serve until their successors are appointed and qualified.

## Section 3. Personnel of Committees

Each committee shall consist of not less than two members of the Board, together with the Chairperson and the Superintendent of Parks and the Superintendent of Recreation acting as ex-officio.

## Section 4. Minutes of Committees

A record of the actions of each committee shall be kept by a member of the said committee and shall be reported in writing to the Board at its next meeting for action by the Board if the Board so desires.

## Section 5. Duties of the Committee on Finance

The Committee on Finance, in cooperation with the Superintendents and the ex-officio Treasurer of the Parks and Recreation Departments shall make a careful study of the anticipated income of the Parks and Recreation Departments from all sources, and shall certify to the Board such amount as needed be raised by taxation or otherwise in order to provide for all anticipated and authorized expenditures of the Parks and Recreation Departments, and shall recommend to the Board the method or methods by which these anticipated needs shall be met.

The Committee on Finance shall have supervision over the financial records of the Parks and Recreation Departments and shall submit to the Board for its consideration and approval a budget for the Parks and Recreation Departments for the year beginning on the first day of January following. The budget submitted shall include an itemized statement of the probably expenses of the Parks and Recreation Department for said year, showing in detail as nearly as possible the salaries to be paid and the persons to whom payable, and set forth in such tabulation as shall be readily compiled according to the requirements of the Statutes of the State of Kansas.

The budget, when approved by the Board of City Commissioners, in accordance with the Statutes of the State of Kansas shall be the authority for incurring expenditures for the various departments included therein. It shall be the duty of the Superintendent of Parks and the Superintendent of Recreation and the Treasurer of the Board, acting as a committee on expenditures, to make a distribution of such budget appropriations as are desired by authorizing expenditures within the limits of such appropriations, subject to the approved procedure.

The Committee on Finance shall review the annual audit of the books and securities of the Parks and Recreation Departments by a firm of public accountants. A copy of the annual audit report shall be made available to each member of the Board, the Superintendents, and the Treasurer.

### Section 6. Duties of Committee on Buildings and Grounds

The Committee on Buildings and Grounds shall exercise supervision over the care and control of all buildings and grounds and equipment of the Parks and Recreation Departments. It shall be the duty of this Committee to see that the buildings and property of the Parks and Recreation Departments are adequately insured.

The Committee shall investigate and determine all major proposals and propositions for the acquisition, construction, maintenance, and improvement of land, facilities and equipment, and beautification of the Parks and Recreation Department's facilities. It shall consider the reports and suggestions of the administration and make its recommendation of specifications and plans for new buildings or improvements by the Board of City Commissioners for consideration; it shall call for bids when directed by the City Commission and shall recommend to the City Commission for approval the contractor or contractors who, in its opinion, should be awarded the contract for construction or major renovation authorized.

### Section 7. Committee on Program

It shall be the duty of this Committee to consider and make recommendations relating to the establishment of any new programs or the modification of existing programs, and any other related matter. The Committee shall report its actions or any recommendations at any meeting of the Board.

### Section 8. Committee on Personnel

It shall be the duty of this Committee to consider and make a recommendation to the Board of City Commissioners for employment of a Superintendent of Recreation and a Superintendent of Parks. This Committee shall also consider the number and organizational structure for full-time professional personnel and all problems, including retirement, tenure, working conditions, etc., and such other matters as may be referred to the Committee for consideration by either the administration of the Board of City Commissioners. The Committee shall report its actions or any recommendations at any meeting of the Board prior to making a recommendation to the Board of City Commissioners.

### Section 9. Advisory Committee

The Board may from time to time appoint advisory committees on departmental welfare and other subjects pertaining to the welfare of the Parks and Recreation Departments. The personnel of such committees may be composed of members of the community, representatives of civic clubs, or others.

## Article IV. Superintendent of Parks and Superintendent of Recreation

### Section 1. Appointment of the Superintendent

The City Commission shall employ as Superintendent of Parks and Superintendent of Recreation a person who gives promise of such leadership as will further the general welfare of the Topeka Parks Department and the Topeka Recreation Department, its service to the city, and the welfare of all those who shall be influenced by it. The Superintendents shall be employed by the City Commission as hereinbefore upon the recommendation of the Board.

### Section 2. Duties of the Superintendent

The Superintendent shall carry out the duties and responsibilities as described in the job description as formulated by the Personnel Department of the City of Topeka.

### Section 3. Reports of the Superintendent

The Superintendent shall make an annual written report to the Board and the City Commission of the work and condition of the Parks and Recreation Departments and from time to time shall give to the Board and to the City Commission such additional reports on the work and condition of the Parks and Recreation Departments as they may require or the welfare of the department may suggest.

## Article V. Board Rules

### Section 1. Rules

The Board may from time to time make and designate the rules and regulations under which the Board shall be operated. These rules shall be kept in suitable records by the Secretary of the Board.

### Section 2. Fiscal Year

The Fiscal Year of the Parks and Recreation Departments shall begin January 1 and continue until midnight, December 31, next following.

### Section 3. Budget: 12-1908

Not later than 20 days prior to the date for the publishing of the budget, the Recreation Department shall certify its budget to the Board of City Commissioners, which as the taxing body shall levy a tax sufficient to raise the amount required by such budget, but *in no event more than three mills as provided by law.*

### Section 4. Tax Collections: 12-1907

A claim voucher should be submitted to the taxing body for the amount of tax funds received by the county treasurer and paid over to the treasurer of the taxing body for the Recreation Department. The total of such claims presented in each fiscal year must not exceed the amount required from the tax levy as shown in the budget submitted to the taxing body and levied for that year.

The taxing body will draw a warrant to the treasurer of the City Commission to set over the City Commission the amount shown by the approved claim voucher. This warrant will be deposited in the official depositary of the City of Topeka.

### Section 5. Other Receipts

All other cash receipts shall be deposited promptly to the credit of the treasurer of the City of Topeka.

The person who receives money should issue triplicate receipts as evidence that payment or settlement has been made, and to provide for a systematic accounting therefore. Receipts should be written at the time monies are received, the original delivered to the party making the payment, first carbon copy to accompany the settlement or reporting of the collection to the treasurer, and the second carbon copy to be retained in the receipt book by the issuing officer.

### Section 6. Disbursements: 12-1907

The disbursements made by the Parks and Recreation Departments shall be paid by voucher, and the claims paid by said Parks and Recreation Departments shall be duly verified.

### Section 7. Purchase Orders

All purchases of materials, supplies, services, and equipment, the cost of which can reasonably be expected to exceed $4,500, shall be by sealed bid. All public improvement projects where the estimated costs thereof shall exceed $4,500, shall be contracted for by formal written contract with the lowest responsible bidder. Formal bids are required on purchase of all motor vehicles and construction equipment where the estimated cost exceeds $2,500. *Purchase orders shall be issued to cover all expenditures.*

### Section 8. Claim Vouchers

Disbursements can be made only upon itemized claim duly verified. Claim vouchers filed by vendors shall be examined by the Superintendents to determine that the calculations and additions are correct, and that it is supported by a proper purchase order and record of delivery, and signed as evidence of his audit and approval. Claim vouchers must bear the approval of the City Commission.

### Section 9. Warrants

Warrants issued in payment of approved claims must be signed by the City Treasurer and the City Auditor and countersigned by the Mayor.

## Article VI. Amendments

### Section 1. Amendments

These by-laws may be amended at any regular meeting of the Board by a majority vote of the entire Board, provided previous notice of the nature of any proposed amendment shall be given at least one regular meeting before the action thereon shall be taken.

Approved, adopted, and recorded by the Topeka Recreation Commission in its minutes of a called meeting dated Thursday, January 21, 1954.

Approved, adopted and recorded as REVISED, by the Topeka Recreation Commission in its minutes of the meeting, December 13, 1960; meeting of January 13, 1965; and at its meeting of July 8, 1976.

Policy & Procedure Manual[1]
Elmhurst Park District
225 Prospect
Elmhurst, Illinois 61026
Michael S. Pope, Director

## 4.00. Organizational & Procedural Functions of the Board

### 4.01. Government

The government of the Elmhurst Park District shall be vested in the Board of Commissioners, duly elected or appointed as provided by law.

### 4.02. Officers

The officers shall be the President and Vice-President and shall be elected from the duly-elected or appointed Commissioners of the Elmhurst Park District. Additional officers shall be the Secretary, Treasurer and Assistant Treasurer. These additional officers may or may not be elected members of the Elmhurst Park Board.

The Director of Parks & Recreation, as well as the Attorney, shall be appointed by the Board of Commissioners, neither of whom shall be a member of the Board.

### 4.03. Election & Appointment of Officers

All officers of the Board shall be elected, and other officers of the Elmhurst Park District shall be appointed by the Commissioners at the annual meeting hereinafter set forth, and at such other times as a vacancy occurs. Each of said officers shall hold office until the next annual meeting and until their successor shall be chosen. In case of temporary absence or inability of any officer to act as such, the Board may fill the office *pro tempore*.

The office of President shall be rotated annually. Each year, at the annual meeting, the Commissioner with the most seniority shall be elected President, and the one with the second most seniority shall be elected Vice-President. Seniority shall be counted from the meeting when a Commissioner was sworn into office (not as a result of re-election) or from the meeting when he or she last served as President, whichever is more recent. When two or more Commissioners were first sworn into office at the same meeting, order of seniority shall be determined by the number of votes cast for them in the election, high number of votes first.

### 4.04. President

The President shall be the executive officer of the Board. It shall be his/her duty to preside at all meetings when present, to sign all contracts and other papers authorized by the Board, to see that all Ordinances of the Board are enforced and

[1] Organizational and Procedural Functions of the Board/no copyright/by permission of Michael Pope, Director, Elmhurst Park District.

that all orders of the Board are faithfully executed, and to exercise general supervision over all officers and employees and over the business and property of the District; all subject, however, to the direction and approval of the Board.

### 4.05. Vice-President

The Vice-President shall be vested with the powers to perform the duties of the President in the absence of the President or in the event of his refusal or inability to act.

### 4.06. Secretary

The Secretary shall have the custody of the Corporate Seal and of all books and papers pertaining to the Secretary's office; shall attest and affix the Corporate Seal to all instruments requiring such action when authorized by Ordinance or vote of the Board; and shall cause all Ordinances, Resolutions and other actions of the Board requiring publication, to be duly published. The Secretary shall attend all meetings of the Board and keep full and true record of its proceedings. The Secretary need not be a member of the Board, but may be a paid employee, and shall act under the general supervision of the Director of Parks & Recreation. Assistant Secretaries may be appointed by the Board, with full or limited powers as specified when the appointment is made.

### 4.07. Treasurer & Assistant Treasurer

The Treasurer shall receive and keep all moneys belonging to the District, depositing same on a timely basis in the name of the District, and shall disburse the same only on warrants authorized by the Board and signed by one Commissioner and countersigned by the Treasurer or Assistant Treasurer. The Treasurer shall make monthly reports to the Board of all receipts and disbursements, and shall submit monthly budget comparison reports. The Treasurer and Assistant Treasurer do not need to be a member of the Board, but may be a paid employee, and shall act under the general supervision of the Director of Parks & Recreation. The Assistant Treasurer may be appointed by the Board with full or limited powers as specified when the appointment is made.

### 4.08. Director of Parks & Recreation

The Director of Parks & Recreation (hereinafter referred to as Director) is directly responsible to the Elmhurst Park District Board of Commissioners. The primary function of the Director shall be to advise the Board on matters pertaining to all functions for which the District is responsible. He shall be the administrative head of all departments of the Park District, in regard to recreation programming, personnel management, facility development, land acquisition and business management. He shall be the official medium of communication between the employees of the District and the Board of Commissioners. He shall have charge of the employment of such employees as are required to operate the Park District and its facilities, subject to employment policies and salary schedules as established by the Board and embodied in the Personnel Policies of the Elmhurst Park District.

### 4.09. Attorney

The Attorney shall have charge of all legal matters and of the prosecution and defense of all litigation as directed by the Board. He shall draft all Ordinances, Resolutions and other instruments required by the Board or any committee thereof, and shall give opinions on all questions referred to him by the Board or any committee or the President of the Board, and shall attend all meetings required by the Board.

### 4.10. Additional Duties of Officers

In addition to the duties hereinbefore specified, each officer shall perform such other duties as may be required of him by law or by the Ordinance or Resolutions of the Board.

### 4.11. Vacancies Declared

Whenever any member of the Board of Commissioners shall die, resign, be declared insane, cease to be a legal voter in the Elmhurst Park District, be convicted of an infamous crime, refuse to take the oath of office after becoming elected to the Board of Commissioners of the Elmhurst Park District, or neglect to attend the duties of his/her office or neglect to attend regular and social meetings of the Board for a period of three consecutive calendar months, or neglect to attend a minimum of twelve regular or special meetings in any 12-month period, said office may be declared vacant by the Board, and may be filled by appointment by a majority of the remaining members of the Board.

### 4.12. Compensation of Officers and Employees

The officers (with the exception of Commissioners serving in such offices) and all employees shall receive such compensation for their services as the Board shall from time to time determine.

### 4.13. Primary Functions of the Park Board

The Park Board's major function is that of establishing policy through the majority vote at duly-called and authorized Park Board meetings. Through the policy-making procedures, the Board determines fiscal procedures, personnel matters, operational procedures, fees and charges, land acquisition and facility development. Specifically included in the above items are the following:

**A.** To provide for the levy of taxes pursuant to the authority granted by State Statute. Such levies shall provide for the various operational concerns by fund so that sufficient revenue is generated to provide for a quality Park & Recreation operation. Decide upon the proper use of funds generated by revenue-producing facilities after the operational and maintenance needs are satisfied and enact periodic adjustments in the operational policies of said revenue facilities to insure proper and meaningful controls for the benefit of the entire District—not just the revenue facility itself.

**B.** The Board should employ a Director of Parks & Recreation as its chief administrative officer to whom the Board places its reliance and authority for the judicious administration of the day-to-day operation of the Park District. The Director of Parks & Recreation shall be charged with the authority to execute the Board's policy, enforce its rules and regulations and act as an advisor to the Board by preparing, or causing to be prepared, written reports for the Board which states the problem, investigates the alternatives, and recommends a course of action. The Board shall adopt and periodically review a set of rules and regulations affecting all full-time personnel in a document known as "Personnel Policies of the Elmhurst Park District".

**C.** The Board shall constantly monitor the operational procedures of the Park District and make additions or alterations to said procedures at duly-called and authorized Board meetings. The individual Board members shall keep themselves informed of the activities and functions of the District by observation, comments from its citizenry, and reports presented by the staff of the Park District. The Board shall act decisively on issues brought before them in the best interest of the District as a whole and will refrain from making decisions which will benefit small factions or special interest groups.

Park Board members should make decisions involving the welfare of the community based on study and evidence rather than feelings, prejudices, personal opinions or other similar subjective factors. Such judgments require mutual considerations of varying points of view before final action is taken.

Park Board members should accept the principle of Board unity and the subordination of personal interests by accepting and supporting majority decisions of the Board and identifying themselves with Board policies and actions.

**D.** The Board shall be responsible to establish the operational philosophy of recreation programming for the Park District and establish the fees and charges which will be approved at a duly-authorized and attended Board meeting. The necessity of assessing fees and charges allows a wide variety of programs and services to the community which would otherwise not be possible without the additional revenue which these fees generate.

**E.** The Board shall realize that land acquisition is of primary importance to the provision of leisure services and the proposition that open space, which is judiciously placed, produces benefits for active and passive use as well as contributing to urban form. A comprehensive and continuously updated land acquisition plan, supported by a set of land acquisition criteria, which will assist in evaluating various parcels, must be the cornerstone of the land acquisition program.

Cooperative ventures with local, county, state, regional and national levels of government should be recognized as important and integral processes toward the orderly acquisition of parcels which may otherwise be too costly for one agency. Various state and federal land grant programs should be viewed as a vehicle for financial assistance. Such grants must be reviewed carefully for their dependencies and provisos which may be considered reasonable or perhaps too restrictive.

**F.** The Board shall concern itself to the establishment and continual care of a well-rounded and broadly-based park system which recognizes the diverse needs and interests of its constituency.

# APPENDIX D

# Rating Records

City of Eugene, Oregon
Parks & Recreation Department
SUPERVISORY EVALUATION

EMPLOYEE ______ EMP. No. ______ DEPT. DIV. ______ ☐ SCHEDULED ☐ UNSCHEDULED ☐ PROBATIONARY

Date employed by City ______ Starting date this class ______

PRESENT CLASS TITLE ______ PRESENT RANGE ______ PRESENT STEP ______

PROPOSED CLASS TITLE ______ PROPOSED RANGE ______ PROPOSED STEP ______ DATE ______

SECTION A — FACTOR CHECK LIST

IMMEDIATE SUPERVISOR MUST CHECK EACH FACTOR IN THE APPROPRIATE COLUMN

| A Exceeds Standards | B Meets Job Requirements | C Needs Improvement | D Unacceptable | Factor | E Does Not Apply |
|---|---|---|---|---|---|
| | | | | 1. Observance of scheduled work hours | |
| | | | | 2. Compliance with established policies | |
| | | | | 3. Safety practices | |
| | | | | 4. Works cooperatively w/other staff | |
| | | | | 5. Relates to & communicated w/public | |
| | | | | 6. Understands responsibilities & functions | |
| | | | | 7. Accepts responsibility | |
| | | | | 8. Work judgments | |
| | | | | 9. Planning and organizing | |
| | | | | 10. Meeting deadlines | |
| | | | | 11. Quality of work | |
| | | | | 12. Effective time management | |
| | | | | 13. Accepts direction | |
| | | | | 14. Adaptability to change (flexibility) | |
| | | | | 15. Effectiveness under stress | |
| | | | | 16. Operation & care of equipment | |
| | | | | 17. Initiative | |
| | | | | 18. Energy and enthusiasm | |
| | | | | 19. Understanding of overall organization | |
| | | | | 20. | |
| | | | | 21. | |
| | | | | 22. | |
| | | | | 23. | |
| | | | | FOR EMPLOYEES WHO SUPERVISE OTHERS | |
| | | | | 24. Training and instructing | |
| | | | | 25. Evaluating | |
| | | | | 26. Scheduling and coordinating | |
| | | | | 27. Productivity, quality of work | |
| | | | | 28. Supervisory control | |
| | | | | 29. Leadership | |
| | | | | 30. Operational economy | |
| | | | | 31. Judgments and decisions | |
| | | | | 32. Fairness and impartiality | |
| | | | | 33. Safety record | |

SECTION B Record job STRENGTHS AND

Record job STRENGTHS and superior performance incidents, and/or DEFICIENCIES or job behavior requiring improvement or correction. (Explain checks in columns C & D.)

SECTION C Record PROGRESS ACHIEVED in attaining previously set goals for improved work performance for personal, or job qualifications.

SECTION D Record specific GOALS or IMPROVEMENT PROGRAMS to be undertaken during next evaluation period.

SUMMARY EVALUATION – Check Overall Performance

1 ☐ Exceeds Standards 3 ☐ Requires Improvement
2 ☐ Effective – Meets Standards 4 ☐ Not Satisfactory

Rater: I certify this report represents my best judgment.

☐ I DO ☐ I DO NOT recommend this employee be granted permanent status. (Final report only)

☐ I DO ☐ I DO NOT recommend a Merit Pay increase for this employee. (Where applicable)

RATED BY: ______ DATE: ______

REVIEWED BY: ______ DATE: ______

EMPLOYEE: I certify that a copy of this report has been given to me. I understand my signature does not necessarily indicate agreement.

COMMENTS: ______

EMPLOYEE SIGNATURE: Date:

(Employee Comments May Be Attached)

## Urbana Park District
## Urbana, Illinois
## Personnel Evaluation Guide[1]

Each employee completing a personnel evaluation form must read thoroughly this personnel evaluation guide. The guide contains definitions of the performance factor ratings, offers general instructions for completing the personnel evaluation form and conducting the evaluation interview, and specifically explains considerations which should be made in assessing each performance factor. The lengthy definitions of performance factors, including suggested questions to be considered, are intended to standardize the evaluation process. For the evaluations to be effective, these definitions must be read carefully and thoughtfully.

The personnel evaluation form has two sections. Both sections must be completed in full. Familiarize yourself with the contents of the form. Before beginning the evaluation you should thoroughly read the job description for the position held by the employee being evaluated. During the evaluation interview you should refer to this job description as necessary.

In completing the evluation form, make every attempt to eliminate personal prejudice. Try to be objective. Do not let subjective feelings blur your perceptions of the employee's competency and effectiveness. Do not assume, for instance, that excellence in one performance factor implies excellence in all factors.

Your evaluation of an employee should be based on performance during the entire rating period. Important single instances of poor or excellent behavior should not be ignored. These instances should, however, be considered in the context of performance during the entire period.

The evaluation report must be typed or written in ink before the evaluation interview. Any changes in the report which might result from the interview must be initialed and dated in ink. To be complete, each evaluation report must contain the signature of the rater and of the employee being evaluated.

Ample space is provided to make additional comments. Some of this space must be used to explain specific examples of poor or excellent performance. Other space is available which may be used at your discretion. This provides you the opportunity, for example, to explain unusual circumstances surrounding the employee's performance such as prolonged illness or other unavoidable adverse conditions. Actual work performance must be evaluated regardless. However, please make use of the space for comment whenever appropriate.

The evaluation process will serve no useful purpose if you are not thoroughly candid in your assessments. You should be generous in rating the employee's best qualities and severe in rating his or her weaknesses. Do not ignore areas of performance which require improvement in order to avoid an uncomfortable or embarrassing evaluation interview. Inevitably, this lack of candor in your ratings does a great disservice to the employee being evaluated. In ignoring the weaknesses you may be robbing him or her of the opportunity to improve their performance.

[1]Urbana Park District Personnel Evaluation Guide/no copyright/by permission of Robin Hall, Director of Parks and Recreation, Urbana, Illinois Park District.

A similar situation arises for those who find it difficult to express honest appreciation and praise. Recognition of a job well done is often a prime motivation for continued good work. If an employee has worked hard with excellent results, the failure of the supervisor to give appropriate praise can have a lasting and demoralizing effect on the employee's work attitude.

If you are a supervisor, a less than candid evaluation will leave you with little or no verification of the need for disciplinary action, dismissal, or promotion. If you are evaluating your supervisor, you will find that a failure to offer constructive criticism will be viewed as a weakness, not a kindness. There is no need to feel your position threatened by the evaluation which you give your supervisor if it is given fairly and honestly. Neither is there any cause to hope that a flattering but less than honest evaluation will enhance your position.

In Section I of the evaluation form, you are asked to rate a variety of performance factors on a scale from unsatisfactory to outstanding. Space for indicating that you feel the performance factor does not apply to the employee is also available under column heading "f." What follows are definitions of the performance factor ratings listed under columns "a" through "e" on the form:

*Outstanding.* Performance which is consistently and distinctly far above the standard expected of a competent worker in the position.

*Above Standard.* Performance which frequently exceeds the standard for a competent worker; sometimes slightly above, sometimes far above.

*Standard.* Performance which consistently meets the standard of a competent worker in the position. On occasion, the employee may exceed or fall below the standard.

*Below Standard.* Performance which too often falls below the standard expected of a competent worker; performance which needs to be improved.

*Unsatisfactory.* Performance which frequently falls below standard; performance which requires corrective action or discipline.

The performance factors upon which each employee is rated are, when taken alone, open to widely different interpretation. For this reason, guidelines for interpreting the scope of each performance factor are provided below in an attempt to standardize the evaluation process. Spaces 16 through 18 (for all personnel) and spaces 30 and 31 (for supervisory and administrative personnel) have been left blank on the evaluation form to give you the opportunity to rate additional factors you believe are essential in evaluating the employee.

You should refer frequently to these performance factor guidelines while completing the evaluation form.

**1.** *Quality of Work.* This pertains to the accuracy and thoroughness, the general degree of excellence of the work performed. Consideration should be given to the consequences of poor quality work. Must the work be redone, or substantially revised? Do errors in the employee's work affect the efforts of others? Does the quality of work too often reflect adversely upon the department or the District?

**2.** *Quantity of Acceptable Work.* This pertains to the accomplishment of work goals as concerns the volume of work produced. Is the volume of work produced sufficient to meet job responsibilities as outlined in the job description.

3. *Dependability.* This pertains to the employee's attitude and attention to work in the absence of supervision, ability to meet deadlines, reliability to satisfactorily complete tasks.

4. *Work Habits.* This pertains to the observance of established policies and procedures, the care of property and materials, attendance, punctuality, and safety practices.

5. *Knowledge of Work.* This pertains to the specific job skills required for the position; to the more general knowledge of the District's operations and policies; and to an understanding of the employee's position in relation to other positions.

6. *Interaction with Employees.* How well does the employee work with fellow employees? Does the employee have a positive influence on employee morale; does he or she contribute to a cooperative work atmosphere? Is the employee sensitive and responsive to those with whom he or she works? The rating would reflect only those contacts which either improve or reduce the effectiveness of the employees involved. It does not apply to an employee's personal popularity.

7. *Interaction with Public.* How well does the employee work with the public? This includes all public contact made through personal or telephone conversation, correspondence, and day-to-day appearance before the public. Does the employee respond attentively, promptly, and courteously to citizens' suggestions, comments, and complaints?

8. *Work Judgement.* Does the employee exercise sound judgement in making decisions? Is the employee's judgement reliable and consistent? Does the quality of judgements decline under pressure?

9. *Planning and Organizing.* Does the employee organize tasks and allocate man-hours efficiently? To what degree does lack of planning decrease the quantity or quality of work performed? Are timetables established and followed in an effort to achieve agreed upon goals? Does the employee take time to plan the sequence of steps required in carrying out tasks?

10. *Accepts Responsibility.* Does the employee readily accept responsibility? Is the employee willing to accept responsibility for his mistakes. Does the employee consistently act in a responsible manner.

11. *Accepts Direction.* This pertains to the acceptance of supervision, training, and instruction. Does the employee show respect for the authority of his or her supervisor? Does the employee often challenge supervision or instruction? Does the employee respond in a constructively critical manner to supervision and instruction, or does he or she passively accept directions he or she thinks may be faulty? Is the employee resentful of supervision?

12. *Adaptability and Flexibility.* Does the employee accept change willingly or display resistance to change? Does the employee adapt well to new work surroundings, new equipment, new policies and procedures, new supervisors or fellow employees? Can the employee adapt to work assignments which are not part of his routine responsibilities?

13. *Effectiveness under Stress.* Does the employee perform well under stress? Is the employee capable of meeting rapidly changing deadlines? Can the employee cope with legitimate but periodically heavy demands on his or her time by other employees? Can the employee produce work of acceptable quality and

quantity in an emergency? Is the employee's work generally organized well enough to accommodate unforeseen circumstances? Consider here to what degree stress is inherent in the position and to what degree it results from the employee's failure to plan and organize his or her work.

**14.** *Initiative.* This pertains to initiation of action by the employee. Does the employee offer valuable suggestions and constructive criticism? Must the employee be prodded into action or does he or she exercise initiative and take independent action within specified policy limitations. Is the employee inventive and creative?

**15.** *Work Coordination.* Does the employee pursue coordination efforts which increase work efficiency? Is the employee able to undertake coordination activities without usurping the authority of other employees? Does the employee perform well in cooperative efforts with other agencies? Does the employee respect the work load and deadlines of other employees in initiating coordination? Has production been slowed because of obstructions caused by the employee?

**16.** Spaces 16 through 18 have been left blank for additional factors the rater may consider necessary in presenting a view of the employee's total job effectiveness.

**19.** *Planning and Organizing.* Does the employee recognize or develop methods of work simplification? How well does the employee analyze and implement improved and more efficient work procedures? Does he or she plan improvements or changes and implement them in a logical and systematic manner?

**20.** *Scheduling and Coordinating.* Does the employee perform the necessary scheduling and rescheduling of work? Does the employee provide the necessary personal coordination of work, not only among subordinates but also between departments? When schedules are changed in a work area does he or she provide for the maintenance or adjustment of related work schedules in other areas?

**21.** *Training and Instructing.* Does the employee plan and carry out a program of orientation and training for new employees? Does he or she provide for the correction of any technical skill deficiencies in new employees? Does he or she provide on-going training for employees in new methods and procedures? Are new ideas, procedures, materials, and techniques presented in such a manner that employees will be receptive to change? Does he or she assist employees in self-development programs?

**22.** *Productivity.* Are assigned functions accomplished completely and on time? Is the quality of work produced by the employee and his or her staff up to standard? Does the employee improvise and find ways to compensate for the failure of others? Does he or she offer excuses instead of reasons for a decrease in productivity? Does the employee keep the supervisor informed of problems and delays, or does he or she wait until these are discovered, or until it is too late for other planning adjustments? Is the employee unwilling to adjust or revise schedules once they are set even though they may impose undue hardships on others?

**23.** *Evaluating Subordinates.* This pertains to the accuracy and manner in which the employee completes evaluations of subordinates. Does the employee strike a balance of constructive criticism and praise in evaluating other employees? Does he or she indicate specifically how an employee's work may be improved? Are his or her evaluations positive contributions to employee development? Are the evaluations consistently objective, fair, and accurate?

**24.** *Leadership.* Is the employee able to obtain the support and cooperation of other employees? Does the employee set a good example for others? Are the employee's methods of supervision excessively authoritarian? Does the employee create a work atmosphere in which work attitudes are positive and employee cooperation is increased? Does the employee fail to give adequate direction when needed?

**25.** *Operational Economy.* This pertains to the conservation of time and material, and the maintenance of adequate fiscal controls. Is the employee conscious of budgetary limitations? Does the employee operate within the budget? Does he or she prepare accurate budget estimates? Does he or she periodically calculate operational costs for units or phases of operational responsibilities? Is the employee able to identify uneconomical procedures and offer more efficient alternatives?

**26.** *Supervisory Control.* This pertains to the maintenance of order in all areas of supervisory jurisdiction. Do the employee's subordinates have a clear understanding of the behavior and performance standards which are expected? Are these standards applied consistently? Is the employee accepted by subordinates and in control of operations at all times? Is the discipline and control excessive? Is discipline too lax?

**27.** *Analytical Skills.* This pertains to the analyzing of data, to problem solving, and to the ability to clearly and accurately report conclusions drawn. Is the employee able to identify the sources of both simple and complex work problems? Is the employee able to analyze a subject clearly and objectively and draw well-organized, accurate conclusions? Does the employee approach problem solving in a systematic manner?

**28.** *Oral and Written Communication.* This pertains to the employee's ability to produce well-organized readable written material, and to speak articulately and tactfully. Is the employee able to produce written reports which do not require extensive revision? Does the employee have knowledge of the principles of business English? Do the employee's written and oral communications reflect positively on the District? Is the employee able to accurately represent the operations and policies of the District when speaking to the public or to other employees?

**29.** *Judgements and Decisions.* This pertains to the practical exercise of authority and responsibility. Is the employee firm and fair in making judgements concerning other employees? Does the employee cause resentment or other adverse reactions to the decisions because of poor timing or the manner in which they are stated?

**30.** Spaces 30 and 31 have been left blank for any additional factors which are considered necessary in assessing the supervisory or administrative effectiveness of the employee.

**Urbana Park District Personnel Evaluation Form Section I**

| Unsatisfactory a | Below Standard b | Standard c | Above Standard d | Outstanding e | Items 1–18 for all personnel | Does Not Apply f |
|---|---|---|---|---|---|---|
| | | | | | 1. Quality of Work | |
| | | | | | 2. Quantity of Acceptable Work | |
| | | | | | 3. Dependability | |
| | | | | | 4. Work Habits | |
| | | | | | 5. Knowledge of Work | |
| | | | | | 6. Interaction with Employees | |
| | | | | | 7. Interaction with Public | |
| | | | | | 8. Work Judgement | |
| | | | | | 9. Planning and Organizing | |
| | | | | | 10. Accepts Responsibility | |
| | | | | | 11. Accepts Direction | |
| | | | | | 12. Adaptability and Flexibility | |
| | | | | | 13. Effectiveness Under Stress | |
| | | | | | 14. Initiative | |
| | | | | | 15. Work Coordination Additional Factors | |
| | | | | | 16. | |
| | | | | | 17. | |
| | | | | | 18. | |
| | | | | | Items 19–31 for supervisory and administrative personnel | |
| | | | | | 19. Planning and Organizing | |
| | | | | | 20. Scheduling and Coordinating | |
| | | | | | 21. Training and Instructing | |
| | | | | | 22. Productivity | |
| | | | | | 23. Evaluating Subordinates | |
| | | | | | 24. Leadership | |
| | | | | | 25. Operational Economy | |
| | | | | | 26. Supervisory Control | |
| | | | | | 27. Analytical Skills | |
| | | | | | 28. Oral and Written Communication | |
| | | | | | 29. Judgements and Decisions Additional Factors | |
| | | | | | 30. | |
| | | | | | 31. | |

Employee Name

Department

Title or Classification

Supervisor's Name

Date

Self Evaluation

Supervisor's Evaluation

Evaluation of Supervisor

Checks in column a must be explained in Section II.

Any alterations in this form must be initialed in ink and dated.

**Urbana Park District Personnel Evaluation Form Section II**

If space for comment is insufficient you may attach additional sheets. These attachments must be signed and dated by both the employee and the rater.

Review the extent to which previously agreed upon objectives for improved work performance or for acquisition of additional job skills have been attained by the employee.

__________________________________________________________

__________________________________________________________

__________________________________________________________

__________________________________________________________

List mutually agreed upon objectives to be attained by the employee during the next evaluative period

__________________________________________________________

__________________________________________________________

__________________________________________________________

__________________________________________________________

Identify job strengths and superior performance incidents (refer to checks in column e of Section I)

__________________________________________________________

__________________________________________________________

__________________________________________________________

__________________________________________________________

Identify specific work performance deficiencies or job behavior requiring improvement or correction (Explain all checks in column a of Section I)

__________________________________________________________

__________________________________________________________

__________________________________________________________

__________________________________________________________

Summary Evaluation (Evaluate overall performance. This summary evaluation reflects your best judgement, not the application of a mathematical formula.)

______ Unsatisfactory    ______ Standard    ______ Outstanding

______ Below Standard    ______ Above Standard

Additional Comments

______________________________________________________________

______________________________________________________________

______________________________________________________________

______________________________________________________________

______________________________________________________________

______________________________________________________________

______________________________________________________________

______________________________________________________________

______________________________________________________________

______ I do ______ I do not recommend this employee be granted permanent status. (For probationary period evaluation only)

Supervisor's Signature ______________________________________

Title ______________________________________________________

Employee's Signature ______________________________________

Title ______________________________________________________

Date ______________________________________________________

Note: The signing of this form indicates that the individual being evaluated has reviewed the contents of the evaluation. Signing of the form does not indicate that the individual being evaluated agrees with the judgements made in the evaluation.

# APPENDIX E

# Job Descriptions

## DIRECTOR OF RECREATION AND PARKS

### Description

The director manages both recreation and park functions, including the development of comprehensive recreation programs and the operational and developmental phases of parks, boulevards, recreation areas and facilities, playgrounds, and specialized areas and structures. Although the basic policies are established by the governing jurisdiction, the director of recreation and parks, in administering the policies, exercises considerable independent professional judgment in directing the work of the department. He or she serves as the technical adviser to the recreation and parks commission, board, or other authority responsible to the public for the recreation and park services.

### Examples of Duties[1]

1. Organizes and directs the services of a recreation and park department in consonance with the general policies established by the public authority.

[1]This classification and many of the remaining descriptions are adapted by permission from Standards for Recreation and Park Personnel, Division of Recreation, Sacramento, Calif.; 1962, Department of Parks and Recreation.

2. Prepares and justifies the budget; controls and supervises departmental expenditures.
3. Recruits, selects, and employs (subject to the personnel policies of the jurisdiction) professional and other personnel needed for the successful operation of the department.
4. Ascertains present and future needs for areas, facilities, and program; prepares long-term plans to meet these needs, including the budgetary requirements.
5. Directs the acquisition, planning, construction, improvement, and maintenance of all areas and facilities that are the responsibility of the department.
6. Develops and administers a broad program of recreation activities for all age groups and interests, taking into consideration the resources and needs of the area.
7. Counsels with community groups and individuals to determine program needs, area and facility requirements, and improvements to interpret scope and purpose of present operations, and to point out program deficiencies and areas in which expansion and improvement are needed.
8. Gives direction and guidance to departmental staff by defining standards and principles of operations, and, together with staff, establishes agency goals and objectives.
9. Confers with other local, regional, state, and national government and voluntary agencies concerned with recreation, parks, conservation, and other recreation resources, so that cooperative planning and working relationships can be developed.
10. Establishes procedures to maintain files, correspondence, and records of the department; provides for a system for reporting, interpreting, and publicizing the work of the agency.

## Required Knowledge, Skills, and Abilities

The director is required to have thorough knowledge of the principles, theory, and philosophy of the recreation and park profession; skill in organizing the services of the department, which would include advance planning for areas and facilities, level of adequacy at which services are to be provided, development of long-term financial plans, and formulation of plans for the development and replacement of personnel; understanding of the recreation needs of the community, and ability to meet these needs with a progressive program; ability to guide and direct the work of the department and its personnel and to delegate tasks with dispatch to particular parts or divisions of the department; administrative skill in supervising the work of specialized personnel in landscape architecture, planning, horticulture, and recreation and park operation; intensive knowlege of the techniques of public administration, and executive capacity to make decisions judiciously; skill in communications; thorough knowledge of the development, maintenance, and operation of recreation and park areas and facilities; and ability to inspire and work harmoniously with peers and employees.

### Minimum Education and Experience

The director is required to have a degree from a college or university with specialization in parks and recreation and leisure services at either the baccalaureate or master's degree level and six years of progressive supervisory or administrative experience in a public park or recreation agency.

## ADMINISTRATIVE ASSISTANT

### Description

The administrative assistant under direction assists the administrator with management operation projects. Assignments may include responsibilities in affirmative action, budgeting, grant writing and administration, long-range planning, office management, personnel, promotion and public relations, purchasing, research, staff education and training, and so on.

### Examples of Duties

The duties would relate to those projects assigned but might include the following among others.

1. Assists the administrator by relieving him or her of administrative detail.
2. Makes special studies and investigations.
3. Prepares administrative reports.
4. Assists operating heads with administrative problems and procedures.
5. Prepares manuals of procedures.
6. Assists in budget analysis.
7. Analyzes the duties and responsibilities of the staff.
8. Assists in the establishment of standard management procedures.
9. Analyzes and makes recommendations on forms, work flow, procedures and the purchasing, storage, and utilization of park and recreation equipment.
10. Tabulates statistical data and summarizes and circulates professional literature.
11. Handles correspondence as directed by the administrator.
12. Assists the administrator with cooperative planning and development of working relationships with local agencies and groups.

### Required Knowledge, Skills, and Abilities

The administrative assistant is required to have knowledge of public administration, personnel management, administrative analysis, budgeting, public finance, modern office methods and procedures, and principles of park and recreation administration; ability to plan, organize, and direct a comprehensive administrative program; ability to train and evaluate the work of staff; skills in collecting and analyzing data and preparing sound recommendations and reports; ability to establish and maintain effective and cooperative relationships; and ability to represent the agency effectively and negotiate with representatives of other agencies.

### Minimum Education and Experience

The administrative assistant is required to have a baccalaureate degree in parks, recreation, and leisure services or a related degree and successful experience in parks and recreation or administration in related fields.

## DIRECTOR (SUPERINTENDENT) OF RECREATION

### Description

The director of recreation formulates and administers a broad public recreation program; plans, organizes, and directs departmental staff and activities, with immediate supervision of staff for major functions or phases of the recreation program usually delegated; and encourages community participation in the program. He or she operates independently, subject to review of policy by the governing authority for recreation.

### Examples of Duties

1. Formulates and executes a broad and varied program of public recreation activities conducted by the public recreation agency for interested individuals and groups.
2. Selects and appoints the recreation personnel in accordance with established procedure and has general supervision of the employees of the department.
3. Prepares, justifies, and administers the budget.
4. Studies and recommends property acquisitions for the expansion of recreation areas.
5. Plans construction, maintenance, and operation of the recreation facilities.
6. Serves as technical adviser to the recreation commission or board in formulating policies of recreation activity, services, and use of facilities; meets with interested individuals and groups to discuss program and facilities; and meets with interested individuals and groups to discuss program and facility requirements and improvements.
7. Recruits, selects, and employs (subject to the personnel policies of the jurisdiction) the professional and other personnel needed for the successful operation of the department.
8. Plans the program, advises and directs supervisors and other staff members, and gives general direction to the departmental staff.
9. Instructs leadership staff in policies and procedures or directs supervisory staff in training activity, by means of staff meetings and conferences and individual direction.

10. Prepares and justifies the budget and administers the expenditure of funds and collection of fees and other revenues.
11. Confers with representatives of public recreation agencies including those of the state and the federal government and with voluntary agencies in effecting cooperative undertaking.
12. Coordinates the work of the department with other government, commercial, or private agencies engaged in the field of recreation and parks and publicly represents the department in varied civic activities.
13. Develops and directs a public relations program and internal personnel policies.
14. Prepares reports on department accomplishments and needs.
15. Prepares agendas for board or commission meetings.
16. May serve as secretary to the governing authority.

### Required Knowledge, Skills, and Abilities

The director of recreation is required to have thorough knowledge of the principles, theory, and philosophy of the recreation profession; skill in organizing the services of the department, which would include advance planning for areas and facilities, level of adequacy at which services are to be provided, development of long-term financial plans, and formulation of plans for the development and replacement of personnel; understanding of the recreation needs of the community, and ability to meet these needs with a progressive program; ability to guide and direct the work of the department and its personnel and to delegate tasks with dispatch to particular parts or divisions of the department; administrative skill in supervising the work of specialized personnel; intensive knowledge of the techniques of public administration, and executive capacity to make decisions judiciously; skill in communications; thorough knowledge of the development, maintenance, and operation of recreation and park areas and facilities; and ability to inspire and work harmoniously with his or her peers and employees.

### Minimum Education and Experience

The director of recreation is required to have a baccalaureate degree in parks, recreation, and leisure services. A master's degree may be required, in addition to four years of progressive supervisory or administrative experience in a public park or recreation agency.

## DIRECTOR (SUPERINTENDENT) OF PARKS

### Description

The director of parks is responsible for the design, construction, and maintenance of recreation and park areas, facilities, and parkways, and the maintenance of street trees. He or she directs the work of employees engaged in the planning, design, construction, landscaping, and maintenance of recreation and park areas and facilities and in the maintenance and repair of equipment and is responsible for operating procedures necessary to accomplish objectives based on administrative directives or department policy.

### Examples of Duties

1. Designs and prepares functional plans for land development related to recreation and park areas.
2. Organizes the work of the department or division in accordance with the general policies or administrative directives established by the agency.
3. Directs and supervises the department employees in the maintenance and care of recreation and park areas, facilities, and equipment.
4. Recruits, selects, and trains the professional, technical, and maintenance employees needed for the successful operation of the department.
5. Prepares and justifies the budget; controls and supervises department or division expenditures.
6. Assigns and supervises the work of highly technical and skilled personnel in planning, forestry, horticulture, landscape architecture, construction, and design.
7. Reviews and evaluates the work of employees; makes field inspections to evaluate the quality and scope of work being performed.
8. Coordinates maintenance and custodial functions with the program of the recreation and other divisions of the jurisdiction.
9. Maintains complete and accurate records of the division's activities.
10. Interprets the work of the division to the administrative authority and local citizenry.
11. Prepares specifications, contracts, documents, and written and oral reports.
12. Ascertains present and future needs for areas and facilities; prepares long-term plans to meet these needs.

### Required Knowledge, Skills, and Abilities

The director of parks is required to have knowledge of present-day concepts of recreation and park functions; ability to formulate detailed plans and specifications; knowledge of plant nomenclature and adaptabilities to environment; knowledge of horticulture, plant pathology, arborculture, and park management; thorough knowledge of the principles, theory, and philosophy of the recreation and park profession; ability to plan, guide, and direct the work of the department or division and its personnel; capacity to make decisions objectively; skill in communications; thorough knowledge of the development, maintenance, and operation of recreation and park areas and facilities; and ability to work harmoniously with fellow employees and the public; knowledge of basic budgeting procedures.

### Minimum Education and Experience

The director of parks is required to have a baccalaureate degree in park administration or an allied field (i.e., landscape architecture, horticulture, forestry, arboriculture), with a minor concentration in planning, conservation, recreation, or public administration, as well as four years of progressive supervisory or administrative experience in a public park agency.

## ASSISTANT DIRECTOR (SUPERINTENDENT)

The position of assistant director of recreation or assistant superintendent of parks is more likely found in the larger departments. These staff assist the administrator in carrying out their duties or acting in the administrator's behalf by assuming responsibility for one or more of the director's duties. The position is a responsible one involving a close working relationship with the director and with line authority over staff members including supervisiors.

## BUSINESS MANAGER[2]

### Description

The business manager is responsible for the planning and controlling of fiscal operations including budget planning, accounting and reporting system, purchasing and inventory control system and supervises the work of the accounting and clerical staff.

### Examples of Duties

1. Coordinates the preparation of the annual budget and submits the final draft proposal to the director.
2. Monitors all income and expenditures and prepares monthly reports of financial information for the director and department heads.
3. Supervises the receipt and deposit of all monies.
4. Establishes with the approval of the director and the agency board all accounting and fiscal controls as deemed necessary.
5. Monitors all insurance programs of the organization and keeps all transactions up to date and the policies in force.
6. Administers the agency payroll in accordance with agency policy.
7. Coordinates purchasing program including bidding and the inventory of assets.
8. Supervises the clerical and accounting staff.
9. Assists independent auditor in the complete examination of the financial operation.

### Required Skills and Abilities

The business manager is required to have thorough knowledge of modern principles and techniques of public finance administration including budgeting, fiscal account-

[2]This classification is adapted by permission from "Park Ridge Recreation and Park District Job Description," unpublished job description, Park Ridge Illinois Park District.

ing, investing, payroll administration and purchasing; ability to communicate effectively; ability to cooperate with and interpret financial matters to the agency administration, other organizations, and the public; and ability to supervise and maintain effective relationships with staff personnel.

### Minimum Education and Experience

The business manager is required to have a baccalaureate degree in finance, accounting, business administration, or related field, in addition to three years of progressive supervisory or administrative experience in industrial or public accounting.

## DISTRICT RECREATION SUPERVISOR

### Description

The district recreation supervisor performs professional recreation work at supervisory level in the development and direction, within an assigned geographical area, of a variety of recreation activities. He or she is responsible for the establishment and evaluation of recreation programs and activities of assigned playgrounds, recreation centers, clubs, and housing projects and for the inspection of facilities to determine adequacy and adherence to established standards. He or she contacts the public in the planning and adaptation of suitable recreation programs. Instructions are usually received in department and staff meetings, and general supervision and review are received from one or more recreation supervisors through periodical reports, conferences, and inspections.

### Examples of Duties

1. Assigns schedules and supervises the work of general and specialist personnel at recreation centers and playgrounds in an assigned geographical district.
2. Inspects and evaluates adequacy of recreation program and administers a communitywide recreation program in an assigned district.
3. Participates in developing training programs for recreation personnel and counsels and instructs such personnel in proper techniques and methods.
4. Assists in the planning and development of communitywide recreation programs.
5. Schedules and conducts district recreation meetings for the purpose of interpreting the general recreation program to the public and to subordinates.
6. Makes recommendations for repairs and improvement of recreation facilities.
7. Prepares news articles and performs other public relations work connected with recreation activities.
8. Maintains records and makes periodic reports on all phases of the recreation program.

### Required Knowledge, Skills, and Abilities

The district recreation supervisor is required to have a knowledge of the basic psychology, techniques, and methodology of organizing groups in a recreation setting; ability to interpret the work of the staff effectively; skills in training procedures; general knowledge of all phases of recreation programs; skills in communications and public relations; and ability to work harmoniously and cooperatively with fellow employees and the public.

## RECREATION SUPERVISOR—FUNCTIONAL

### Description

The recreation supervisor plans and formulates an assigned major function or phase of the recreation program and supervises playgrounds, recreation centers, and/or assigned special staff in the conduct of such assigned phases. He or she may assist the director of recreation or assistant director in details of administration; or may, on the other hand, devote a portion of time to leadership, organization, or recreation activities at a playground or center. He or she is concerned with the orientation, training, and supervision of staff members, particularly seasonal and part-time workers and conducts supervisory staff meetings and conferences, subject to general direction and coordination of the director (superintendent).

### Examples of Duties

#### Recreation Centers and Playgrounds.

1. Plans and promotes a variety of activities and events for various age and special interest groups.
2. Coordinates schedules for recreation centers and playgrounds.
3. Orients, trains, and instructs recreation leaders, particularly part-time and volunteer employees.
4. Conducts regular staff meetings to discuss and evaluate program content, techniques, and special problems.
5. Inspects operations at each center and playground.
6. Confers individually with staff members in the exercise of supervisory authority and leadership.

#### Sports and Athletics.

1. Organizes or encourages the formation of teams and leagues of participants in popular competitive athletic games.
2. Plans and stimulates interest in organized competition including leagues and tournaments.
3. Cooperates with, and advises, center directors on the development of sports activities for children.
4. Directly promotes and organizes teenage and adult groups, and arranges schedules and facilities for competition among industrial or commercial teams and leagues.

**Arts and Crafts.**

1. Plans, coordinates, and supervises the arts and crafts program offered by the organization.
2. Directs the orientation and training of part-time and full-time special activity leaders and observes and supervises their work.
3. Confers with staff members on program and course content, advises on problems of techniques, and evaluates and discusses effectiveness.
4. Purchases supplies and materials for art programs.
5. Plans and supervises special exhibitions or competitions.

**Other Specialties.** Depending on local requirements and interests, additional supervisory positions of this type are not uncommon in the following specialized fields of recreation.

1. Aquatics, golf, tennis, or other sports activity.
2. Camping and outdoor education and nature programs.
3. Music and drama.
4. Special groups (e.g., teens, handicapped, aging, housebound).

### Required Knowledge, Skills, and Abilities

The recreation supervisor is required to have a knowledge of the basic psychology, techniques, and methodology of organizing groups in a recreation setting; ability to use volunteers effectively; skills in training procedures; thorough knowledge of various special recreation programs, and ability to lead participants in a variety of activities; skills in communications and public relations; and ability to work harmoniously and cooperatively with fellow employees and the public.

## PARK FOREPERSON OR SUPERVISOR

### Description

Under general direction, the park foreperson supervises the immediate performance of park maintenance and construction work; supervises and may perform the more skilled work in the planting and care of plants and lawns and the installation, repair, and maintenance of park structures and equipement; and does other work as required.

### Examples of Duties

1. Supervises, and assists in the performance of, a variety of tasks relating to installation, repair, and maintenance work in park development and maintenance of park structures.
2. Supervises and performs the more skilled work in the planting, propagation, cultivation, care, and treatment of shrubs, flowers, and lawns.
3. Supervises the cleaning of park buildings and grounds.

4. Plans and/or coordinates all construction projects including work with consultants estimating and obtaining materials and supervision of the project.
5. Plans, organizes, and directs the care and maintenance of park sites and facilities including buildings, vehicles, equipment, and landscape materials.
6. Supervises and instructs in the use and care of vehicles, power equipment, and hand tools and keeps records and reports.
7. Conducts safety inspections for park facilities and equipment and develops work safety programs.
8. Hires, trains, and supervises landscape and maintenance personnel and sees that proper personnel practices are carried out.

### Required Knowledge, Skills, and Abilities

The park foreperson is required to have knowledge of park and building maintenance including the planting, cultivating, pruning, propagating, and care of shrubs, plants, flowers, and turf; skill in the use of tools and equipment used in park maintenance and development and in building custodial and maintenance work; knowledge of the operation, repair, and use of trucks, tractors, and other equipment used for park maintenance; ability to supervise a group of employees performing a wide variety of park maintenance and construction tasks; ability to instruct staff in good job procedures and methods; ability to meet and deal tactfully and effectively with the public, keep proper records, and follow oral and written instructions; skill in evaluating the productivity of staff, developing work schedules, and making assignments and decisions judiciously; and skill in coordinating the work of the staff with that of the recreation department employees.

## PARK SUPERVISOR—FUNCTIONAL

### Description

The park supervisor plans and formulates an assigned major function or phase of the park operations program. He or she may assist the superintendent of parks in details of administration or devote the majority of time to one phase of the park operation.

### Examples of Duties

#### Horticultural Supervisor[3].

1. Directs and coordinates the grading, planting, renovating, and modifying of landscaped areas connected with recreation facilities and buildings.
2. Inspects work when it is in progress and on its completion; plans and schedules work priorities.

[3]Adapted from "Horticultural Supervisor," *Job Description and Specifications* (Los Angeles, Calif.: Department of Recreation and Parks).

3. Estimates material, equipment, personnel, and time needed for specific work projects.
4. Approves requisitions for flowers, shrubs, and trees, including requests from other departments.
5. Directs the preparation and maintenance of specialized horticultural displays.
6. Directs the nursery and other horticulture areas including the propagation of a wide variety of flowers, shrubs, and trees.

***Required Knowledge, Skills, and Abilities.*** The horticultural supervisor is required to have a thorough knowledge of propagation, planting, cultivation, pruning, and maintenance of plants, shrubs, flowers, trees, and lawns, including proper soil mixtures and fertilization; thorough knowledge of insects and diseases affecting plants and trees and of measures for their control; thorough knowledge of the requirements for propagation and growing of flowers, shrubs, and trees in a nursery, greenhouse, or bathhouse; thorough knowledge of the common and botanical names and characteristics of a wide variety of plant life, particularly those forms common to the local region; thorough knowledge of the methods of transplanting and balling shrubs and trees; knowledge of the principles of landscape planning and design; knowledge of methods used in maintaining landscaped areas, including equipment and watering systems and their care and use; ability to plan, organize, and direct the work of employees in various locations; ability to instruct employees in skilled propagation, planting, cultivation, pruning, and balling techniques; ability to prepare correspondence, keep records, and make reports; and ability to meet and deal tactfully and effectively with employees, the public, and personnel of wholesale nurseries.

**Other Specialties.** Depending on the size of the organization and local requirements and interests other supervisory positions might include:

1. Agronomist.
2. Building Maintenance Supervisor.
3. Forester.
4. Groundskeeper.
5. Landscape Supervisor.
6. Shop Maintenance Supervisor.

## FACILITY DIRECTOR

One or more major recreation complexes are not uncommon in park and recreation departments today. These facilities, which might be an ice skating complex, tennis and racquetball center, cultural arts center, zoo, golf complex, or auditorium or stadium, usually are major revenue-producing facilities and special leadership is required. The description and duties of the facility director vary according to the type of facility but the basic requirements are the following.

### Description

The facility director is responsible for planning, organizing, operating, and promoting the activities and program of the special facility or complex.

### Examples of Duties

1. Develops and implements programs for the facility.
2. Provides for training, assignment, and supervision of staff.
3. Maintains records of activities and prepares reports of facility programs.
4. Prepares recommended budget for facility including capital items and directs expenditures after approval.
5. Provides for collecting revenue and accounting for revenue for activities.
6. Prepares brochures, flyers, and other public information items promoting the facility.
7. Reviews and evaluates operations and recommends policy changes for operation of the facility.

### Required Knowledge Skills and Abilities

The facility director is required to have considerable knowledge of the program opportunities and maintenance requirements of the facility; skill in efficient business methods for revenue producing operations; ability to work effectively with staff and the general public; background and skill in effective record keeping and evaluation procedures; and skill in public relations and promotional activities.

## RECREATION LEADER

The designation recreation leader can relate to many positions within an organization, ranging from senior recreation leader; recreation leader I, II, or III; to recreation leader trainee. The basic description of these positions is about the same with additional duties added according to the level of the position and as the responsibilities for the position increase.

### Description

The recreation leader does responsible professional recreation work in directing a wide variety of activities at a recreation center or other program site, with responsibility for planning, coordination, and supervision of recreation activities. Work requires the application of specialized skills and training in the conduct of recreation activities and the meeting of particular recreation needs of the area or of various age groups. He or she develops and expands programs to meet specific area needs. General instructions are received from a supervisor concerning the overall recreation program to be administered. Work is supervised through field visits, staff conferences, and review of activity reports. Senior recreation leaders will normally supervise other recreation leaders and subordinates.

### Examples of Duties

1. Plans, organizes, and directs a wide variety of recreation activities at a recreation center or other facility.
2. Supervises the issuance, use, care, and maintenance of recreation supplies and equipment.

3. Determines the recreation needs and desires of the public in the area of service and initiates the programs required.
4. Works with community and special interest groups on matters of civic and recreation interest.
5. Advises and assists individuals and groups on social and recreation problems in the assigned recreation area.
6. Instructs and supervises recreation leaders in the lower classifications in the performance of assigned duties; checks on proper completion of work.
7. Schedules, coordinates, and conducts programs in drama, arts and crafts, and dancing, with the assistance of visiting special activity leaders.
8. Attends staff conferences and professional meetings; cooperates with voluntary agencies in the area.
9. Maintains recreation activity and progress records and prepares periodic reports.

## RECREATION ATTENDANT

### Description

The recreation attendant does routine work involving the performance of a variety of manual, clerical, and routine leadership tasks at a community recreation facility. He or she may assist a recreation leader at a center, playground, or camp, or may serve as an attendant in a swimming pool locker room. Work requires an ability to perform specific tasks that are readily learned after appointment and that involve considerable routine contact with the public and work is usually performed in accordance with specific instructions and reviewed for compliance with instructions. (The position may be filled primarily by part-time or seasonal workers.)

### Examples of Duties

1. Assists recreation leaders with the issuing and collecting of play equipment such as balls, bats, gloves, and horseshoes.
2. Assists recreation leaders with routine recreation activities including simple types of leadership of children, in accordance with specific instructions.
3. Moves and sets up equipment, cleans areas and facilities, and performs light maintenance and custodial works.
4. Assigns swimming pool or golf lockers to patrons, maintains order in locker rooms and showers, and enforces rules and regulations regarding the use of facilities.

## SPECIAL ACTIVITY LEADER—ARTS AND CRAFTS

### Description

The special activity leader (arts and crafts) does technical recreation work involving the organization and instruction of various age-level groups in arts and crafts. Develops courses of study and conducts classes in one or more fields of arts and crafts, such as painting, drawing and sketching, sculpture, ceramics, leatherworking, silk-screen work, metalworking, or woodcraft. He or she receives instructions on the scheduling of classes and on course content from a supervisor, but must exercise independent judgment in the application of specialized knowledge and skills in the conduct of the work. He or she may also counsel and train other staff members in crafts work. Work is reviewed through periodic inspections, staff conferences, and analyses of activity reports. Work may include the supervision of one or more assistants.

### Examples of Duties

1. Plans, schedules, and conducts classes in arts and crafts.
2. Develops arts and crafts classes for groups of all age levels.
3. Stimulates interest and appreciation in one or more fields of arts and crafts.
4. Maintains facilities and materials for class use.
5. Consults with recreation supervisor on the scheduling of classes to meet the needs of various groups in specific districts and for the city as a whole.
6. Visits playgrounds and centers or, in staff conferences, advises on arts and crafts activities, group organization, and instruction and leadership techniques.

### Required Knowledge, Skills, and Abilities

The special activity leader (arts and crafts) is required to have ability to plan, organize, and conduct a wide range of arts and crafts activities; technical knowledge of the field; ability to train and supervise others; skills in a variety of arts and crafts and a knowledge of a number of additional cultural activities; skills in the use of volunteers and the ability to work cooperatively with other staff members; and knowledge of both the basic philosophy of recreation and individual and group behavior.

## SPECIAL ACTIVITY LEADER—DANCE

### Description

The special activity leader (dance) does technical recreation work involving the organization and instruction of dance groups and classes in one or more of the major types of dancing, such as social, folk, square, and modern. He or she teaches the techniques of dancing to groups of various age levels; receives general instructions on the scheduling of classes from a supervisor, but exercises independent judgment in the application of specialized techniques and skills. He or she may also counsel and train other staff members on dance activities. Work is reviewed by a supervisor through staff conferences, inspections, and analyses of activity reports. Work may include the supervision of one or more assistants.

### Examples of Duties

1. Plans, organizes, schedules, and conducts classes in dancing.
2. Schedules classroom activities and coordinates them with other recreation activities.

3. Maintains group and individual discipline.
4. Maintains facilities and materials for class activities.
5. Operates record player and public address system in connection with class activities.
6. Plans logical sequence of dances to fit abilities and interests of participants.
7. Acts as host to participants in the classes.
8. Leads training conferences or consults with, and advises, playground and center directors on problems related to dance instruction techniques or dance activities.

### Required Knowledge, Skills, and Abilities

The special activity leader (dance) is required to have technical knowledge of the field and skill in planning, organizing, and conducting a dance program; skills in a variety of dance activities; understanding of the dance as related to other cultural programs; ability to train and supervise others; skills in the use of volunteers and the ability to cooperatively work with other staff members; and ability to plan and organize festivals, special events, and pageants in cooperation with other specialists.

## SPECIAL ACTIVITY LEADER—DRAMA

### Description

The special activity leader (drama) does technical recreation leadership work involving the direction and production of plays and similar dramatic activities for stage and radio. He or she plans, organizes, schedules, and produces dramatic activities on a district- or communitywide basis and receives instructions on the scheduling of classes and course content from a supervisor, but exercises independent judgment in the application of dramatic skills. He or she may also counsel and train other staff members on drama activities. Work is reviewed through occasional inspection visits and by analysis of attendance records and general community acceptance of the activity. Work may include the supervision of one or more nonspecialized assistants.

### Examples of Duties

1. Plans, organizes, schedules, and produces plays, festivals, pageants, and other dramatic activities.
2. Consults with supervisors of recreation on the need, in specific districts and the community as a whole, for drama programs for various age groups.
3. Selects types of plays to be produced and maintains script library.
4. Furnishes technical guidance on costumes, wardrobes, and props.
5. Writes radio scripts and produces radio skits and plays.
6. Studies and analyzes radio and dramatic materials in the development and collection of scripts for the dramatic library of the department.
7. Demonstrates methods, conducts training conferences, and discusses special problems with playground or center staff members whose program of activities includes dramatics.
8. Stimulates interest and appreciation in the field of dramatics.

### Required Knowledge, Skills, and Abilities

The special activity leader (drama) is required to have professional knowledge of all phases of dramatic productions; ability to promote, organize, and direct a specialized drama program including children's creative drama, puppetry, community or little theater, radio and TV, dramatics, drama institutes and workshops, and formal dramatic productions; skills in working with volunteers and the ability to work cooperatively with other staff members; personal skills in a variety of dramatic activities; and ability to work cooperatively with other specialists in the cultural field to produce major productions, pageants, festivals, and special events.

## PARK STAFF

The description of park or operations staff includes positions of many levels and with a variety of technical skills. The basic description and responsibilities are the same; however, the descriptions for the technical skill positions must include many additional requirements and responsibilities.

### Description

The park staff does responsible professional work in the area of park operations including construction, care and maintenance of facilities, park areas, and plant materials. Work requires the application of specialized skills and training in specific areas including landscape management and care, turf management, carpentry, masonry, plumbing, mechanics, welding, painting, electrical work, and so on. General instructions are received from a supervisor concerning the overall park operations and maintenance program. Work is supervised through field visiting and conferences and a review of project reports. Senior park staff will normally supervise other staff and work crews.

### Examples of Duties

1. Performs a variety of tasks relating to installation, repair, and maintenance work in park development and facility operation.
2. Operates trucks, tractors, and other equipment safely, efficiently, and responsibly.
3. Safeguards tools and equipment and maintains them in safe, efficient working order.
4. Inspects parks and facilities regularly and reports to supervisor.

### Required Knowledge, Skills, and Abilities

The park staff is required to have knowledge of park and building maintenance and care of equipment and landscape materials with special skills in specific areas and ability to keep proper records and follow oral and written instructions.

# APPENDIX 

# Agreements for Cooperation Between School and Recreation Authorities

AGREEMENT OF COOPERATIVE OPERATION
OF A
PUBLIC EDUCATION AND RECREATIONAL PROGRAM
IN THE
CITY OF SOUTH BEND
AND IN THE SOUTH BEND COMMUNITY SCHOOL CORPORATION[1]

This agreement is executed by and between the Board of Park Commissioners of the City of South Bend, hereinafter called Park Commissioners, and the Board of School Trustees of the South Bend Community School Corporation, hereinafter called the Board of Education, under and in pursuance to the authority granted under Chapter 118 of the acts of the Indiana General Assembly, 1957 (Burns Ind. Stat. Annot. Sec. 53-1101 to 53-1107), commonly called the Interlocal Cooperation Act.

Now, therefore, it is agreed by and between the parties hereto as follows:

**1.** *Provision for a joint board to be known as the Public Recreation Commission.* The joint or cooperative undertaking to be accomplished hereunder shall be administered by a joint board to be known as the Public Recreation Commission for the City of South Bend, Indiana, and the South Bend Community School

[1]Agreement of Cooperative Operation of a Public Education and Recreational Program in the City of South Bend and in the South Bend Community School Corporation, permission by Robert Goodrich, Director, South Bend Public Recreation Commission.

Corporation. It shall be composed of two (2) members of the Board of Park Commissioners, two (2) members of the Board of Education, and one (1) person to be appointed by the mayor who will serve as the Chairman of the Commission. Of the members of said Commission first selected by the Board of Education and the Park Commissioners, one shall serve for a term of one year and one for a term of two years. The chairman of the Commission, appointed by the mayor, shall serve for a term of two years. Thereafter, the members of the Commission shall be selected in the same manner as the member he is succeeding and the term of office shall be for two years. All terms shall expire on the first Monday in January, but an appointee shall continue in office until his successor is appointed. Except that no person will continue to serve as a member of the Public Recreation Commission after his term as a member of either board has expired. In case of death or resignation a member's unexpired term shall be filled according to the by-laws of the Board he represents.

**2.** *Purpose of joint or cooperative undertaking.* The purpose of this joint or cooperative undertaking is to provide for the cooperative administration of a public educational and recreational program in the City of South Bend, Indiana, that lies within the school district of the South Bend Community School Corporation through the creation of a Public Recreation Commission; such cooperation is on a basis of mututal advantage and will provide services and facilities in a manner and pursuant to forms of governmental organization that will accord best with geographic, economic, population, and other factors influencing the public educational and recreational needs of said units of government.

The educational program to be undertaken by the Public Recreation Commission is in no way to be construed as being that which is the normal responsibility of the said school corporation: Instead it is that educational program which customarily is associated with a recreation program involving the teaching of hobbies, crafts, games, skills, and similar activities which adults and children seek to learn or acquire during their leisure hours.

**3.** *Financing, Staff, and Supplying.*

**a.** Financing, establishing and maintaining a budget. On or before June 15 of each year, the Public Recreation Commission shall prepare and submit to each of the parties for approval, a proposed annual budget for the ensuing year. The budget for the ensuing year shall be the final adjusted budget as formally approved by both the parties hereto.

The parties shall contribute to the budget on an equal basis to that portion of the expenditures allocable to the program within the city limits, and the School Corporation shall bear the entire expense for the portion of the program which is allocable to that area outside the city limits but within the boundaries of the said School Corporation. Said contribution and expense shall be determined from and according to funds derived from tax sources only and exclusive of any funds from any other source.

Within the respective amounts so appropriated, each party shall employ such personnel, shall make such purchases and shall expend such funds as the Recreation Commission shall recommend in the manner prescribed by law; provided, however, that as to the employment of personnel, each of the parties hereto may, in its sole discretion, discharge any person employed by it whom it believes to be an improper or unsatisfactory employee. Either party may also, in its sole discretion, at any time refuse to make any purchases or to expend any funds on behalf of said Recreation Commission out of its respective share of the budget if it deems such purchases or expenditures to be improper or unnecessary.

The disbursement of all funds shall be upon such written warrant, voucher, requisition or claim as may be required by law and properly executed by the person or persons required by law. There shall be a Director of Recreation who shall endorse his written approval upon every such warrant, voucher, requisition or claim before presentation for payment.

Except for funds budgeted and to be paid by the parties to this agreement, as herein otherwise provided, all income from all activities under the sole and complete direction of the Recreation Commission shall be paid over to the secretary of the Board of Park Commissioners for deposit in the Park General Fund.

Funds which are collected from members by various clubs and organizations for program activities requiring entry fees and carrying special operation costs are to be deposited in a Recreation Activities Fund and may be disbursed without appropriation but on the proper order and according to the provisions set out and approved by the State Board of Accounts.

In the event either the Public Recreation Commission or the Comptroller of the City of South Bend shall, at any time, have reason to believe that any moneys are not being deposited as herein provided, immediate notice thereof shall be given to each of the parties hereto.

All receipts and disbursements made pursuant to this agreement shall at all times be in full compliance with the regulations of the State Board of Accounts of Indiana and the laws of the State of Indiana.

The South Bend School Corporation shall also make available for use of the Recreation Commission in the conduct of the program contemplated by this agreement, such of its buildings, grounds, equipment and facilities as the South Bend School Corporation, in its sole discretion, may determine can be so used for such purpose without interference or detriment to the principal educational functions of the South Bend School Corporation.

The Board of Park Commissioners shall also make available for use of the Recreation Commission in the conduct of the program contemplated by this agreement, such of its buildings, ground, equipment and facilities as the Board of Park Commissioners, in its sole discretion, may determine can be so used for such purpose without interference or detriment to its principal functions namely, the maintenance of public parks and grounds.

**b.** Staffing and Supplying. The Recreation Commission shall have the authority to recommend that the Parties hereto hire such personnel and obtain such supplies as are necessary to carry out the purposes of this agreement and as are authorized and provided by the budget which is provided by the parties hereto in the manner hereinabove set out.

**4.** *Duration of Agreement.* This agreement shall operate for the calendar year 1966, and for each year thereafter, unless terminated as herein provided. This agreement may be terminated at any time by either party giving one (1) year's written notice of termination to the other party.

**5.** *Methods of Complete or Partial Termination of the Agreement and Disposition of Property Thereupon.* This agreement may be terminated at any time by either party giving one (1) year's written notice of termination to the other party. In the event that such termination involves the disposition of property which has been acquired by the Commission, any such property shall be returned to that party hereto which provided the funds for its purchase.

**6.** *Acquisition, Holding and Disposition of Personal Property.* No real property shall be acquired by the Commission. Personal Property may be acquired by the Recreation Commission on behalf of the parties hereto to the extent authorized and provided for by the budget as appropriated each fiscal year and shall be under the jurisdiction of the Commission for the purposes herein set forth. No such property shall be acquired unless it is reasonably necessary and proper for the accomplishment of the purpose of this agreement. When the Commission ceases to have a use of any such personal property, the same shall be disposed of it by returning any such property to that Party hereto which provided the funds for its purchase.

**7.** *Powers and Authority.* The Public Recreation Commission shall have all power not prohibited by law that shall be necessary or proper in order to accomplish the purposes of this agreement. By way of illustration and not by way of limitation upon the powers of the Commission, the Commission shall have the following powers:

**a.** The power to adopt by-laws for its own government and management, which by-laws shall be adopted or amended by two-thirds' vote of all members of the Commission.

**b.** The Commission shall plan and conduct an educational and recreational program within the corporate limits of the City of South Bend, Indiana, and in that area outside the City of South Bend that lies within the school district of the South Bend Community School Corporation for the benefit of all citizens living therein and to the extent authorized by the purpose of this agreement. The Commission shall have full power and authority over the conduct of these programs and shall be accountable to the parties to this agreement for such programs.

**c.** The Commission may request either of the parties hereto to supply such personnel as it shall determine to be necessary in order to accomplish its purpose and to perform its duties set forth herein. The Commission shall have full power and authority over the direction of the efforts and work of such personnel as come within its supervision.

**d.** The Commission shall have no authority whatsoever to expend budget funds to make major repairs, improvements, or structural alterations to either buildings or real estate.

**e.** Said Commission shall keep complete financial records pertaining to the programs and activities contemplated by this agreement, including the contributions of the parties hereto, receipts from activities, all funds referred to in Paragraph 4 above, and any and all other moneys from every other source whatsoever, together with a complete record of all expenditures and disbursements made for or on behalf of the activities and programs contemplated by this agreement. Said Commission shall also keep a permanent record of all its proceedings and shall in all respects comply with the general law regulating the conduct of public agencies and applicable to the City of South Bend, Indiana. Reports of the activities of said Commission and also financial reports shall be made to the parties hereto at such times as requested.

**8.** *Cancellation of Prior Agreements.* This agreement cancels and supersedes all prior agreements by and between the parties hereto.

## Agreement Between Board of School Directors—Common Council City of Milwaukee[1]

The following statement of policy shall govern the actions of the Common Council and the Board of School Directors in the execution of their plan of cooperation for the joint planning and separate use of combined properties in the financing, purchasing, construction, reconstruction, and maintenance of neighborhood playgrounds and playfields in the City of Milwaukee. (Board Proceedings—April 6, 1955)

### 1. Location of Neighborhood Playgrounds

**a.** So far as possible, playgrounds should be a part of or adjoining the grounds of elementary schools or located in neighborhood parks. When this is not possible, there is a less desirable alternative—the construction of playground segregated from a school or park.

**b.** Neighborhood playgrounds should, whenever possible, extend to street or at least to alley lines and not abut private property.

### 2. Financing and Purchasing

**a.** Provisions should be made for the purchase and improvement of neighborhood playgrounds and playfields through the City of Milwaukee's Capital Improvement Program and funds. The Commissioner of Public Works or his representative of the Board of School Directors, submit annually to the Capital Improvements Committee a recommended program based as far as possible upon the recommendations contained in the most recent Ten-Year Playground Sites Survey and Program.

[1]Milwaukee School and Common Council Agreement.

**b.** The Common Council of the City of Milwaukee, whenever it contemplates the purchase of property for new neighborhood playgrounds or playfields not already approved in the most recent Ten-Year Playground Sites Survey and Program, shall notify, consult with and obtain the approval of the Board of School Directors before the purchase is made if the Board of School Directors is to assume the maintenance and operations costs.

**c.** The Board of School Directors, whenever it contemplates the purchase of property for new play areas not already approved in the most recent Ten-Year Playground Sites Survey and Program, or in the most recent Five-Year School Building and Future Sites Program, shall take into consideration the cost of playground construction because of terrain, drainage, and accessibility in order that future costs to be borne by the City shall be kept to a minimum.

**d.** Purchase of land needed for future neighborhood playgrounds or playfields should, when possible, be made before private or public buildings are erected thereon or any real estate development is started, even though the area is not to be developed immediately.

## Construction and Reconstruction

**a.** Construction and reconstruction and the erection of buildings on all neighborhood playgrounds and playfields used in connection with public recreation as distinguished from areas used in connection with the educational program shall be under the control of the Commissioner of Public Works; work to be done according to plans and specifications submitted by him and such representative of the Board of School Directors as shall be designated by it, subject to the approval of the Common Council.

**b.** The Board of School Directors shall include the surfacing and development of areas and facilities adjacent to new school buildings to be used in connection with the educational program (for recess activities) as part of the new school building project cost.

**c.** Reconstruction means changes in layout or grade necessitating and involving a complete restudy of the grounds and facilities.

**d.** The Common Council shall supply funds for neighborhood playgrounds and playfield reconstruction. The plans for reconstruction shall be made by the Department of Public Works in cooperation with the Board of School Directors as for a new neighborhood playground or playfield, and such reconstruction plans shall be submitted to the Common Council for approval.

**e.** No major change in layout of a recreation area shall be made by the Common Council or Board of School Directors, except upon mutual agreement.

**f.** Salvageable material and equipment resulting from such reconstruction shall revert to the department which originally supplied it.

## 4. Maintenance

**a.** After neighborhood playgrounds and playfields are completed and equipped according to the plans and specifications, the same shall be under the supervision, both as to maintenance and play organization, of the department of municipal recreation of the Board of School Directors.

**b.** Maintenance means resurfacing of grounds, replacement of lighting, equipment, minor changes in dimensions or grades, minor remodeling and general repairs.

**c.** The Board of School Directors shall provide in its Extension Fund budget sufficient funds to maintain neighborhood playgrounds and playfields transferred to it in a reasonable manner to prevent deterioration which might require major reconstruction or rehabilitation.

### 5. Cooperative Development

**a.** Where the Board of School Directors has purchased a school site sufficiently large to provide for a school and a neighborhood playground or playfield, the Common Council will provide for the development of the neighborhood play area portion of such site in the Capital Improvement Program and in accordance with 3a. The Board of School Directors shall surface an area of 100 square feet per child, on the basis that each school room in the building has an enrollment of 35 children; that an area of 50 square feet shall be surfaced for each child in the kindergarten; that each kindergarten has an average enrollment of 30 children.

**b.** In such development, a representative of the Department of Public Works shall be invited to participate from the very start in the planning of the play area portion of the site in order that there will be maximum coordination of playground and building functions.

**c.** Plans for such neighborhood playgrounds and playfields to be constructed with funds supplied by the Common Council shall be submitted to the Common Council for approval.

### 6. Partially Completed Grounds

**a.** Because of the length of time required to complete the development of the larger playgrounds or playfields and the pressing need for the use of the facilities, the Board of School Directors will accept for operation and maintenance the usable improved portions and facilities prior to completion of the entire project.

**b.** The Common Council shall do everything within reason to complete the construction as soon as possible of partially completed neighborhood playgrounds or playfields according to the plans approved by the Common Council and the Board of School Directors, so that they may be fully and finally accepted by the Board of School Directors.

### 7. Program and Operation

The program, the type of leadership and supervision, the days and hours of operation, and the use of all facilities on all playgrounds and playfields shall be determined by the Board of School Directors.

### 8. Abandonment

**a.** When it is deemed advisable to abandon any playground or playfield, the initiative may come from either the Common Council or the Board of School Directors, but such abandonment shall be only by mutual consent.

**b.** Upon abandonment, the control of the land shall revert to the entity which acquired it; salvageable equipment and accessories shall become the property of the agency which purchased them.

## Board of School Directors—Milwaukee County Park Commission

The following statement of policy was accepted by the School Board and the Park Commission: (Board Proceedings—July 2, 1957)

Any community to be successful and permanent must provide for its people:

1. A place to work
2. A place to live
3. Facilities for transportation
4. Opportunities for recreation, inspiration, and relaxation

The Master Plan of the community acknowledges these factors and makes adequate provision for them in number and in locations, so integrated as to produce a most efficient, beautiful and economic whole.

In order that the opportunities for recreation, inspiration and relaxation might fully meet the demand of the residents of the City and County of Milwaukee a committee was formed some ten years ago.

This committee designated as the Joint City-County Committee on Park Recreation Coordination and Policy was comprised of members of the City of Milwaukee School Board, City of Milwaukee Land Commission and the Milwaukee County Park Commission and its Planning Department. Innumerable meetings were held by the committee and in addition thereto the staff studies were undertaken to properly evaluate and to integrate the different recreational activities of the various agencies. Studies were also made of the geographic distribution of facilities so as to avoid duplication of function.

In May of 1950 "Recommendations" were submitted to the Policy Boards by the staff subcommittee, comprised of two parts, namely:

Part I. Program of Activities
Part II. Facilities

These recommendations were approved by the Milwaukee County Park Commission on June 2, 1950, and by the City of Milwaukee School Board on June 6, 1950.

During the ensuing years, that is, since 1951, experience has indicated the desirability of enlarging this basic agreement. Within the past few months the Recreation Directors and other staff members of the Milwaukee County Park Commission and of the School Board have met on a number of occasions. These conferences held in a spirit of complete cooperation and understanding, have resulted in the following being respectfully submitted to the policy-level agencies for consideration:

**1.** The Municipal Recreation Department is to conduct the organized playground programs in the parks at such park locations as [will be] mutually agreed upon.

**2.** The Municipal Recreation Department has a staff of professional recreation workers to plan, organize, and supervise such programs.

**3.** The conduct of such a program is to be carried on by the local unit because duplication could arise if a similar program were carried on by another public recreation agency within the local unit, and to provide an equally satisfactory program by two public recreation agencies within the same local unit would entail additional and unnecessary costs to the community.

**4.** In locations, under the jurisdiction of the Milwaukee County Park Commission, where a study indicates the need for an organized supervised program, the local municipality is to conduct and supervise this program, if mutually agreed upon.

**5.** The Park Commission is to approve the type of activities to be presented and to have the final authority to control the use of its facility by the program director.

**6.** The Park Commission Recreation Department and the Municipal Recreation Department are directed to work out a joint plan to advertise and promote the activities and programs of the other department, and are directed to jointly plan and conduct programs which would require the use of facilities and personnel of both departments.

It is, therefore, respectfully recommended that the various policy agencies, responsible for public recreation programs, such as Municipal Planning Boards, School Boards, and the Milwaukee County Park Commission direct their staffs to immediately establish a joint-planning procedure in order to provide a most complete and well-balanced program to be conducted on a cooperative basis to produce the most efficient service to the Metropolitan Community.

## Statement of Policy

### The Joint Acquisition, Development, and Operation Of School Park Areas By the City of Eugene and School District 4–J Lane County, Oregon

WHEREAS, it is in the public interest to maximize the use of both park and school facilities, and

WHEREAS, there is considerable overlapping of interest in the operation of these facilities by the City and the District, and

WHEREAS, there has been an occasional development of school playground and city park areas as a joint operation by the City and the School District, and

WHEREAS, Oregon Revised Statutes, Section 190.010, to Section 190.040, recognize and authorize local governments, including school districts, to make agreements for joint performance of functions,

It shall be the policy of the City and the District to cooperate in the acquisition, development, planning, and operation of integrated School/Park sites, and the operation of separate facilities where these already exist, subject to the conditions and regulations of the local budget laws.

**I. Acquisition**

**A.** The City and the District will locate new park and school facilities as centrally as possible in the neighborhoods as defined in the 1990 Development Plan of the Eugene-Springfield Metropolitan Area.

**B.** Neither the City nor the District will purchase additional land without conferring with the other agency as to its needs in the area of the land being acquired.

**C.** If both a park and school are needed in a neighborhood, every effort will be made by the City and the District to acquire sufficient land for the appropriate integrated use. It may be necessary or more economical for the City or District to acquire all of the land necessary for an integrated site until the other agency can acquire its share.

**D.** Where a school already exists and a park is needed, sufficient additional land will be acquired by the City to create an integrated site, when and if this is economically and physically possible. And conversely, the District will locate school sites adjacent to existing park sites when this is economically and physically possible.

**II. Development**

**A.** Whenever possible, development of school and park facilities on an integrated site shall proceed concurrently, with full consultation between the City and the District, before any construction begins.

**B.** If concurrent development is not possible, the School Administration shall be consulted in the event the park area is being developed first or the City Administration shall be consulted in the event the school is being built first, in order to ensure orderly and economical development of the integrated site and thus not building duplicate facilities.

**C.** The Architect of a school or the Landscape Architect of a park shall be instructed to maximize the joint use of certain specified facilities (e.g., play equipment, gymnasiums, swimming pools, locker rooms, craft and hobby rooms, rest rooms, etc.) by locating them carefully so that they may be conveniently used by the patrons of park or school personnel.

**III. Operation**

**A.** In the joint use of facilities, the liability of the City and the District and the responsibility for maintenance and upkeep shall be carefully spelled out in contracts between the City and the District.

**B.** There shall be a separate contract for each integrated site development and operation, stating each agency's responsibilities. (See attached.)

**C.** A schedule shall be established, setting forth the exact hours that specified school facilities shall be reserved for use by the City and specified City Recreation facilities by the District. Any use not set up in the schedule must be requested in writing in order to maintain clear lines of responsibility and liability.

**D.** The City and the District shall explore the possibility of joint support of supervisory personnel, with a view to year-round after-school and vacation supervision, if funds permit. Such a joint support shall be set forth in a contract between the City and the District.

**E.** The City and the District shall provide an agreement of understanding on the maintenance of school grounds, with a view to minimizing the duplication of maintenance equipment and staff.

**IV. Planning and Coordination**

The City of Eugene and School District 4J will appoint staff members on a joint committee to review and recommend courses of action necessary to implement this agreement or to submit progress reports as necessary to keep the general public and governing agencies aware of status and progress.

**V. Financial Participation**

**A.** Degree or amount of financial participation must be decided prior to implementation of a school-park project, taking into consideration budget limitations and project scheduling.

**B.** Equal participation (50-50) by both agencies is the ultimate goal. The degree of financial participation will vary from year to year, so equal sharing will eventually work out.

## Agreement for Community Recreation Services

THIS AGREEMENT, is made and entered into this 7th day of February, 1977, by and between the CITY OF SAN MATEO, a municipal corporation of the State of California, hereinafter called "City," and the SAN MATEO UNION HIGH SCHOOL DISTRICT, hereinafter called "District."

RECITAL:

WHEREAS, it is to the best interests of City, District and the people thereof, that joint use of City and District recreation facilities be organized, promoted and conducted to improve the health and general welfare of the people of City and District and to cultivate the development of good citizenship therein;

WITNESSETH:

NOW, THEREFORE, in consideration of the mutual covenants herein contained, the parties hereto agree that they shall organize, promote and conduct joint use of recreational facilities and equipment upon the following basis:

**1.** *City Facilities and Equipment.* City shall furnish and maintain for the use of District all City park and recreation facilities and equipment mutually agreed upon.

**2.** *District Facilities and Equipment.* District shall furnish and maintain for use of City all District facilities and equipment mutually agreed upon.

**3.** *Costs of Use.* District and City agree to pay actual and necessary costs incurred in use of facilities belonging to the opposite entity.

**4.** *District Policy—Employee Use.* When District policy requires the presence of a District employee before a facility may be used, District agrees to charge employee costs only for that part of the employee's time that cannot be used to advantage by District. It is understood that because a District employee is on a school site it may not always be possible to use the employee to the advantages of District.

**5.** *District Employee's Costs to Be Apportioned to Use.* When use of District facilities does not require the constant attendance of the responsible District employee, and the employee would otherwise be employed by District, only that portion of the employee's time that he is required to be in attendance shall be charged as a cost of using the facility.

**6.** *District Employee to Be Available.* When use of District facilities does require the constant attendance of the responsible District employee, the employee shall remain in or immediately about the facility and on call at all times except for normal breaks.

**7.** *Priority in Use.* It is understood, under terms of this agreement, that District use of District facilities herein described shall be given first priority. Only after all dates requested by District have been scheduled will City be granted use of the facilities. It is further understood, under terms of this agreement, that District shall be given second priority on the use of City facilities after City's first priority needs have been fulfilled. Once schedules have been approved, they shall not be altered without mutual consent.

**8.** *District Insurance.* District shall carry public liability insurance to protect its officers and employees against claims for damage to person and/or property arising from City's participation in the Community Recreation Program, the form of such insurance to be satisfactory to the District Attorney of the County of San Mateo.

**9.** *City Insurance.* City shall carry public liability insurance to protect its officers and employees against claims for damage to person and/or property arising from City's participation in the Community Recreation Program, the form of such insurance to be satisfactory to the City Attorney of the City of San Mateo.

**10.** *Enforcement of Rules and Regulations.* When District facilities are used by City, all District and State rules and regulations submitted by District must be enforced by City. When City facilities are used by District, all City and State rules and regulations submitted by City must be enforced by District.

**11.** *Termination.* This agreement may be terminated by either City or District upon giving 90 days' written notice of intention to terminate to the other party, which notice, if given, shall specify the date on which this agreement, and the terms thereof, is to terminate.

**12.** *Coordination Representatives.* For implementation of this agreement, the representative of District shall be the Director of Maintenance, Operations and Engineering and of City shall be the Director of Parks and Recreation.

IN WITNESS WHEREOF, the parties hereto have affixed their signatures on behalf of their respective entities to be effective the date hereinabove first set forth.

SAN MATEO UNION HIGH SCHOOL DISTRICT

By ______________________________
Assistant Superintendent-Operations

______________________________
Clerk

CITY OF SAN MATEO, a municipal corporation

By ______________________________
Mayor

ATTEST:

______________________________
City Clerk

## Recreation Agreement Between the City of Corvallis and the Corvallis School District No. 509J[1] Corvallis, Oregon

### Recitals

The City of Corvallis, through its Park and Recreation Department and the Corvallis School District 509J are mutually interested in and concerned with the provision of adequate facilities for the recreation and physical well-being of the people of the City of Corvallis and the people of District 509J.

Corvallis School District No. 509J has certain play areas, toilets, auditoriums, gymnasiums, community rooms and other educational facilities under its jurisdiction owned and operated for school purposes, but suitable, incidentally, for a community recreation program. The City, in its Park and Recreation Department has in its employ certain personnel qualified to supervise, direct and conduct such a community recreation program.

For many years past, the City of Corvallis and the Corvallis School District 509J have maintained a cooperative working arrangement whereunder many school grounds and facilities have been and are being used for general recreational purposes, thus affording to the community greatly increased recreational opportunities at costs much below what would otherwise be necessary.

This agreement supersedes all prior agreements on this same subject and is made to continue and improve the cooperative efforts of Corvallis Parks and Recreation Department and the Corvallis School District 509J.

Corvallis School District 509J, referred to as the "District" and the city of Corvallis referred to as the "City" agree that:

**1.** *Recreational Use of School Facilities.* The District will make available to the City for community recreation activities:

**a.** All permanently operated playground areas which are suitable for community recreation activities; these areas are to be selected by the Park and Recreation Director of the City and approved by the City Manager and the Superintendent of the District.

**b.** The District will allow the City to use other selected school facilities which are suitable for community recreation programs. These facilities are to be selected by the Park and Recreation Director, approved by the City Manager and the District Superintendent.

**c.** The use of such selected school facilities shall be in accordance with the regular procedures of the District in granting permits for use of school facilities as provided for by laws of Oregon and the policies and procedures of the District Board.

[1]Recreation Agreement: Between the City of Corvallis and the Corvallis School District No. 509J, permission by Rene D. Moye, Director of Parks and Recreation, Corvallis, Oregon.

**d.** Schedules shall be established for use of selected school facilities by designated representatives of the District Superintendent and the City Manager.

**e.** Any City program which is operated in or on District facilities shall be open and available to residents of the District equally with residents of the City.

**2.** *School Use of City Facilities.* Consistent with usage by the general public, the City will make available to the District selected City-owned park and recreation facilities for school use. These areas and facilities are to be selected by the District Superintendent and approved by the City Manager. Schedules shall be established for the use of established park facilities by the District Superintendent and the City Manager.

**3.** *Scheduling.* A schedule of dates for the use of the City and School District facilities will be worked out in advance. This schedule will be so arranged as to avoid any conflict between School and City use. In the scheduling of school facilities, school events and regularly scheduled school-related programs shall have the first priority and the recreation program established by the City shall have second priority and any other events by other groups or agencies shall have third priority. In the scheduling of City facilities the City shall have first priority, the District second priority, and others third.

**a.** The schedule may be changed at the request of either party by mutual consent.

**b.** School principals shall advise in scheduling of an approved recreation program to be conducted by the City on or in the facilities under the principal's jurisdiction.

**c.** School properties and facilities are intended primarily for school purposes and for the benefit of children of school age. It is, therefore, agreed that in planning programs and scheduling activities on school grounds, the recreation needs and opportunities of children, will be well provided for and adequately protected.

**d.** School properties and facilities are not required for school purposes during certain evening hours and on certain days of the week and during summer vacation periods. It is, therefore, agreed that at such times suitable facilities will be made available for recreation use by the City and that designated school playgrounds will be open for public recreation use.

**4.** *Rental and Other Charges.* The City promises to pay a basic rental charge to the District each year to cover utility costs and general wear and tear. Basic rental charges for fiscal years' 1976–1977 and 1977–78 shall be $5,000 per year. The charge is due and payable on December 31 of each year. Basic rental charge for the present and subsequent years shall be based on the current percentage (38%) of adult use of school facilities for City recreation programs. The rental charge, based on this percentage, shall be renegotiated annually, prior to budget preparation, but not later than January 31. Basic rental charges do not cover custodial time or overtime required for special meetings, hearings or any activities which may require furniture and equipment arrangements for such special meetings, hearings or activities. Such custodial time or overtime will be charged as required. This agreement does not preclude the City from charging for the use of its facilities if and when appropriate and if agreed by both the School District and the City.

**5.** *Supervision.* The City agrees to provide adequate personnel to supervise the recreation activities which take place after school hours and during holiday and vacation periods.

> **a.** It shall be permissible where such activities are beneficial to both school and recreation programs to allow the working hours of the City personnel to be integrated with school hours. In the event such activities are conducted during school hours with school children, the employee of the City shall be subject to the administrative authority and supervision of the principal of the school.
>
> **b.** The personnel employed by the City shall be under the supervision of the City Park and Recreation Department except as outlined above.

**6.** *Development by School District.* The School District will install and maintain all fences, play apparatus and facilities necessary, in its judgment, for its school programs. That equipment, apparatus, and facilities may be used by the City for community recreation purposes.

> **a.** The District may participate and assist in recreation facility development on City lands where those facilities are designed to supplement normal school facilities and equipment necessary for the community recreation program which are not included as a requirement for the school program. The facilities and equipment so furnished may be used by the District for school purposes.

**7.** *Maintenance.* The maintenance of facilities on City-owned land shall be the responsibility of the City while the maintenance of facilities on District land shall be the responsibility of the District.

**8.** *Disputes and Differences.* In the event any dispute or difference arises as a result of the recreation program being conducted on the sites jointly used and selected as above outlined or as to the use of a District facility, the dispute or difference shall be settled by the respective department heads of the District and the City.

**9.** *Supplementary Agreements and/or Termination.* It may be necessary from time to time to enter into supplementary agreements to deal with any proposed joint development of facilities on City and District lands. This agreement may be terminated by either party upon ninety (90) days written notice to the other party.

**10.** *Responsibility for Damage.* The City shall be responsible for any damages to school property which may be incurred as a proximate result of any recreation activity being conducted by the City Park and Recreation Department. The School District shall be responsible for any damages to City property which may be incurred as a proximate result of any school activity being conducted by the District.

**11.** *Liability to Third Parties.* The City shall hold the District, its officers, employees, and agents harmless on account of any claims, suits or actions for personal injuries to third persons or damage to the property of third persons which may result from any recreation activity being conducted by the City on District facilities, and the District shall hold the City, its officers, employees, and agents harmless on account of any claims, suits or actions resulting from District use of City Park and Recreation facilities. To satisfy this mutual responsibility, each of the parties, in their policies of public liability insurance covering the premises to be used under this agreement, shall cause the other party and its officers, employees, and agents to be named as insureds.

# APPENDIX G

# Hints for Obtaining Informal Bids and Quotes and Sample Price Quote Form

1. Initially contact as many firms as possible and from experience develop a list of the firms that submit bids. Do not send information to firms not interested.
2. Develop a simple form for price quotes and send duplicate copies so the firm can retain a copy.
3. Provide a complete description of items desired including, quantity desired, size, number, colors, and the like.
4. List exact time and place that bids are to be returned and do not make exceptions. One minute late is late, and the bid cannot be considered. Your best bet is to refuse to accept the bid, or if it is left, record the time received on the bid envelope and return it unopened. (Note to avoid these problems it is sometimes helpful to set your clocks back a few minutes. Then a bidder cannot complain that the bids were opened early.)
5. Establish the price FOB your city and state that shipping charges must be included. A seemingly low quote might not be low when shipping charges are included.
6. Inform all firms submitting quotes or bids of the action taken. There is nothing that will cut down on a firm's willingness to submit bids or price quotes than the failure to inform them of the action taken.

# QUOTATION REQUEST

Inquiry No ____________

Date ____________

Classification ____________

P. O. BOX 3161 COUNTRY FAIR STATION ▲ CHAMPAIGN, ILLINOIS 61820 ▲ PHONE (217) 352-0071

To

Address replies to: ______________________________

Quotations must be in this office by: ______________________________

| Item No. | QUANTITY | Description | Unit Price | Total |
|---|---|---|---|---|
| | | | | |
| | | | | |
| | | | | |
| | | | | |
| | | | | |
| | | | | |
| | | | | |
| | | | | |
| | | | | |

QUOTATIONS TO BE F.O.B. CHAMPAIGN

Shipping method ______________________________

Shipment can be made in ________ days after receipt of order.

This quotation expires ______________

Signed ______________________________ per ______________________________
Name of Company                                                    Authorized agent

**THIS IS AN INQUIRY—NOT AN ORDER**

Permission by Champaign Park District, Champaign, Illinois.

# APPENDIX 

# A Bond Prospectus[1]

## Official Statement

Willamalane Park and Recreation District
Lane County, Oregon
$1,300,000 General Obligation Bonds Series 1978

*Notice* is hereby given that sealed proposals for the purchase of $1,300,000 Senior Adult Activity Center General Obligation Bonds, Series 1978, of the Willamalane Park & Recreation District, will be received by the District at the place and up to the time specified.

***Time.*** All proposals for the purchase of the bonds must be received by 8:00 P.M. Pacific Standard Time, on January 26, 1978.

***Place.*** Mailed bids should be addressed to Willamalane Park & Recreation District, Gary Walker, Superintendent, PO Box 153, Springfield, OR 97477. Bids may be delivered in person to Gary Walker, Superintendent, Willamalane Park & Recreation District at 765 North "A" Street, Springfield, OR 97477.

***Place and Time of Consideration.*** The bids will be opened at 8:00 P.M. Pacific Standard Time, on January 26, 1978 at the Springfield Utility Board Meeting Room at 250 North "A" Street, Springfield, OR 97477. At that time all bids that

[1]Bond Prospectus, permission by Roger E. Delles, Assistant Superintendent, Willamalane Park and Recreation District, Springfield, Oregon.

have been received will be considered and acted upon by the Board of Directors of Willamalane Park & Recreation District.

***Rating.*** Moody's rating has been applied for.

***Issue.*** The total amount of the bonds issued shall be $1,300,000 consisting of 260 bonds, numbered 1 to 260, "Senior Adult Activity Center General Obligation Bonds, Series 1978", both inclusive, of the denomination of $5,000, and each issued February 1, 1978. All bonds issued hereunder shall be in accordance with ORS Chapter 266.

***Maturities.*** The bonds will mature in consecutive numerical order on February 1 in the amounts for each of the several years as follows:

| *Year* | *Principal Amount* | *Year* | *Principal Amount* |
|---|---|---|---|
| 1979 | $45,000 | 1989 | $60,000 |
| 1980 | 45,000 | 1990 | 70,000 |
| 1981 | 50,000 | 1991 | 70,000 |
| 1982 | 50,000 | 1992 | 70,000 |
| 1983 | 50,000 | 1993 | 75,000 |
| 1984 | 50,000 | 1994 | 80,000 |
| 1985 | 50,000 | 1995 | 85,000 |
| 1986 | 55,000 | 1996 | 90,000 |
| 1987 | 55,000 | 1997 | 100,000 |
| 1988 | 60,000 | 1998 | 100,000 |

***Interest.*** The bonds shall be dated February 1, 1978 and shall bear interest at the rate to be fixed upon the sale thereof but not to exceed a net effective rate of seven percent (7%) per annum; initial payment on July 1, 1978, and semi-annual payments on February 1 and August 1 of each year.

***Payment.*** The bonds and the interest thereon are payable in lawful money of the United States and are payable at the office of the treasurer of Lane County, Lane County Courthouse, Eugene, OR 97401.

***Redemption.*** Bonds maturing after February 1, 1988 are callable on February 1, 1988 and on any interest payment date thereafter, as a whole or in part, in numerical order of maturity and by lot within a maturity, at one-hundred percent (100%).

***Registration.*** The bonds are issued in bearer form with no option for registration.

***Purpose and Security.*** The bonds are to be issued by the District as general obligation bonds to finance a Senior Adult Activity Center in the District.

***Interest Rate.*** No proposal for less than all of the said bonds or for less than one-hundred percent (100%) of the par value thereof and accrued interest to the date of delivery thereof will be considered. Bidders are requested to name the rate or rates of interest which the bonds are to bear in multiples of one-eighth (1/8) or one-twentieth (1/20) of one percent (1%). Each coupon rate of interest for a bond must be within two percent (2%) of the coupon rate of interest of any other bond. The net effective interest rate shall not exceed seven percent (7%). Each bidder must specify in his bid the amount and maturity of the bonds of each rate named. All bonds maturing on the same date must bear interest at the same coupon rate.

***Award.*** The bonds will be awarded to the responsible bidder who offers the lowest net interest cost to the District, to be determined by computing the total interest on all of the bonds from February 1, 1978 to the respective maturities thereof, and deducting therefrom the premium bid, if any. Each bid must specify the total interest cost to the District under the terms of the bid and the effective interest rate. Each installment of interest will be represented by only one (1) coupon, and any bids specifying additional or supplemental coupons will be rejected. In the event that there are several identical bids which are the lowest, then the Board of Directors of Willamalane Park & Recreation District shall determine in their discretion the responsible bidder who shall be awarded the bonds.

***Right of Rejection.*** The Willamalane Park & Recreation District reserves the right in its discretion, to reject any and all bids and, to the extent not prohibited by law, to waive any irregularity or informality in any bid.

***Delivery and Payment.*** Delivery of the bonds will be made to the successful bidder at Portland, Oregon, or Springfield, Oregon, at the expense of Willamalane Park & Recreation District or at any other place at the expense of the buyer. Payment for the bonds must be made in Federal Reserve Bank funds or other funds immediately available to the Willamalane Park & Recreation District in Springfield, Oregon.

***Form of Bid.*** No bid form will be furnished. All bids must be unconditional. Each bid, together with the bid check, must be in a sealed envelope, addressed to the Willamalane Park & Recreation District with the envelope and bid clearly marked "Proposal for Bonds".

***Bid Check.*** Each proposal must be accompanied by a good faith deposit evidenced by a certified check, cashier's check, drawn on a solvent commercial bank doing business in the State of Oregon, payable to the order of the Willamalane Park & Recreation District, for $26,000.00, to secure the District from any loss resulting from the failure of the successful bidder to comply with the terms of the proposal. The check of the successful bidder will be deposited by the District and credited on the purchase price of the bonds. No interest will be allowed upon the check. If the successful bidder fails to take up and pay for the bonds, the check will be retained by the District as the District's liquidated damages. The check accompanying each unaccepted bid will be returned promptly.

***Legal Opinions.*** The unqualified opinion of Rankin, McMurry, Osburn & Gallagher, Bond Counsel, Portland, Oregon, approving the validity of the bonds and stating that interest on the bonds is exempt from income taxes of the United States of America under present federal income tax laws, and that such interest is also exempt from personal income taxes of the State of Oregon under present state income tax laws, will be provided to the successful bidder. A copy of the opinion of Rankin, McMurry, Osburn & Gallagher, certified by an officer of Willamalane Park & Recreation District, by signature or facsimile signature, will be printed on the back of each bond.

***Cusip Numbers.*** It is anticipated that CUSIP numbers will be printed on the bonds at the expense of the District. However, failure to print numbers on bonds shall not constitute grounds for refusal to accept bonds.

***Official Statement.*** The Official Statement and additional information are available from: Willamalane Park & Recreation District, Gary Walker, Superintendent, PO Box 153, Springfield, OR 97477, (503) 746-1669 and United States Consulting Co., Inc., 1149 Ferry Street, Eugene, Oregon 97401, (503) 485-0490.

***Adopted*** this 8th day of December, 1977.

WILLAMALANE PARK &
RECREATION DISTRICT

______________________________

President
Board of Directors

______________________________

Secretary
Board of Directors

# APPENDIX I

# EEI: A Survival Tool[1]

by Robert L. Wilder

Mr. Wilder is administrator of the Washington State Interagency Committee for Outdoor Recreation in Olympia, Washington.

PARK AND RECREATION departments around the nation are suffering a great deal of anguish over their financial plight, for this is a time of economic stress. Leisure services are hard to quantify and even harder to justify because the values involved are socially oriented and intrinsic and subjective in nature. As a result, the case for parks and recreation is often lost to a degree in the battle for budgets and for the shrinking dollar.

One of the challenges that has been facing park and recreation professionals is the development of a tool or measure which would allow them to report in quantifiable terms the value of recreation and leisure services—a tool which allows them to be more sophisticated and scientific in their approach to justifying expenditures for park and recreation facilities and services.

Many attempts have been made, time and time again, to develop some form of quantification that would appear to be useful and represent some of the values and importance of recreation facilities and leisure services.

The state of the art regarding quantification and economic determination in recreation is sadly lacking. Public recreation is not generally sold for a price.

[1]Reprinted with permission of Robert L. Wilder and National Recreation and Park Association, 1601 North Kent Street, Arlington, Va. 22209.

Since there is no adequate dollar measure of the worth of recreation experiences at public facilities, there therefore is great difficulty in judging the direct economic values that accrue to people engaging in recreation.

As David E. Gray, speaking at the 1972 banquet of the California Association of Park and Recreation Commissioners, put it so well: "We desperately need a method of planning that permits social cost-benefit analysis. Lacking such a system, we are turning control of our social enterprises over to the accounting mind. The accounting mind reaches decisions by a method in which short-range fiscal consequences are the only criterion of value. . . . Recreation and park services will not survive in that kind of an environment. Most of the great social problems that disfigure our national life cannot be addressed in a climate dominated by that kind of value system."

The modern day name of the game seems to be quantification, justification, competition, and cost-benefit analysis. This is where the economic equivalency index (EEI) comes into prominence. More than ever we need a tool to justify, to quantify—a tool for survival.

One criterion for any survival tool is that it have practical applicability so that it is used. It must be functional so it is not shoved in a corner because of its complexity and its users' inability to interpret it to their constituents and legislative bodies. Such a criterion demands that the tool utilize some form of language and units of measurement that are easily understood by all. The units must have some common meaning and must have a quality that will get attention and hold the attention of the legislative decision makers. The tool should be such that it can be used nationwide, thereby being judged as a recognized and acceptable standard. Yet it must be able to be modified to meet local conditions or changing conditions. It also must be flexible enough to have a multiplicity of uses in justifications of leisure services, including but not limited to budgets, programs, areas, facilities, and other essential elements of the park and recreation system.

The economic equivalency index (EEI) could be just such a tool. It can be used to quantify the value of leisure experience to the citizen or to the participant in the activity. And it is stated in dollar terms for the purpose of quantification and interpretation.

The economic equivalency index is based upon the concept that time is money, and that every waking hour has a value. If the hour is used one way, that hour is given up some place else. This hour is characterized as an "opportunity cost" to perform some other activity, such as work for remuneration. The EEI is based on the fact the participant has leisure time which is discretionary and can allocate it as he sees fit. He makes a choice of using a certain amount of time for leisure activities rather than investing it to accumulate income at a minimum wage.

In short, the economic equivalency index is an attempt to place a monetary value on a social service or good. It is a measure of the value of the leisure experience to the individual, and in turn the program, area, or facility. The common denominator for this index is the participant hour (PH), which is the unit of time that can be measured and recorded. Participation at recreation areas and facilities in various leisure time activities is recorded, aggregated, and converted to a conservative dollar value (EEI). For the sake of being conservative, the federally accepted minimum wage (MW) is used in all cases for adult participants. Youth of 13 to 18 are evaluated at one-half minimum wage and children 12 and under are evaluated at one-third the minimum wage rate. Therefore, the minimum wage in the EEI formula would be divided by an age factor (AF) which would equal 1 for adults, 2 for youth, and 3 for children.

The formula itself is represented by the following equation: PH × NP × MW ÷ AF = EEI, or participant hour × number of participants × minimum wage ÷ age factor = economic equivalency index. Take an example: if citizens participate in an activity for 50 hours, the PH would equal 50; if that leisure activity had 10 participants, the NP would equal 10. Assume the minimum wage is $2 an hour and the users are adults, 19 years or older. Then the computation would be as follows: 50 participant hours × 10 participants × $2 per hour = $1,000. The EEI for the leisure experience of the citizen participants in that particular activity would then be $1,000.

Another example might be the participation by youths in an athletic league an average of 100 hours each, counting practice and games during the season. The league has 300 participants. Again, using the $2 as the illustrative minimum wage, the formula would be as follows: 100 participant hours × 300 participants × $2 minimum wage ÷ age factor of 2 = $30,000 EEI. Therefore, $30,000 would be the economic equivalency index for the leisure experience of the youths in the athletic league. This social value to the community, expressed in economic terms, is assumed to be the equivalent of what the citizens would be willing to pay or give up for the recreation program.

It is also proposed that an agreed upon multiplier could be used for special services or aimed at special groups, such as the needy, handicapped, or aged. The multiplier would vary according to the value system of the area in which the EEI was to be applied. Here is how the multiplier works: Assume that the legislative body of a governmental entity has determined that certain social programs and/or populations should be given a priority over others. This would lead to the adoption of multipliers that could be used to weight the EEI to reflect those priorities. For example, assume that the following populations are assigned the corresponding multipliers: senior citizens, 5; handicapped, 5; children, 3; youth, 3; economically underprivileged, 4.

These values are then used in the appropriate place in the equation which yields an adjusted economic equivalency index: PH × NP × MW ÷ AF = EEI × multiplier = adjusted EEI.

Using the adjusted EEI formula and the above multiplier values, the computation for a 100-hour program for 25 handicapped youth would be: 100 participant hours × 25 participants × $2 minimum wage ÷ age factor of 2 = $2,500 EEI × multiplier of 5 = $12,500 adjusted EEI.

For a 100-hour program for 100 senior citizens, the computation would be: 100 × 100 × $2 ÷ 1 = $20,000 × 5 = $100,000 adjusted EEI.

For a two-hour "music in the park" program attended by different age groups—say, 1,000 adults, 500 youth, and 100 children—there would be a computation for each age group. The three adjusted EEI values would then be added to reach the total adjusted EEI for the program. For the adults the computation would be: 2 × 1,000 × \$2 ÷ 1 = \$4,000 × 1 = \$4,000 adjusted EEI. For youth: 2 × 500 × \$2 ÷ 2 = \$1,000 × 3 = \$3,000. For children: 2 × 100 × \$2 ÷ 3 = \$134 × 3 = \$402. To find the total adjusted EEI, the computation would be: \$4,000 + \$3,000 + \$402 = \$7,402 total adjusted EEI.

For a 25-hour crafts program for 10 adults, the computation would be: 25 × 10 × \$2 ÷ 1 = \$500 × 1 = \$500.

The aggregate economic value to a community of all these recreation programs would then be the sum of the services offered the citizens, or: \$12,500 for handicapped youth + \$100,000 for senior citizens + \$7,402 for "music in the parks" + \$500 for adult crafts = \$120,402 total value. This figure could be presented as either a gross value or as a net value after the costs have been computed. It represents the social value to the community in terms of dollars that the citizens would be willing to pay or give up for recreation services.

The economic equivalency index could also be used for a tool in cost-benefit analysis. The benefit (EEI) could be related to the cost of providing the program or service so that rankings and comparisons could be made between programs or services and certain standards of service and performance which could be adopted.

Following is an example of how a cost-benefit application of the economic equivalency index could be computed for a justification of dollar expenditures and for management analysis.

Assume the previous program examples for which adjusted EEIs were calculated have direct and indirect *costs* amounting to the following: handicapped youth program, \$2,500; senior citizen program, \$50,000; "music in the parks" program, \$500; adult crafts program, \$750.

Using the above figures, cost-benefit ratios can be determined by placing the costs over the EEI (the benefit) in fraction form or in decimal equivalents. For the handicapped youth program \$2,500/\$12,500 = 1/5 = .200. For the senior citizen program: \$50,000/\$100,000 = ½ = .500. For music in the parks: \$500/\$7,402 = 1/15 = .068. For the adult crafts program: \$750/\$500 = 3/2 = 1.500. After the ratios have been determined, standards can be established for measurement or comparisons.

For example, in ranking the above programs, those with the smaller ratio or decimal equivalent would be judged to be of greater economic return for each dollar invested.

A policy decision could then be made as to what cost-benefit ratios are acceptable to the agency. For example, the decision might be that any cost-benefit ratio that exceeds .50 must either be dropped or income must be raised to reduce the ratio to .50 or less.

In the example above, the adult crafts program would not be supported unless the ratio could be reduced to .50 or less. To do this, an additional \$1,000 value would have to be generated to support the program by: (1) increasing participant hours, (2) offsetting departmental costs, or (3) a combination of 1 and 2.

For the purpose of quantification and applicability, the EEI is a very simple approach. It is important for a tool such as this to be highly simplified in order for it to be used consistently. It is also essential that it be expressed in terms that everyone will understand—hence the use of dollar values. This seems to get everyone's attention. Remembering the saying "money talks," park and recreation administrators can use the EEI to give an economic equivalency to the social value of an individual's recreation experience and the aggregate sums to show the conservative social value of the recreation experiences to the community.

Certainly the EEI is not an answer in itself, and it will be open to skepticism and a certain amount of criticism. The assumptions can and will be debated forever, but its value is that it is a simplified tool that can be applied uniformly, can be modified to meet local conditions, has a multiplicity of uses, can be expressed in dollar terms, and will have the ability to get the attention of the decision makers who have such a drastic impact on park and recreation budgets. Park and recreation professionals must start to put together a tool kit for survival and the economic equivalency index is one tool that certainly should be in the toolbox.

# Index